T0350134

# FOURIER RESTRICTION, DECOUPLING, AND APPLICATIONS

The last 15 years have seen a flurry of exciting developments in Fourier restriction theory, leading to significant new applications in diverse fields. This timely text brings the reader from the classical results to state-of-the-art advances in multilinear restriction theory, the Bourgain–Guth induction on scales and the polynomial method. Also discussed in the second part are decoupling for curved manifolds and a wide variety of applications in geometric analysis, PDEs (Strichartz estimates on tori, local smoothing for the wave equation) and number theory (exponential sum estimates and the proof of the Main Conjecture for Vinogradov's Mean Value Theorem).

More than 100 exercises in the text help reinforce these important but often difficult ideas, making it suitable for graduate students as well as specialists. Written by an author at the forefront of the modern theory, this book will be of interest to everybody working in harmonic analysis.

**Ciprian Demeter** is Professor of Mathematics at Indiana University, Bloomington. He is one of the world's leading experts in Fourier restriction theory and its applications to number theory, which he teaches regularly at the graduate level. He received the Sloan fellowship in 2009 and was an invited speaker at the 2018 International Congress of Mathematicians in Rio de Janeiro.

# Fourier Restriction, Decoupling, and Applications

CIPRIAN DEMETER

*Indiana University*

CAMBRIDGE
UNIVERSITY PRESS

# CAMBRIDGE
## UNIVERSITY PRESS

University Printing House, Cambridge CB2 8BS, United Kingdom

One Liberty Plaza, 20th Floor, New York, NY 10006, USA

477 Williamstown Road, Port Melbourne, VIC 3207, Australia

314–321, 3rd Floor, Plot 3, Splendor Forum, Jasola District Centre,
New Delhi – 110025, India

79 Anson Road, #06–04/06, Singapore 079906

Cambridge University Press is part of the University of Cambridge.

It furthers the University's mission by disseminating knowledge in the pursuit of
education, learning, and research at the highest international levels of excellence.

www.cambridge.org
Information on this title: www.cambridge.org/9781108499705
DOI: 10.1017/9781108584401

First published 2020

A catalogue record for this publication is available from the British Library.

ISBN 978-1-108-49970-5 Hardback

# Contents

# Preface

Fourier restriction is a subfield of harmonic analysis that, broadly speaking, studies the properties of Fourier transforms of measures supported on curved manifolds. Since its inception in the late 1960s, it has established deep connections to geometric measure theory and incidence geometry, and has led to important developments in dispersive partial differential equations (PDEs) and number theory.

There are two main threads in the book. The first part is concerned with the restriction conjecture for hypersurfaces and its relation to various aspects of the Kakeya phenomenon. The discussion here includes some classical topics such as the Stein–Tomas and Córdoba–Fefferman arguments, but is mostly focused on the developments from the last three decades. The modern era of Fourier restriction was born in the early 1990s, with a flurry of landmark papers by Bourgain that have shaped many of the subsequent advances in the field. Different chapters of the book are dedicated to introducing fundamental tools such as wave packets, bilinear and multilinear restriction estimates, the Bourgain–Guth induction on scales, and the polynomial method.

The second part of the book is focused on decoupling. This Fourier analytic tool introduced by Wolff and later refined by Bourgain, the author, and others, is a culmination of many of the techniques presented earlier in the book. Due to its many applications, decoupling is continuing to broaden the scope of Fourier restriction and to connect it with new areas of mathematics.

This book is certainly not aimed to be an encyclopedia. It only tries to follow the major developments in the field, with the main focus on the most recent advances that have not previously appeared in book format. It is in part a reflection of the author's taste. The following are a few examples.

In Chapter 1, we present the details for two "folklore results". Proposition 1.6 quantifies the meaning of restricting the Fourier transform of an $L^p$

function to a manifold. In Section 1.4, we present a lesser-known proof of the endpoint Stein–Tomas theorem in $\mathbb{R}^3$, which does not make use of oscillatory integrals. Near the end of the chapter, we introduce constructive interference and square root cancellation, and describe their connection to restriction estimates.

Chapter 3 is a long reflection on biorthogonality. A lot of effort is spent in building the intuition behind the role of $L^4$ and the significance of diagonal behavior for caps and tubes. The main result proved here is the bilinear restriction estimate for the paraboloid. We start with an argument that relies on finding good bounds for bilinear incidences of tubes. To ease the exposition of this difficult result, Section 3.7 analyzes the warm-up scenario where tubes are replaced with lines. Near the end of the chapter, we also present an alternative argument that essentially avoids incidence geometry.

Chapters 5 and 6 investigate the connections between restriction estimates, square functions, and the hierarchy of Kakeya-type conjectures, first in the linear, then in the multilinear setting. The Bourgain–Guth method of deriving linear restriction estimates from their multilinear counterparts is described in Chapter 7, in a variety of contexts. Chapter 8 introduces the polynomial method and shows how it leads to the state-of-the-art results on the restriction conjecture for the paraboloid.

Chapters 9 through 12 are concerned with presenting decoupling in a more unified fashion. Some of the results that have not appeared in an article format include the decoupling for the anisotropic neighborhood of the moment curve (Chapter 11) and a slightly different, more pedagogical, proof of the decoupling for the parabola due to Guth (Section 10.4).

The last chapter contains applications of decoupling to a large variety of topics in number theory (exponential sums and Diophantine systems), PDEs (Strichartz estimates on tori, local smoothing for the wave equation), combinatorics (estimates for additive energies), and spectral theory ($L^p$ estimates for eigenfunctions of the Laplacian on tori). The crowning achievement of decoupling is the resolution of Vinogradov's mean value theorem. This was a long-standing conjecture in number theory whose final answer came, to the surprise of the mathematical community, via purely harmonic analytic methods.

This book was written to be accessible to graduate students with a moderate background in harmonic analysis. It can be used to teach graduate-level classes, and it can be covered completely in a two-semester sequence. It is the author's hope that, in addition to harmonic analysts, the book will also be accessible to other groups of mathematicians, including (but not restricted to) number theorists and people working in PDEs. If the intention is to cover decoupling in a one-semester course, the only serious prerequisites are contained in Chapters 6 and 7. The book contains more than 100 exercises, many of which are rather substantial and come with generous hints. In order to keep things both simple enough and sufficiently rigorous, most of the detailed arguments are presented in the low-dimensional setting ($\mathbb{R}^2$ or $\mathbb{R}^3$), followed by brief discussions of the higher-dimensional case.

# Acknowledgments

The idea of writing this book came in the summer of 2017, motivated by the flurry of recent developments in the area of harmonic analysis concerned with the Fourier restriction phenomenon. The material presented here was polished during the two-semester course that I taught at Indiana University in the academic year 2018–2019. I would like to thank my students Shival Dasu, Hongki Jung, Dominique Kemp, and Aric Wheeler for their contributions to this process, which included additional weekly meetings aimed at discussing the exercises from the book.

My special thanks go to the dedicated readers Bartosz Langowski, Zane Li, and Po-Lam Yung, whose many comments have led to significant improvements of the presentation. I am also grateful to Nets Katz for using an earlier version of this manuscript to run a reading seminar at Caltech, with participants including Maksym Radziwill, Zachary Chase, and Liyang Yang. The book has benefited from feedback offered by Joshua Zahl, Jonathan Bennett, and Maxim Gilula. My colleague Hari Bercovici has taught me a few things about how to write a book and offered technical support when needed.

When writing a book, there are always concerns about the time investment and the emotional toll that inevitably lead to sacrificing other areas of one's life. The patience and the support of my wonderful wife Iulia have been invaluable throughout.

This book owes a lot to my collaborators, especially to Jean Bourgain and Larry Guth. I would like to thank Larry for allowing me to include his insight on the decoupling for the parabola in Section 10.4. The book also contains other examples that I learned about from the inspiring conversations with him.

Jean Bourgain and Elias Stein passed away on consecutive days of December 2018. The mathematical contributions of these two giants, in particular those to the topics discussed here, could not be overestimated. Stein has set up the foundations of Fourier restriction, while Bourgain has taken this field to unimaginable heights over the last 30 years. This book is a testimony to their extraordinary legacy.

My collaboration with Jean started in December 2012. It was largely due to reading his remarkable paper [20], on applications of decoupling to exponential sums, that I decided to work on this type of problem. Throughout our six-year collaboration, which has resulted in 11 papers, he never ceased to strike me with his enthusiasm and generosity. I would often start my day by breathtakingly reading his latest emails, sent to me around 3 am. Some of them would contain notes, some others would ask questions. Most of his messages were short, yet incredibly rich.

There is hardly any chapter in Fourier restriction where Jean's work did not play a transformative role. In all these years, he changed tremendously not only the field, but also the lives of so many mathematicians, myself included. I dedicate this book to his memory.

*Ciprian Demeter*
Bloomington

# Background and Notation

We will denote by $e(z)$ the quantity $e^{2\pi i z}$, $z \in \mathbb{R}$.

The Fourier transform of an $L^p$ function on $\mathbb{R}^n$ will be denoted by either $\widehat{F}$ or $\mathcal{F}(F)$. In particular, when $F \in L^1(\mathbb{R}^n)$ we will use the definition

$$\widehat{F}(\xi) = \int_{\mathbb{R}^n} F(x) e(-x \cdot \xi) \, dx.$$

More generally, if $d\mu$ is a finite complex Borel measure on $\mathbb{R}^n$ we will write

$$\widehat{d\mu}(\xi) = \int_{\mathbb{R}^n} e(-x \cdot \xi) \, d\mu(x).$$

The space of complex-valued Schwartz functions on $\mathbb{R}^n$ will be denoted by $\mathcal{S}(\mathbb{R}^n)$.

We will make repeated use of both the Fourier inversion formula, valid for each $F \in \mathcal{S}(\mathbb{R}^n)$

$$\mathcal{F}(\mathcal{F}(F))(x) = F(-x),$$

and Plancherel's identity, valid for $L^2$ functions (with natural generalizations for measures)

$$\int_{\mathbb{R}^n} F(x) \overline{H(x)} \, dx = \int_{\mathbb{R}^n} \widehat{F}(\xi) \overline{\widehat{H}(\xi)} \, d\xi.$$

Given a subset $U$ of $\mathbb{R}^n$, we define the Fourier projection operator $\mathcal{P}_U$ via the formula

$$\mathcal{P}_U(F)(x) = \int_U \widehat{F}(\xi) e(x \cdot \xi) \, d\xi.$$

The function $\mathcal{P}_U(F)$ will be referred to as the Fourier projection of $F$ onto $U$.

We will write

$$B_n(c, R) = \{y \in \mathbb{R}^n : |c - y| < R\},$$

and we will drop the lower index $n$ whenever there is no confusion about the dimension of the ambient space. We will often denote by $B_R$ a ball with radius $R$ and arbitrary center.

For two quantities $A, B$, we denote by either $A \lesssim_{par} B$ or $A = O_{par}(B)$ the fact that the inequality $A \leq CB$ holds with an implicit constant $C$ depending on one or more parameters $par$ (most often a scale or $\epsilon$). Typically, the implicit constants will depend on the dimension $n$ of the Euclidean space, on the manifold we are investigating, and on the exponents of the Lebesgue spaces under consideration.

We will write $A \sim B$ if $A \lesssim B$ and $B \lesssim A$.

The slightly less rigorous notation $A \approx B$ will sometimes be used to indicate that $A$ and $B$ are essentially similar in size, if small, unimportant error terms are neglected.

We will use $\lesssim$ to denote logarithmic dependence of the implicit constant on a certain parameter $P \geq 1$ (typically a scale) that is clear from the context. Thus, $A \lesssim B$ will mean that $A \leq (\log P)^C B$ for some $C = O(1)$.

Also, $A \ll B$ and $B \gg A$ will refer to the fact that $A \leq cB$, for some sufficiently small absolute constant $c$.

Given a cube $Q$ or a ball $B$, we will denote by $CQ$ and $CB$ their dilation by a factor of $C$ around their centers.

We will use $|S|$ to denote the Lebesgue measure of a measurable set $S$, or its cardinality, if the set is finite. For a vector $\mathbf{v} \in \mathbb{R}^n$, we will write $|\mathbf{v}|$ to denote its length.

Given $S \subset \mathbb{R}^n$ (typically a ball or a cube), we write

$$\|F\|_{L^p_\sharp(S)} = \left( \frac{1}{|S|} \int_S |F|^p \right)^{\frac{1}{p}}.$$

For $d$ vectors $\mathbf{v}_1, \ldots, \mathbf{v}_d$ in $\mathbb{R}^n$, $d \leq n$, the quantity

$$|\mathbf{v}_1 \wedge \cdots \wedge \mathbf{v}_d|$$

is the $d$-dimensional volume of the parallelepiped determined by $\mathbf{v}_1, \ldots, \mathbf{v}_d$.

# 1

## Linear Restriction Theory

We will start by motivating part of the forthcoming investigation with a problem from partial differential equations (PDEs). Let $u(\bar{x},t)$ be the solution of the free Schrödinger equation with initial data $\phi$

$$\begin{cases} 2\pi i u_t(\bar{x},t) = \Delta_{\bar{x}} u(\bar{x},t), & (\bar{x},t) \in \mathbb{R}^{n-1} \times \mathbb{R} \\ u(\bar{x},0) = \phi(\bar{x}), & \bar{x} \in \mathbb{R}^{n-1}. \end{cases} \tag{1.1}$$

The solution $u(\bar{x},t)$ is obtained by applying the following operator $E$ to the function $f = \widehat{\phi}$

$$Ef(\bar{x},t) = \int_{\mathbb{R}^{n-1}} f(\xi) e(\xi \cdot \bar{x} + |\xi|^2 t) \, d\xi. \tag{1.2}$$

In the literature, $\bar{x}$ is often referred to as the spatial variable, $t$ stands for time, while $\xi$ is called frequency. We will almost always denote the time variable $t$ by $x_n$, and will only rarely treat it differently from the spatial variable $\bar{x}$. In fact, more often than not, the expression *spatial localization* will refer to the support of a function with respect to the $(\bar{x}, x_n)$ variable.

We will call $E$ the *extension operator*. This terminology will soon be motivated, once we realize that $E$ is the adjoint of a certain *restriction operator*. More generally, let $U$ be an open and bounded subset of $\mathbb{R}^d$, $1 \leq d \leq n - 1$ (typically a cube, ball, or annulus).

**Definition 1.1** Given a smooth function $\psi : U \to \mathbb{R}^{n-d}$, we define the $d$-dimensional manifold in $\mathbb{R}^n$

$$\mathcal{M} = \mathcal{M}^{\psi} = \{(\xi, \psi(\xi)) : \xi \in U\} \tag{1.3}$$

1

and its associated extension operator acting on functions $f \in L^1(U)$

$$E^\psi f(x) = E^{\mathcal{M}} f(x) = \int_U f(\xi) e(\bar{x} \cdot \xi + x^* \cdot \psi(\xi)) \, d\xi,$$

where $x = (\bar{x}, x^*) \in \mathbb{R}^d \times \mathbb{R}^{n-d}$.

For a subset $S \subset U$, we will denote $E^{\mathcal{M}}(f 1_S)$ by $E_S^{\mathcal{M}} f$.

Thus, the operator introduced in (1.2) corresponds to the case $\psi(\xi) = |\xi|^2$, which gives rise to the elliptic paraboloid. We will prefer to work with its truncated version

$$\mathbb{P}^{n-1} = \{(\xi_1, \dots, \xi_{n-1}, \xi_1^2 + \dots + \xi_{n-1}^2) \colon |\xi_i| < 1\}.$$

We will drop the superscript $\psi$ whenever we deal with $\psi(\xi) = |\xi|^2$ and will implicitly assume that the operator $E$ is associated with $\mathbb{P}^{n-1}$.

A significant part of this book will be concerned with hypersurfaces. Other than the paraboloid, examples of interest will include the hemispheres

$$\mathbb{S}_\pm^{n-1} = \left\{ \left( \xi, \pm\sqrt{1 - |\xi|^2} \right), \ |\xi| < 1 \right\},$$

the (truncated) cone

$$\mathbb{Cone}^{n-1} = \{(\xi, |\xi|) \colon 1 < |\xi| < 2\},$$

and the parabolic cylinder

$$\mathbb{Cyl}^{n-1} = \{(\xi_1, \dots, \xi_{n-1}, \xi_1^2 + \dots + \xi_{n-2}^2) \colon |\xi_i| < 1\}.$$

The sphere

$$\mathbb{S}^{n-1} = \{(\xi_1, \dots, \xi_n) \colon \xi_1^2 + \dots + \xi_n^2 = 1\}$$

is, strictly speaking, not of type (1.3). However, the topological distinction between the sphere and the hemispheres will bear no significance for the topics discussed in this book.

We have described the connection between the extension operator for $\mathbb{P}^{n-1}$ and the Schrödinger equation. A similar relation exists between the cone and the wave equation and also between the sphere and the Helmholtz equation.

Besides hypersurfaces, of particular interest are the low-dimensional manifolds – most notably the curves – due to their connection to the theory of exponential sums. Note for example that if $E^\psi$ is applied (somewhat informally) to a measure such as the sum of Dirac deltas $\sum_i \delta_{\xi_i}$ ($\xi_i \in U$), the result is an exponential sum

$$E^\psi \left( \sum_i \delta_{\xi_i} \right)(x) = \sum_i e((\xi_i, \psi(\xi_i)) \cdot x).$$

In the second part of the book, we will see that many counting problems related to systems of Diophantine equations can now be solved by embedding them into a more general (continuous) framework.

A major theme throughout our investigation will be to find various quantitative estimates for the extension operators associated with such manifolds. In addition to the motivation coming from PDEs and number theory, there is also considerable intrinsic interest in studying these operators. We will see that there is a wide range of approaches that will naturally lead us to explore fascinating connections to geometric measure theory and incidence geometry.

## 1.1 The Restriction Problem for Manifolds

We turn each manifold (1.3) into a measure space by equipping it with the induced Borel $\sigma$-algebra from $\mathbb{R}^n$ and with the pullback $d\sigma^{\psi}$ of the $d$-dimensional Lebesgue measure $d\xi$ from $\mathbb{R}^d$, under the canonical projection map. In the case of the hemispheres, this measure will occasionally be denoted by $d\sigma^{\mathbb{S}^{n-1}_{\pm}}$.

Given $f : \mathcal{M}^{\psi} \to \mathbb{C}$, we define $f_{\psi} : U \to \mathbb{C}$ via $f_{\psi}(\xi) = f(\xi, \psi(\xi))$. We thus have

$$\|f\|_{L^p(\mathcal{M}^{\psi})} = \|f_{\psi}\|_{L^p(U)}.$$

Note that when $\mathcal{M}^{\psi}$ is a hypersurface that is not a horizontal hyperplane, $d\sigma^{\psi}$ is not the same as the surface measure, which in turn is the pullback of $\sqrt{1 + |\nabla\psi|^2}\,d\xi$. However, the problems we will investigate are not sensitive to this difference.

**Definition 1.2** Given a manifold $\mathcal{M}^{\psi}$ we define its restriction operator acting on Schwartz functions $F \in \mathcal{S}(\mathbb{R}^n)$ as follows:

$$\mathcal{R}^{\psi} F = \widehat{F}|_{\mathcal{M}^{\psi}}.$$

Thus, $\mathcal{R}^{\psi} F$ is a function on $\mathcal{M}^{\psi}$, which coincides with $\widehat{F}$ on this space. Note that this definition makes in fact sense whenever $\widehat{F}$ is a continuous function, in particular for $F \in L^1(\mathbb{R}^n)$.

Let us also define the operator $\mathcal{R}^{\psi,*}$ acting on functions $f \in L^1(\mathcal{M}^{\psi})$ via

$$\mathcal{R}^{\psi,*} f(x) = E^{\psi} f_{\psi}(x).$$

As an application of Plancherel's identity, $\mathcal{R}^{\psi,*}$ can be seen to be a formal adjoint for $\mathcal{R}^{\psi}$, in the sense that

$$\int_{\mathbb{R}^n} F(x)\overline{\mathcal{R}^{\psi,*} f(x)}\,dx = \int_{\mathcal{M}^{\psi}} (\mathcal{R}^{\psi} F(\xi))\bar{f}(\xi)\,d\sigma^{\psi}(\xi).$$

**Definition 1.3** Given $1 \leq p, q \leq \infty$, we will say that the restriction estimate $R_{\mathcal{M}^\psi}(p \mapsto q)$ holds if the operator $\mathcal{R}^\psi$ is bounded from $\mathcal{S}(\mathbb{R}^n)$ equipped with the $L^p$ norm to $L^q(\mathcal{M}^\psi, d\sigma^\psi)$. For example, when $q < \infty$ this amounts to

$$\left( \int_U |\widehat{F}(\xi, \psi(\xi))|^q \, d\xi \right)^{\frac{1}{q}} \lesssim \|F\|_{L^p(\mathbb{R}^n)},$$

with implicit constant independent of $F \in \mathcal{S}(\mathbb{R}^n)$.

We will say that $R^*_{\mathcal{M}^\psi}(q' \mapsto p')$ holds if the operator $\mathcal{R}^{\psi,*}$ is bounded from $L^{q'}(\mathcal{M}^\psi, d\sigma^\psi)$ to $L^{p'}(\mathbb{R}^n)$. When $p > 1$, this can be written as follows:

$$\|E^\psi f\|_{L^{p'}(\mathbb{R}^n)} \lesssim \|f\|_{L^{q'}(U)},$$

with implicit constant independent of $f \in L^{q'}(U)$.

Recall that we have assumed $U$ to be bounded. This will guarantee the inclusion $L^{q'}(U) \subset L^1(U)$, and thus the fact that $E^\psi f$ is well defined for $f \in L^{q'}(U)$. Write

$$\|\mathcal{R}^\psi\|_{p \mapsto q} := \sup_{\substack{F \in \mathcal{S}(\mathbb{R}^n): \, \|F\|_p = 1 \\ \|g\|_{L^{q'}(U)} = 1}} \left| \int_U \mathcal{R}^\psi F(\xi, \psi(\xi)) \bar{g}(\xi) \, d\xi \right|.$$

It is easy to check that $\|\mathcal{R}^\psi\|_{p \mapsto q} = \|\mathcal{R}^{\psi,*}\|_{q' \mapsto p'}$, and thus

$$R_{\mathcal{M}^\psi}(p \mapsto q) \iff R^*_{\mathcal{M}^\psi}(q' \mapsto p').$$

We will refer to either of these as *linear restriction estimates*, or simply as *restriction estimates*.

It is easy to check that if $\mathcal{M}_2$ is a nonsingular affine image of $\mathcal{M}_1$, then the two manifolds obey identical restriction estimates. For example, this is the case with $\mathbb{S}^{n-1}_+$ and $\mathbb{S}^{n-1}_-$. The restriction estimates $R_{\mathbb{S}^{n-1}}(p \mapsto q)$, $R^*_{\mathbb{S}^{n-1}}$ $(q' \mapsto p')$ for the sphere $\mathbb{S}^{n-1}$ will be understood as the corresponding restriction estimates for the upper (or lower) hemisphere.

Let us take note of the identity

$$E^\psi f(x) = \widehat{f^\psi d\sigma^\psi}(-x),$$

where $f^\psi(\xi_1, \ldots, \xi_n) = f(\xi_1, \ldots, \xi_d)$. Since the measure $f^\psi d\sigma^\psi$ is singular with respect to the Lebesgue measure, this identity shows that, unless $f$ is identically equal to zero, the function $E^\psi f$ is never in any $L^s(\mathbb{R}^n)$ with $1 \leq s \leq 2$. Thus, there cannot be any restriction estimate $R_{\mathcal{M}^\psi}(p \mapsto q)$ when $p \geq 2$. We will unravel a rather different story for smaller values of $p$.

The following example describes the most elementary and universal restriction estimate.

**Example 1.4** The Fourier transform of each $F \in L^1(\mathbb{R}^n)$ is continuous, and moreover, we have a global estimate $\|\widehat{F}\|_{L^\infty(\mathbb{R}^n)} \leq \|F\|_{L^1(\mathbb{R}^n)}$. This implies that $R_{\mathcal{M}^\psi}(1 \mapsto q)$ holds for each manifold $\mathcal{M}^\psi$ and each $1 \leq q \leq \infty$.

The next example shows that there is no nontrivial restriction estimate for completely flat manifolds.

**Example 1.5** Consider the extension operator $E^\psi$ associated with a $d$-dimensional affine space in $\mathbb{R}^n$. We choose $\psi \equiv 0$, but this example can be adapted to the case of an arbitrary affine map. Then for each $g \in L^1(U)$ and each $(\bar{x}, x^*) \in \mathbb{R}^d \times \mathbb{R}^{n-d}$, we have $E^\psi g(\bar{x}, x^*) = \widehat{g}(-\bar{x})$. Note that $E^\psi g \notin L^{p'}(\mathbb{R}^n)$, if $p > 1$ and $g \not\equiv 0$. We conclude that no estimate $R_{\mathcal{M}^\psi}(p \mapsto q)$ holds if $p > 1$.

If $p > 1$, the Fourier transform of an $L^p$ function is measurable, but not always continuous. In particular, its values may not be well defined on the Lebesgue null set $\mathcal{M}^\psi$. The restriction estimates $R_{\mathcal{M}^\psi}(p \mapsto q)$ quantify the extent to which the Fourier transform $\widehat{F}$ can be given a meaning when restricted to arbitrary translates of the manifold $\mathcal{M}^\psi$. We will make this more precise as follows. For each $\mathbf{v} \in \mathbb{R}^n$, let us equip the set $\mathcal{M}^{\psi, \mathbf{v}} := \mathbf{v} + \mathcal{M}^\psi$ with the translate $d\sigma^{\psi, \mathbf{v}}$ of the measure $d\sigma^\psi$.

**Proposition 1.6** *Assume that $p < 2$. If $R_{\mathcal{M}^\psi}(p \mapsto q)$ holds, then for each $F \in L^p(\mathbb{R}^n)$ there is an (everywhere defined) function $G_F \colon \mathbb{R}^n \to \mathbb{C}$, which is Borel measurable, so that $\widehat{F}$ and $G_F$ coincide almost everywhere with respect to the Lebesgue measure, and moreover, for each $\mathbf{v} \in \mathbb{R}^n$, we have*

$$\|G_F\|_{L^q(\mathcal{M}^{\psi, \mathbf{v}}, d\sigma^{\psi, \mathbf{v}})} \lesssim \|F\|_{L^p(\mathbb{R}^n)}, \tag{1.4}$$

*with an implicit constant independent of $F$ and $\mathbf{v}$.*

*Proof* We start by noting that if $G_F$ is Borel measurable, the restriction $G_F|_{\mathcal{M}^{\psi, \mathbf{v}}}$ will be measurable with respect to the Borel $\sigma$-algebra on $\mathcal{M}^{\psi, \mathbf{v}}$. As a result, the quantity $\|G_F\|_{L^q(\mathcal{M}^{\psi, \mathbf{v}}, d\sigma^{\psi, \mathbf{v}})}$ is well defined.

For $F \in L^p(\mathbb{R}^n)$, we let $F_m$ be an arbitrary sequence of Schwartz functions such that $\lim_{m \to \infty} \|F - F_m\|_{L^p(\mathbb{R}^n)} = 0$. By invoking the Hausdorff–Young inequality, we find that $\lim_{m \to \infty} \|\widehat{F} - \widehat{F_m}\|_{L^{p'}(\mathbb{R}^n)} = 0$. This will allow us to extract a subsequence of $F_m$, which for reasons of simplicity will also be denoted by $F_m$, such that $\lim_{m \to \infty} \widehat{F_m}(x) = \widehat{F}(x)$ for a.e. $x \in \mathbb{R}^n$. Let $S$ be the set of all $x \in \mathbb{R}^n$ for which $\widehat{F_m}(x)$ converges. A standard exercise shows that $S$ is a Borel set. Define $G_F(x) := \lim_{m \to \infty} \widehat{F_m}(x)$ when $x \in S$, and $G_F(x) := 0$ when $x \notin S$.

The only thing left to be checked is (1.4). Fix $\mathbf{v}$. Note that

$$\|\widehat{F_m}\|_{L^q(\mathcal{M}^{\psi, \mathbf{v}}, d\sigma^{\psi, \mathbf{v}})} = \|\widehat{F_{m, \mathbf{v}}}\|_{L^q(\mathcal{M}^\psi, d\sigma^\psi)}$$

where

$$F_{m,\mathbf{v}}(\xi) = F_m(\xi)e(-\mathbf{v} \cdot \xi).$$

Thus, an application of our hypothesis $R_{\mathcal{M}^\psi}(p \mapsto q)$ combined with Fatou's lemma (or a variant of it for $q = \infty$) leads to

$$\begin{aligned}
\|G_F\|_{L^q(\mathcal{M}^{\psi,\mathbf{v}},d\sigma^{\psi,\mathbf{v}})} &\leq \left\|\liminf_{m\to\infty} |\widehat{F_m}|\right\|_{L^q(\mathcal{M}^{\psi,\mathbf{v}},d\sigma^{\psi,\mathbf{v}})} \\
&\leq \liminf_{m\to\infty} \left\|\widehat{F_m}\right\|_{L^q(\mathcal{M}^{\psi,\mathbf{v}},d\sigma^{\psi,\mathbf{v}})} \\
&= \liminf_{m\to\infty} \left\|\widehat{F_{m,\mathbf{v}}}\right\|_{L^q(\mathcal{M}^\psi,d\sigma^\psi)} \\
&\lesssim \liminf_{m\to\infty} \|F_{m,\mathbf{v}}\|_{L^p(\mathbb{R}^n)} \\
&= \liminf_{m\to\infty} \|F_m\|_{L^p(\mathbb{R}^n)} = \|F\|_{L^p(\mathbb{R}^n)}. \qquad \square
\end{aligned}$$

Throughout this book, we will see that there are plenty of restriction estimates in the range $1 < p < 2$, assuming $\mathcal{M}^\psi$ is sufficiently curved. Whenever this happens, Proposition 1.6 states the rather nonintuitive fact that (the merely measurable) $\widehat{F}$ can be modified on a set of zero Lebesgue measure, in such a way that (1.4) holds on each translate of $\mathcal{M}^\psi$ (also Lebesgue null). The reader may consult [86] for a different angle on the significance of restriction estimates.

## 1.2 The Restriction Conjecture for the Sphere and the Paraboloid

Our main focus throughout this book will be the restriction problem for the sphere $\mathbb{S}^{n-1}$ and the paraboloid $\mathbb{P}^{n-1}$. The methods used to study these manifolds are very similar. The relevant common feature is the fact that they have positive definite second fundamental form. However, we will see that from a technical point of view it is a bit easier to work with the paraboloid, since it is well behaved under rescaling.

The following conjecture records the expected range of estimates for the restriction operator.

**Conjecture 1.7** (Restriction Conjecture for $\mathbb{S}^{n-1}$ and $\mathbb{P}^{n-1}$) *The estimates* $R_{\mathbb{S}^{n-1}}(p \mapsto q)$ *and* $R_{\mathbb{P}^{n-1}}(p \mapsto q)$ *hold if and only if*

$$\frac{n-1}{q} \geq \frac{n+1}{p'} \tag{1.5}$$

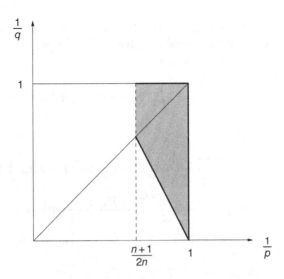

Figure 1.1 Restriction range for $\mathbb{S}^{n-1}$ and $\mathbb{P}^{n-1}$.

*and*

$$p' > \frac{2n}{n-1}. \tag{1.6}$$

The expected range of estimates is depicted in Figure 1.1, using a $(1/p, 1/q)$ coordinate system. It consists of the closure of the trapezoid with vertices $((n+1)/2n, (n+1)/2n), ((n+1)/2n, 1), (1,1), (1,0)$, minus the edge containing the points with first coordinate $(n+1)/2n$.

Conjecture 1.7 was first proved for $\mathbb{S}^1$ in [**49**] and [**128**]. In Section 1.3, we will present a proof for general curves in $\mathbb{R}^2$ with nonzero curvature (this includes the circle and the parabola). Two additional arguments for $\mathbb{P}^1$ will be discussed in Corollary 4.7 and Remark 5.28. The case $n \geq 3$ of the conjecture is still open, though a lot of progress has been made in recent years.

Let us explain why restrictions (1.5) and (1.6) are necessary. The first one applies to arbitrary hypersurfaces $\mathcal{M}^{\psi}$ and has behind it the so-called Knapp example.

**Example 1.8** (Knapp example) Consider a hypersurface as in (1.3). Let $\Upsilon \not\equiv 0$ be a nonnegative Schwartz function supported on $[-1,1]^{n-1}$. Given a cube $\omega \subset U$ with side length $R^{-1/2}$ and center $c_\omega$ and a cube $q \subset \mathbb{R}^{n-1}$ with side length $R^{1/2}$ and center $c_q$, we let

$$\Upsilon_{q,\omega}(\xi) = R^{\frac{n-1}{2}} \Upsilon(R^{1/2}(\xi - c_\omega))e(-c_q \cdot (c_\omega - \xi)).$$

While working with $c_q = c_\omega = 0$ would certainly suffice here, we will carry out the computations in the general case, for future reference.

A simple computation reveals that

$$|E^\psi \Upsilon_{q,\omega}(x)| = \left| \int \Upsilon(\eta) e(\varphi_{x,q,\omega}(\eta)) \, d\eta \right|,$$

where for $x = (\bar{x}, x_n) \in \mathbb{R}^n$ we write

$$\varphi_{x,q,\omega}(\eta) = \eta \cdot \frac{(\bar{x} - c_q)}{R^{1/2}} + \left( \psi\left(\frac{\eta}{R^{1/2}} + c_\omega\right) - \psi(c_\omega) \right) x_n$$

$$= \eta \cdot \frac{(\bar{x} - c_q) + \nabla\psi(c_\omega)x_n}{R^{1/2}} + \frac{x_n}{R} O(|\eta|^2).$$

Note that $\sup_{\eta \in \mathrm{supp}\,(\Upsilon)} |\varphi_{x,q,\omega}(\eta)| \ll 1$ whenever $x$ is in the tube

$$\{(\bar{x}, x_n) \colon |(\bar{x} - c_q) + \nabla\psi(c_\omega)x_n| \ll R^{1/2}, \ |x_n| \ll R\}.$$

Writing

$$\left| \int \Upsilon(\eta) e(\varphi_{x,q,\omega}(\eta)) \, d\eta - \int \Upsilon \right| \le \int \Upsilon(\eta) |e(\varphi_{x,q,\omega}(\eta)) - 1| \, d\eta$$

$$\ll \int \Upsilon$$

shows that for each $x$ in this tube, we have

$$|E^\psi \Upsilon_{q,\omega}(x)| \sim \int \Upsilon. \tag{1.7}$$

Since the tube has volume $\sim R^{(n+1)/2}$, we conclude that for each $s$

$$\|E^\psi \Upsilon_{q,\omega}\|_{L^s(\mathbb{R}^n)} \gtrsim R^{\frac{n+1}{2s}}.$$

On the other hand,

$$\|\Upsilon_{q,\omega}\|_{L^s(\mathbb{R}^{n-1})} \sim R^{\frac{n-1}{2s'}}.$$

Testing the hypothesis $R^*_{\mathcal{M}^\psi}(q' \mapsto p')$ with this example leads to (1.5).

Let us now explain the requirement (1.6) for $\mathbb{S}^{n-1}$. A similar argument would apply to $\mathbb{P}^{n-1}$. Exercises 1.12 and 6.31(b) explore simpler constructions that show the necessity of the slightly weaker restriction $p' \ge 2n/(n-1)$ for arbitrary hypersurfaces $\mathcal{M}^\psi$. Also, the correct restriction estimate at the endpoint $p = 2n/(n+1)$ is discussed in Exercise 1.32.

**Example 1.9** (Fourier transform of full surface measure) Let $d\sigma_\pm$ be the surface measure on the hemisphere $\mathbb{S}^{n-1}_\pm$, more precisely the pullback of the weighted Lebesgue measure $d\xi/\sqrt{1-|\xi|^2}$ from $B_{n-1}(0,1)$. Then $d\sigma := d\sigma_+ + d\sigma_-$ is the surface measure on $\mathbb{S}^{n-1}$. A standard oscillatory integral estimate (see for example corollary 6.7 from *[117]*) shows that

$$\widehat{d\sigma}(x) = C|x|^{-\frac{n-1}{2}} \cos\left(2\pi\left(|x| - \frac{n-1}{8}\right)\right) + O\left(|x|^{-\frac{n+1}{2}}\right). \tag{1.8}$$

Note that $\|\widehat{d\sigma}\|_{L^{p'}(\mathbb{R}^n)} = \infty$ whenever $p' \le 2n/(n-1)$. We prove that the same holds for $\widehat{d\sigma^{\mathbb{S}^{n-1}_\pm}}$. Let $\theta_1,\ldots,\theta_m : \mathbb{S}^{n-1} \to [0,\infty)$ be bounded functions on $\mathbb{S}^{n-1}$ such that

$$\theta_1 + \cdots + \theta_m = 1_{\mathbb{S}^{n-1}}$$

and each supp$(\theta_i)$ fits inside the interior of $\mathbb{S}^{n-1}_+$, after a possible rotation. It follows that for each $p' \le 2n/(n-1)$ there is some $1 \le i \le m$ such that $\|\widehat{\theta_i d\sigma}\|_{L^{p'}(\mathbb{R}^n)} = \infty$. By applying a rotation, we may assume that supp$(\theta_i)$ lies inside $\mathbb{S}^{n-1}_+$. Thus, with $\psi(\xi) = \sqrt{1-|\xi|^2}$ and $f(\xi) = \theta_{i,\psi}(\xi)/\sqrt{1-|\xi|^2}1_{U'}(\xi)$ (see the definition of $\theta_{i,\psi}$ at the beginning of the previous section), we have

$$\widehat{\theta_i d\sigma}(x) = \int_{U'} e((\xi,\psi(\xi))\cdot x)\frac{\theta_{i,\psi}(\xi)}{\sqrt{1-|\xi|^2}}\,d\xi$$
$$= E^\psi f(x),$$

for some $U'$ whose closure lies inside the open ball $B_{n-1}(0,1)$. Since $f$ is in $L^{q'}(\mathbb{R}^{n-1})$ for each $1 \le q' \le \infty$, the restriction (1.6) follows.

Conjecture 1.7 has an equivalent formulation, which we will prefer in our investigation. A detailed presentation of the equivalence between the two formulations of the Restriction Conjecture for $\mathbb{S}^{n-1}$ appears in section 19.3 of **[83]**.

**Conjecture 1.10** (Restriction Conjecture for $\mathbb{S}^{n-1}$ and $\mathbb{P}^{n-1}$) *The restriction estimates*

$$R^*_{\mathbb{S}^{n-1}}(\infty \mapsto p'), \quad R^*_{\mathbb{P}^{n-1}}(\infty \mapsto p')$$

*hold for each $p' > 2n/(n-1)$.*

The range of the restriction estimates for a hypersurface is sensitive to the number of its zero principal curvatures. We illustrate this by indicating that the expected range for the cone $\mathbb{C}\text{one}^{n-1}$ is the same as the one for the sphere $\mathbb{S}^{n-2}$

of smaller dimension. The range of restriction estimates for conical surfaces with more zero principal curvatures will experience further drops in dimension.

**Conjecture 1.11** (Restriction Conjecture for $\mathbb{C}\text{one}^{n-1}$) *The estimate* $R_{\mathbb{C}\text{one}^{n-1}}(p \mapsto q)$ *holds if and only if*

$$\frac{n-2}{q} \geq \frac{n}{p'}$$

*and*

$$p' > \frac{2(n-1)}{n-2}.$$

It was proved in [87] that Conjecture 1.7 for either $\mathbb{S}^{n-2}$ or $\mathbb{P}^{n-2}$ implies Conjecture 1.11 for $\mathbb{C}\text{one}^{n-1}$. The reverse implication is not known. In fact, the Restriction Conjecture for the cone has been verified not only for $n = 3$ (Barcelo [3]), but also for $n = 4$ (Wolff [118]) and $n = 5$ (Ou–Wang [88]). See Exercise 4.8.

**Exercise 1.12** Let $\mathcal{M}^\psi$ be a hypersurface as in (1.3). Show that $\|\widehat{d\sigma^\psi}\|_{L^{p'}(\mathbb{R}^n)} = \infty$ for each $p' < 2n/(n-1)$. Conclude that $R^*_{\mathcal{M}^\psi}(q' \to p')$ cannot hold if $p' < 2n/(n-1)$.

Hint: Choose an appropriate $\phi \in \mathcal{S}(\mathbb{R}^n)$ and let $\phi_R(\xi) = R^n \phi(R\xi)$. Show that

$$\|(d\sigma^\psi) * \phi_R\|_{L^2(\mathbb{R}^n)} \sim R^{\frac{1}{2}}.$$

Then use Plancherel's identity and Hölder's inequality.

**Exercise 1.13** Prove the "only if" part of Conjecture 1.11.
Hint: Refine the Knapp example with the cube $\omega$ replaced with an annular sector of length $\sim 1$ and angular width $\sim R^{-1/2}$.

## 1.3 Proof of the Restriction Conjecture for Curves in $\mathbb{R}^2$

Let $\phi \colon [-1, 1] \to \mathbb{R}$ be a $C^2$ function with $\inf_{|\xi| \leq 1} |\phi''(\xi)| \geq \nu > 0$. For $f \colon [-1, 1] \to \mathbb{C}$, we let

$$E^\phi f(x_1, x_2) = \int f(\xi) e(x_1 \xi + x_2 \phi(\xi)) \, d\xi$$

be the extension operator associated with the curve $(\xi, \phi(\xi))$. In this section, we verify Conjecture 1.7 for all these curves, in particular for $\mathbb{P}^1$ and $\mathbb{S}^1$. The argument we present is from [68].

**Theorem 1.14** *We have*

$$\|E^\phi f\|_{L^q(\mathbb{R}^2)} \lesssim_\nu \|f\|_{L^r([-1,1])}$$

*for each $q > 4$ and $3/q + 1/r \leq 1$.*

*Proof* It suffices to verify the inequality with $3/q + 1/r = 1$. Let $1 \leq p < 2$ with $2p' = q$ and $2p/(3-p) = r$. We first write

$$\|E^\phi f\|_{L^q(\mathbb{R}^2)}^2 = \|E^\phi f E^\phi f\|_{L^{p'}(\mathbb{R}^2)}.$$

Using the change of variables $(t,s) = T(\xi_1, \xi_2)$ with $(\xi_1, \xi_2) = (t+s, \phi(t) + \phi(s))$, we may write

$$(E^\phi f)^2(x_1, x_2) = 2 \iint_{t>s} f(t)f(s)e(x_1(t+s) + x_2(\phi(t) + \phi(s))) \, dt \, ds$$

$$= \iint (f \otimes f)(T(\xi_1, \xi_2)) |\det[T'(\xi_1, \xi_2)]| e(x_1\xi_1 + x_2\xi_2) \, d\xi_1 d\xi_2.$$

This expression is the Fourier transform of the function

$$F(\xi_1, \xi_2) = (f \otimes f)(T(\xi_1, \xi_2)) |\det[T'(\xi_1, \xi_2)]|.$$

Using the Hausdorff–Young inequality, we find

$$\|(E^\phi f)^2\|_{L^{p'}(\mathbb{R}^2)} \lesssim \|F\|_{L^p(\mathbb{R}^2)}.$$

It thus remains to prove that

$$\|F\|_{L^p(\mathbb{R}^2)} \lesssim_\nu \|f\|_{L^r([-1,1])}^2. \tag{1.9}$$

Changing back to the variables $t, s$, then using that $|\phi'(t) - \phi'(s)| \geq \nu|t-s|$ and Hölder's inequality we get

$$\|F\|_{L^p(\mathbb{R}^2)}^p = \iint |f(t)f(s)|^p \frac{1}{|\phi'(t) - \phi'(s)|^{p-1}} \, dt \, ds$$

$$\lesssim_\nu \iint |f(t)f(s)|^p \frac{1}{|t-s|^{p-1}} \, dt \, ds$$

$$\leq \| |f|^p \|_{\frac{r}{p}} \left\| |f|^p * \frac{1}{|t|^{p-1}} \right\|_{(\frac{r}{p})'}.$$

The Hardy–Littlewood–Sobolev inequality on fractional integration (section 8.4 in [**102**]) asserts that

$$\left\| |f|^p * \frac{1}{|t|^{p-1}} \right\|_{(\frac{r}{p})'} \lesssim \| |f|^p \|_{\frac{r}{p}}.$$

The proof of (1.9) now follows by combining the last two inequalities.    $\square$

## 1.4 The Stein–Tomas Argument

Consider a hypersurface as in (1.3) and let us assume for a moment that $\nabla \psi(\xi_0) = 0$, for some $\xi_0 \in U$. The eigenvalues of the $(n-1) \times (n-1)$ matrix $(\partial^2 \psi / \partial \xi_i \partial \xi_j)(\xi_0)$ are called the *principal curvatures* of $\mathcal{M}^\psi$ at $\xi_0$. Their product is called the *Gaussian curvature* of $\mathcal{M}^\psi$ at $\xi_0$. These quantities can be defined at any other point $\xi_0 \in U$, by rewriting $\mathcal{M}^\psi$ as the graph of a function with zero gradient at $\xi_0$.

We record the following fundamental result; see theorem 1, section 8.3.1 from [**102**]. This decay rate is sharp for the sphere as shown by (1.8).

**Theorem 1.15** *Assume that $w$ is in $C^\infty(U)$, and compactly supported on $U$. If $\psi \in C^\infty(U)$ and $\mathcal{M}^\psi$ has nonzero Gaussian curvature at every point, then*

$$|\widehat{d\mu}(x)| \lesssim_{\psi,w} (1 + |x|)^{-\frac{n-1}{2}},$$

*where $d\mu$ is the pullback to $\mathcal{M}^\psi$ of the measure $w(\xi)\,d\xi$ from $\mathbb{R}^{n-1}$.*

The following theorem represents the first significant progress on the Restriction Conjecture in arbitrary dimension. It was first proved in [**115**] for the sphere $\mathbb{S}^{n-1}$ by Tomas, except for the endpoint $p = 2(n+1)/(n+3)$. This result followed some unpublished partial results by Stein from the late 1960s. It was Stein who also proved the endpoint result in [**103**]. Thus, the estimate $R_{\mathbb{S}^{n-1}}(p \mapsto 2)$ is called the Stein–Tomas theorem. A similar result has been proved by Strichartz [**105**] in the case of the paraboloid. See also Exercise 1.24. The version we present appears in [**53**]. Example 1.8 shows that the range $1 \le p \le 2(n+1)/(n+3)$ is sharp.

**Theorem 1.16** *Let $U \subset \mathbb{R}^{n-1}$ be a ball or a cube. Assume that $\psi : U \to \mathbb{R}$ has an extension $\tilde{\psi}$ to $2U$, which is in $C^\infty(2U)$, and moreover that $\mathcal{M}^{\tilde{\psi}}$ has nonzero Gaussian curvature at every point inside $2U$. Then $R_{\mathcal{M}^\psi}(p \mapsto 2)$ holds for each $1 \le p \le 2(n+1)/(n+3)$.*

*Proof* We only present the proof in the range $1 \le p < 2(n+1)/(n+3)$. The endpoint is more delicate; it involves complex interpolation. There is an easier alternative argument for $\mathbb{S}^2$, which we will present at the end of this section.

The proof will rely on the classical $T^*T$ method. This method exploits the fact that the square root of the norm of the operator $T^*T$ equals the norm of $T$, the former quantity often being easier to evaluate. The operator $T^*T$ is well defined in our case, since $T$ takes values in Hilbert space.

We start the proof with a technicality that is enforced by the presence of the cut-off $w$ in the proof of Theorem 1.15. Let $w \in C^\infty(\mathbb{R}^n)$ be such that

$1_U \leq w \leq 1_{2U}$. We equip $\mathcal{M}^{\tilde{\psi}}$ with the measure $\widetilde{d\sigma}^{\tilde{\psi}}$ equal to the pullback of $w(\xi)\,d\xi$ from $\mathbb{R}^{n-1}$. Note that it suffices to prove that

$$\|\mathcal{R}^{\tilde{\psi}}\|_{L^p(\mathbb{R}^n,dx)\mapsto L^2(\mathcal{M}^{\tilde{\psi}},\widetilde{d\sigma}^{\tilde{\psi}})} < \infty.$$

The adjoint of $\mathcal{R}^{\tilde{\psi}}$ is easily seen to be the operator defined for $f: \mathcal{M}^{\tilde{\psi}} \to \mathbb{C}$ via

$$\widetilde{\mathcal{R}^{\tilde{\psi},*}} f(x) = \int_{2U} f(\xi, \tilde{\psi}(\xi)) w(\xi) e((\xi, \tilde{\psi}(\xi)) \cdot x)\,d\xi.$$

It suffices to show that $\|\widetilde{\mathcal{R}^{\tilde{\psi},*}\mathcal{R}^{\tilde{\psi}}}\|_{L^p(\mathbb{R}^n)\mapsto L^{p'}(\mathbb{R}^n)} < \infty$. A simple computation reveals that

$$\widetilde{\mathcal{R}^{\tilde{\psi},*}\mathcal{R}^{\tilde{\psi}}} F(x) = \int_{2U} \widehat{F}(\xi, \tilde{\psi}(\xi)) w(\xi) e((\xi, \tilde{\psi}(\xi)) \cdot x)\,d\xi$$
$$= F * \widehat{d\mu}(x),$$

where $d\mu$ is the image of $\widetilde{d\sigma}^{\tilde{\psi}}$ under the reflection map $\mathbf{v} \mapsto -\mathbf{v}$ in $\mathbb{R}^n$. We will see that the only properties of a measure $d\mu$ on $\mathbb{R}^n$ that we need in order to prove the estimate

$$\|F * \widehat{d\mu}\|_{L^{p'}(\mathbb{R}^n)} \lesssim \|F\|_{L^p(\mathbb{R}^n)} \tag{1.10}$$

are

$$d\mu(B(\xi,r)) \lesssim r^{n-1}, \text{ uniformly over } \xi \in \mathbb{R}^n, r > 0 \tag{1.11}$$

and

$$|\widehat{d\mu}(x)| \lesssim (1 + |x|)^{-\frac{n-1}{2}}. \tag{1.12}$$

Our measure satisfies (1.11) since it is supported on a smooth hypersurface, and it satisfies (1.12) due to Theorem 1.15. Let us first note that (1.12) guarantees that $\widehat{d\mu} \in L^q(\mathbb{R}^n)$ for $q > 2n/(n-1)$. Using this and Young's inequality implies that (1.10) holds for $p' > 4n/(n-1)$. To bridge the gap between $2(n+1)/(n-1)$ and $4n/(n-1)$ we will also need to use (1.11).

Write 1 as a sum of $C^\infty$ functions

$$\sum_{j\geq 0} \phi_j \equiv 1,$$

with $\phi_0$ supported near 0 and $\phi_j$ supported on $\{x \in \mathbb{R}^n : 2^{j-1} < |x| < 2^{j+1}\}$ for $j \geq 1$. Moreover, we may arrange that $\phi_j(x) = \varphi(x/2^j)$ for $j \geq 1$, with $\varphi \in C^\infty$.

Write

$$K_j = \phi_j \widehat{d\mu}.$$

Since $\|K_j\|_\infty \lesssim 2^{-j(n-1)/2}$, an application of Young's inequality shows that

$$\|F * K_j\|_\infty \lesssim 2^{-\frac{j(n-1)}{2}} \|F\|_1. \tag{1.13}$$

A simple computation involving (1.11) shows that

$$\|\widehat{K_j}\|_\infty \lesssim 2^j.$$

This leads to the estimate

$$\|F * K_j\|_2 = \|\widehat{F}\widehat{K_j}\|_2 \lesssim 2^j \|F\|_2. \tag{1.14}$$

Interpolating (1.13) and (1.14) leads to the following estimate, uniform over $j \geq 0$

$$\|F * K_j\|_{L^{p'}} \lesssim 2^{-\epsilon_p j} \|F\|_{L^p},$$

for some $\epsilon_p > 0$, whenever $p < 2(n + 1)/(n + 3)$. Summing this over $j$ and using the triangle inequality leads to the desired conclusion. $\qquad\square$

Note that $\mathbb{P}^{n-1}$ satisfies the requirement in Theorem 1.16. The sphere $\mathbb{S}^{n-1}$ will be discussed in Exercise 1.23.

**Corollary 1.17** (Strichartz estimate, [**105**]) *Assume that $u$ solves the Schrödinger equation (1.1) with initial data $\phi$ compactly supported in frequency. Then for each $p \geq 2(n + 1)/(n - 1)$, we have*

$$\|u\|_{L^p(\mathbb{R}^n)} \lesssim \|\phi\|_{L^2(\mathbb{R}^{n-1})}.$$

The argument from Theorem 1.16 can be easily extended to produce the following generalization, in the range $p < 2(n + \beta - \alpha)/(2n + \beta - 2\alpha)$. The endpoint result was proved in [**2**]. This paper also contains a nice introduction describing the various contributors to this problem.

**Theorem 1.18** *Let $d\mu$ be a finite, compactly supported Borel measure on $\mathbb{R}^n$ satisfying*

$$d\mu(B(\xi, r)) \lesssim r^\alpha, \text{ uniformly over } \xi \in \mathbb{R}^n, r > 0$$

*and*

$$|\widehat{d\mu}(x)| \lesssim (1 + |x|)^{-\beta},$$

*for some $0 < \alpha < n$ and $\beta > 0$. Then*

$$\|F * \widehat{d\mu}\|_{L^{p'}(\mathbb{R}^n)} \lesssim \|F\|_{L^p(\mathbb{R}^n)}$$

*for $1 \leq p \leq 2(n + \beta - \alpha)/(2n + \beta - 2\alpha)$.*

We will next present a proof of the endpoint Stein–Tomas estimate $R^*_{\mathbb{S}^2}(2 \mapsto 4)$ for the two-dimensional sphere. This is different from the original proof of Stein. It does not rely on the estimate (1.12) for the decay of the Fourier transform of the surface measure $\widehat{d\sigma}$. Instead, it uses information about the convolution $d\sigma * d\sigma$, which can be obtained using purely geometric considerations. The reader may consult [50] for an application of this type of argument.

Recall that given two positive, finite Borel measures $dv_1, dv_2$ on $\mathbb{R}^n$, the convolution $dv_1 * dv_2$ is a Borel measure, whose value on a Borel set $A$ is given by

$$dv_1 * dv_2(A) = \int_{\mathbb{R}^n} \int_{\mathbb{R}^n} 1_A(\xi + \eta) \, dv_1(\xi) \, dv_2(\eta).$$

**Proposition 1.19** *Let $d\mu$ be a positive, finite Borel measure on $\mathbb{R}^n$ such that the measure $d\mu * d\mu$ is absolutely continuous with respect to the Lebesgue measure, $d\mu * d\mu = F d\xi$, and moreover $F \in L^\infty(\mathbb{R}^n, d\xi)$. Then we have the estimate*

$$\|\widehat{g \, d\mu}\|_{L^4(\mathbb{R}^n)} \lesssim \|g\|_{L^2(d\mu)}.$$

*Proof* Let $\phi$ be a smooth, positive function on $\mathbb{R}^n$ with integral equal to 1. Define $\phi_\epsilon(\xi) = \epsilon^{-n}\phi(\xi/\epsilon)$ and

$$\mu_\epsilon(\xi) := \phi_\epsilon * d\mu(\xi) = \int \phi_\epsilon(\xi - \eta) \, d\mu(\eta).$$

An application of the Fubini–Tonelli theorem shows that $\mu_\epsilon \in L^1(\mathbb{R}^n, d\xi)$ and moreover $\mu_\epsilon d\xi$ converges weakly to $d\mu$ as $\epsilon \to 0$. It will thus suffice to prove the following estimate for all $G \in L^2(\mathbb{R}^n, d\xi)$, with implicit constant independent of $\epsilon$

$$\int_{\mathbb{R}^n} |\widehat{G\mu_\epsilon}|^4 \lesssim \left( \int_{\mathbb{R}^n} |G|^2 \mu_\epsilon \right)^2.$$

The left-hand side is

$$= \int_{\mathbb{R}^n} |(G\mu_\epsilon) * (G\mu_\epsilon)|^2$$

$$= \int_{\mathbb{R}^n} \left| \int_{\mathbb{R}^n} G(\xi - \eta) G(\eta) \mu_\epsilon(\xi - \eta) \mu_\epsilon(\eta) \, d\eta \right|^2 d\xi$$

$$\leq \int_{\mathbb{R}^n} \left[ \int_{\mathbb{R}^n} |G|^2(\xi - \eta)|G|^2(\eta)\mu_\epsilon(\xi - \eta)\mu_\epsilon(\eta)\, d\eta \right]$$

$$\times \left[ \int_{\mathbb{R}^n} \mu_\epsilon(\xi - \eta)\mu_\epsilon(\eta)\, d\eta \right] d\xi$$

$$\leq \|\mu_\epsilon * \mu_\epsilon\|_{L^\infty(\mathbb{R}^n)} \int_{\mathbb{R}^n} \int_{\mathbb{R}^n} |G|^2(\xi - \eta)|G|^2(\eta)\mu_\epsilon(\xi - \eta)\mu_\epsilon(\eta)\, d\eta d\xi$$

$$= \|\mu_\epsilon * \mu_\epsilon\|_{L^\infty(\mathbb{R}^n)} \left( \int_{\mathbb{R}^n} |G|^2 \mu_\epsilon \right)^2.$$

Finally, two applications of Young's inequality lead to the following estimate:

$$\|\mu_\epsilon * \mu_\epsilon\|_{L^\infty(\mathbb{R}^n)} \leq \|d\mu * d\mu\|_{L^\infty(\mathbb{R}^n)} \|\phi_\epsilon * \phi_\epsilon\|_{L^1(\mathbb{R}^n)}$$

$$= \|F\|_{L^\infty(\mathbb{R}^n)}. \qquad \square$$

**Lemma 1.20** *Let $d\sigma$ be the surface measure on $\mathbb{S}^{n-1}$. Then for each $n \geq 2$ the measure $d\sigma * d\sigma$ is absolutely continuous with respect to the Lebesgue measure, $d\sigma * d\sigma = F d\xi$. Moreover, $F$ is zero outside $B(0,2)$ and satisfies for a.e. $\xi$*

$$|F(\xi)| \lesssim \upsilon(\xi) := \begin{cases} \frac{1}{|\xi|}, & 0 < |\xi| \leq 1 \\ (2 - |\xi|)^{\frac{n-3}{2}}, & 1 \leq |\xi| \leq 2. \end{cases}$$

*Proof* Let $\mathbb{S}_\epsilon^{n-1}$ be the $\epsilon$-neighborhood of $\mathbb{S}^{n-1}$ and let $\sigma_\epsilon = \epsilon^{-1} 1_{\mathbb{S}_\epsilon^{n-1}}$. Then the measures $\sigma_\epsilon d\xi$ converge weakly to $d\sigma$, as $\epsilon \to 0$. Note that

$$\sigma_\epsilon * \sigma_\epsilon(\xi) = \frac{1}{\epsilon^2} |\mathbb{S}_\epsilon^{n-1} \cap (\xi + \mathbb{S}_\epsilon^{n-1})|.$$

Since $\mathbb{S}_\epsilon^{n-1} \cap (\xi + \mathbb{S}_\epsilon^{n-1})$ is a body of revolution, its volume is at most a constant multiple of the area of the cross section $\mathbb{S}_\epsilon^1 \cap ((r,0) + \mathbb{S}_\epsilon^1)$, with $r = |\xi|$.

Assume that $r \leq 1$. Note that any $y = (y_1, y_2) \in \mathbb{S}_\epsilon^1 \cap ((r,0) + \mathbb{S}_\epsilon^1)$ satisfies

$$1 - 2\epsilon \leq y_1^2 + y_2^2 \leq 1 + 3\epsilon,$$

$$1 - 2\epsilon \leq (y_1 - r)^2 + y_2^2 \leq 1 + 3\epsilon,$$

and thus also

$$|2y_1 - r| \leq \frac{5\epsilon}{r}.$$

This means that the horizontal projection of $\mathbb{S}_\epsilon^1 \cap ((r,0) + \mathbb{S}_\epsilon^1)$ sits inside an interval of length $5\epsilon/r$. Since $r \leq 1$, the vertical slices of $\mathbb{S}_\epsilon^1 \cap ((r,0) + \mathbb{S}_\epsilon^1)$ have length $\lesssim \epsilon$. Using Fubini's theorem, we find that $|\mathbb{S}_\epsilon^1 \cap ((r,0) + \mathbb{S}_\epsilon^1)| \lesssim \epsilon^2/r$. We conclude that if $|\xi| \leq 1$,

$$\sup_{\epsilon < 1} \sigma_\epsilon * \sigma_\epsilon(\xi) \lesssim \frac{1}{|\xi|}.$$

A similar computation shows that if $1 \leq |\xi| \leq 2$,

$$\sup_{\epsilon < 1} \sigma_\epsilon * \sigma_\epsilon(\xi) \lesssim (2 - |\xi|)^{\frac{n-3}{2}}.$$

Since $(\sigma_\epsilon * \sigma_\epsilon)\, d\xi$ converges weakly to $d\sigma * d\sigma$, the result of the lemma follows. $\qquad\square$

Exercise 1.22 shows that when $n = 3$, the magnitude of the density $F$ from the previous lemma is in fact comparable to $1/|\xi| 1_{B(0,2)}(\xi)$. Since this is not in $L^\infty(\mathbb{R}^3)$, Proposition 1.19 is not immediately applicable. The singularity of $d\sigma * d\sigma$ at 0 is explained by the symmetry of $\mathbb{S}^2$, which allows 0 to be represented in multiple ways as $\xi + \eta$, with $\xi, \eta \in \mathbb{S}^2$. This issue can be resolved by partitioning $\mathbb{S}^2$ into finitely many regions $S_i$ with 0 in the interior of the complement of each $S_i + S_i$. Writing $d\sigma_{S_i} = 1_{S_i} d\sigma$, it is clear that $d\sigma_{S_i} * d\sigma_{S_i} = F_i d\xi$ with $0 \leq F_i \leq \upsilon$, and moreover $F_i$ is zero near the origin. This in fact guarantees that $F_i \in L^\infty(\mathbb{R}^3)$. Now, invoking Proposition 1.19 for each $d\sigma_{S_i}$ we may write

$$\|\widehat{g d\sigma}\|_{L^4(\mathbb{R}^3)} \leq \sum_i \|\widehat{g d\sigma_{S_i}}\|_{L^4(\mathbb{R}^3)}$$

$$\lesssim \sum_i \|g\|_{L^2(d\sigma_{S_i})} \lesssim \|g\|_{L^2(d\sigma)}.$$

This concludes the proof of the endpoint of the Stein–Tomas theorem in three dimensions.

**Remark 1.21** This argument actually proves the estimate $R^*_{\mathbb{S}^{n-1}}(2 \mapsto 4)$ for each $n \geq 3$. When $n = 2$, the function $\upsilon$ has an additional singularity at each point on the circle $|\xi| = 2$ that cannot be removed as before. And in fact, this is not surprising, as we know $R^*_{\mathbb{S}^1}(2 \mapsto 4)$ is in fact false (cf. Example 1.8). However, it seems likely that the estimate $R^*_{\mathbb{S}^1}(2 \mapsto 6)$ could be proved using the strategy we just described, this time by understanding the triple convolution $d\sigma * d\sigma * d\sigma$.

**Exercise 1.22** Show that if $d\sigma$ is the surface measure for $\mathbb{S}^2$, then

(i) $\widehat{d\sigma}(x) = 2\sin(2\pi|x|)/|x|$.

(ii) $d\sigma * d\sigma = F d\xi$ with $F(\xi) = C/|\xi| 1_{[0,2]}(|\xi|)$, for some constant $C$.

**Exercise 1.23** Note that $\psi(\xi) = \sqrt{1 - |\xi|^2}$ does not have a $C^2$ extension outside $B(0, 1)$. Modify the proof of Theorem 1.16 to argue that $R_{\mathbb{S}^{n-1}}(p \mapsto 2)$ still holds for each $p < 2(n + 1)/(n + 3)$. Do so by first showing the slightly stronger estimate

$$\left( \int_{|\xi| < 1} |\widehat{F}(\xi, \psi(\xi))|^2 \sqrt{1 + |\nabla \psi(\xi)|^2} \, d\xi \right)^{\frac{1}{2}} \lesssim \|F\|_{L^p(\mathbb{R}^n)}.$$

Use (1.8) along the way.

**Exercise 1.24** (Scale-invariant Strichartz estimate) Assume that the restriction estimate $R^*_{\mathbb{P}^{n-1}}(q' \mapsto p')$ holds for the compact paraboloid and some $p, q$ satisfying $(n - 1)/q = (n + 1)/p'$. Show that the same restriction estimate also holds true for the infinite paraboloid, in other words,

$$Ef(x) = \int_{\mathbb{R}^{n-1}} f(\xi) e(\xi \cdot \bar{x} + |\xi|^2 x_n) \, d\xi$$

satisfies

$$\|Ef\|_{L^{p'}(\mathbb{R}^n)} \lesssim \|f\|_{L^{q'}(\mathbb{R}^{n-1})}$$

for each $f : \mathbb{R}^{n-1} \to \mathbb{C}$. In particular, deduce that the following scale-invariant Strichartz estimate holds, where $u$ is the solution of (1.1)

$$\|u\|_{L^{\frac{2(n+1)}{n-1}}(\mathbb{R}^n)} \lesssim \|\phi\|_{L^2(\mathbb{R}^{n-1})}.$$

## 1.5 Constructive Interference

This short section describes the situation where distinct waves are combined to give rise to a wave of maximum possible amplitude.

**Definition 1.25** (Constructive interference) For $1 \le i \le N$, let $z_i \in \mathbb{C}$ and $\xi_i \in \mathbb{R}^n$. We will say that there is constructive interference for the waves $z_i e(x \cdot \xi_i)$ on a set $S \subset \mathbb{R}^n$ if for each $x \in S$ we have

$$\left| \sum_{i=1}^N z_i e(x \cdot \xi_i) \right| \gtrsim \sum_{i=1}^N |z_i|. \tag{1.15}$$

Throughout this book we will use constructive interference to build (almost) extremizers in various contexts. Typically, we will have $|\xi_i|, |z_i| = O(1)$. In this case, if (1.15) holds for $x = x_0$, it automatically holds for all $x$ with $|x - x_0| \ll 1$, since

$$\left| \sum_{i=1}^{N} z_i e(x \cdot \xi_i) \right| \geq \left| \sum_{i=1}^{N} z_i e(x_0 \cdot \xi_i) \right| - \sum_{i=1}^{N} |z_i| |e((x - x_0) \cdot \xi_i) - 1|$$

$$\geq \left| \sum_{i=1}^{N} z_i e(x_0 \cdot \xi_i) \right| - 2|x - x_0| \sum_{i=1}^{N} |z_i|.$$

We will also encounter the situation where constructive interference holds on sets other than balls, such as boxes and half-spaces.

## 1.6 Local and Discrete Restriction Estimates

We will denote by $B_R$ an arbitrary ball of radius $R$ in $\mathbb{R}^n$. Given a hypersurface as in (1.3) and $\alpha \geq 0$, we will write $R_{\mathcal{M}^\psi}(p \mapsto q; \alpha)$ if the inequality

$$\|\mathcal{R}^\psi F\|_{L^q(\mathcal{M}^\psi)} \lesssim R^\alpha \|F\|_{L^p(B_R)}$$

holds for each $R \geq 1$, each ball $B_R$ in $\mathbb{R}^n$, and each $F \in \mathcal{S}(\mathbb{R}^n)$ that is compactly supported on $B_R$. Note that $R_{\mathcal{M}^\psi}(p \mapsto q; 0)$ is equivalent to the global restriction estimate $R_{\mathcal{M}^\psi}(p \mapsto q)$ introduced earlier.

Similarly, we write $R^*_{\mathcal{M}^\psi}(q' \mapsto p'; \alpha)$ if

$$\|\mathcal{R}^{\psi,*} f\|_{L^{p'}(B_R)} \lesssim R^\alpha \|f\|_{L^{q'}(\mathcal{M}^\psi)}$$

holds for each $R \geq 1$, each ball $B_R$ in $\mathbb{R}^n$, and each $f \in L^{q'}(\mathcal{M}^\psi)$.

We will repeatedly use small variations of the following result.

**Lemma 1.26** *There is a Schwartz function $\phi \in \mathcal{S}(\mathbb{R}^n)$ with $\mathrm{supp}\,(\phi) \subset B(0,1)$ and with nonnegative Fourier transform satisfying $1_{B(0,1)} \leq \widehat{\phi}$.*

*Proof* Let $\phi_1$ be an even, nonnegative Schwartz function supported on $B(0, 1/2)$. Then $\phi_2 := \phi_1 * \phi_1$ is supported on $B(0, 1)$ and has nonnegative Fourier transform. Moreover, $\widehat{\phi_2}(0) = (\int \phi_1)^2 > 0$. Choose $C_1 > 1, C_2 > 0$ large enough so that $\min_{|x| \leq 1} C_2 \widehat{\phi_2}(x/C_1) \geq 1$. Then $\phi(\xi) := C_1^n C_2 \phi_2(C_1 \xi)$ satisfies all the desired properties. $\square$

Define the $\delta$-neighborhood of $\mathcal{M}^\psi$

$$\mathcal{N}_{\mathcal{M}^\psi}(\delta) = \mathcal{N}_\psi(\delta) = \{(\xi, \psi(\xi) + t) \colon \xi \in U, |t| < \delta\}. \tag{1.16}$$

**Proposition 1.27** *The following are equivalent:*

*(L1)* $R_{\mathcal{M}^\psi}(p \mapsto q; \alpha)$.
*(L2)* $R^*_{\mathcal{M}^\psi}(q' \mapsto p'; \alpha)$.

*(L3)* $\|\widehat{F}\|_{L^q(\mathcal{N}_\psi(R^{-1}))} \lesssim R^{\alpha-1/q}\|F\|_{L^p(B_R)}$ *holds for each* $R \geq 1$, *each ball* $B_R$ *in* $\mathbb{R}^n$, *and each* $F \in L^p(B_R)$ *supported on* $B_R$.

*(L4)* $\|\widehat{F}\|_{L^{p'}(B_R)} \lesssim R^{\alpha-1/q}\|F\|_{L^{q'}(\mathcal{N}_\psi(R^{-1}))}$ *holds for each* $R \geq 1$, *each ball* $B_R$ *in* $\mathbb{R}^n$, *and each* $F \in L^{q'}(\mathcal{N}_\psi(R^{-1}))$ *supported on* $\mathcal{N}_\psi(R^{-1})$.

*Proof*  The equivalences *(L1)* $\Longleftrightarrow$ *(L2)* and *(L3)* $\Longleftrightarrow$ *(L4)* are consequences of duality.

We start by proving that *(L1)* $\Rightarrow$ *(L3)*. It suffices to prove (L3) for Schwartz functions. Note that (L1) implies that $R_{\mathcal{M}^{\psi_t}}(p \mapsto q; \alpha)$ also holds, uniformly over $t \in \mathbb{R}$, where $\psi_t(\xi) = t + \psi(\xi)$. Write

$$\int_{\mathcal{N}_\psi(R^{-1})} |\widehat{F}|^q = \int_{-R^{-1}}^{R^{-1}} dt \int_U |\widehat{F}|^q(\xi, t + \psi(\xi))\, d\xi$$

$$\lesssim R^{\alpha q} \int_{-R^{-1}}^{R^{-1}} \|F\|_{L^p(B_R)}^q$$

$$= 2R^{\alpha q-1}\|F\|_{L^p(B_R)}^q.$$

It remains to prove that *(L4)* $\Rightarrow$ *(L2)*. It suffices to check (L2) with $B_R$ centered at the origin. Let $\phi \in \mathcal{S}(\mathbb{R})$ be as in Lemma 1.26. This function is easily seen to have an even Fourier transform. Let us define $\phi_R(x_n) = R\phi(Rx_n)$.

Given $f: U \to \mathbb{C}$, we consider the function $F(\xi, \xi_n) = f(\xi)\phi_R(\xi_n - \psi(\xi))$. Note that $F$ is supported on $\mathcal{N}_\psi(R^{-1})$ and its Fourier transform is

$$\widehat{F}(x) = \widehat{\phi_R}(x_n)E^\psi f(-x).$$

We conclude the argument with the following sequence of inequalities:

$$\|E^\psi f\|_{L^{p'}(B(0,R))} \leq \|\widehat{\phi_R} E^\psi f\|_{L^{p'}(B(0,R))}$$

$$= \|\widehat{F}\|_{L^{p'}(B(0,R))}$$

$$\lesssim R^{\alpha-1/q}\|f\|_{L^{q'}(U)}R^{1/q}, \quad \text{by (L4) and Fubini.}$$

$\square$

We will call (L1) through (L4) *local restriction* estimates. The parameter $\alpha$ measures the extent to which a global restriction estimate fails to hold.

The local restriction estimates for the sphere $\mathbb{S}^{n-1}$ will be understood as the corresponding local restriction estimates for the upper (or lower) hemisphere.

**Example 1.28**  For each $\mathcal{M}^\psi$, the local estimate $R^*_{\mathcal{M}^\psi}(2 \mapsto 2; \alpha)$ holds if $\alpha \geq 1/2$. This is a consequence of the conservation of mass

$$\int_{\mathbb{R}^{n-1}} |E^\psi f(\bar{x}, x_n)|^2 \, d\bar{x} = \|f\|_2^2, \quad \text{for each } x_n \in \mathbb{R}. \tag{1.17}$$

We illustrate the usefulness of local inequalities, by rephrasing the restriction estimates in a discrete language. Let Discres $(\mathcal{M}^\psi, p, q)$ be the smallest constant such that the following inequality holds for each $R \geq 1$, each collection $\Lambda \subset \mathcal{M}^\psi$ consisting of $1/R$-separated points $\lambda := (\xi, \psi(\xi))$ with $\text{dist}(\xi, \partial U) > 1/R$, each sequence $a_\lambda \in \mathbb{C}$, and each ball $B_R$:

$$\left\| \sum_{\lambda \in \Lambda} a_\lambda e(\lambda \cdot x) \right\|_{L^{p'}(B_R)} \leq \text{Discres}(\mathcal{M}^\psi, p, q) R^{\frac{n-1}{q}} \|a_\lambda\|_{l^{q'}}.$$

We will refer to this inequality as a *discrete restriction* estimate.

**Proposition 1.29** *For each* $1 \leq p, q \leq \infty$ *we have*

$$\text{Discres}(\mathcal{M}^\psi, p, q) \lesssim \|\mathcal{R}^{\psi,*}\|_{L^{q'}(\mathcal{M}^\psi, d\sigma^\psi) \mapsto L^{p'}(\mathbb{R}^n)}.$$

*Also, if* $p \geq q$ *we have*

$$\|\mathcal{R}^{\psi,*}\|_{L^{q'}(\mathcal{M}^\psi, d\sigma^\psi) \mapsto L^{p'}(\mathbb{R}^n)} \lesssim \text{Discres}(\mathcal{M}^\psi, p, q).$$

*Proof* To prove the first inequality, let us fix a collection $\Lambda \subset \mathcal{M}^\psi$ of $1/R$-separated points. Let also $\phi_R(\xi) = R^n \phi(R\xi)$, with $\phi \in \mathcal{S}(\mathbb{R}^n)$ as in Lemma 1.26. Then, using Proposition 1.27 we may write

$$\left\| \sum_{\lambda \in \Lambda} a_\lambda e(\lambda \cdot x) \right\|_{L^{p'}(B(x_0, R))}$$

$$\leq \left\| \sum_{\lambda \in \Lambda} b_\lambda \widehat{\phi_R}(x) e(\lambda \cdot x) \right\|_{L^{p'}(B(0, R))}, \quad (b_\lambda := a_\lambda e(\lambda \cdot x_0))$$

$$\lesssim R^{-\frac{1}{q}} \|\mathcal{R}^{\psi,*}\|_{L^{q'}(\mathcal{M}^\psi, d\sigma^\psi) \mapsto L^{p'}(\mathbb{R}^n)} \left\| \sum_{\lambda \in \Lambda} b_\lambda \phi_R(\xi + \lambda) \right\|_{L^{q'}(\mathbb{R}^n)}$$

$$\sim R^{-\frac{1}{q}} \|\mathcal{R}^{\psi,*}\|_{L^{q'}(\mathcal{M}^\psi, d\sigma^\psi) \mapsto L^{p'}(\mathbb{R}^n)} \|b_\lambda\|_{l^{q'}} \|\phi_R\|_{L^{q'}(\mathbb{R}^n)}$$

$$\sim R^{\frac{n-1}{q}} \|\mathcal{R}^{\psi,*}\|_{L^{q'}(\mathcal{M}^\psi, d\sigma^\psi) \mapsto L^{p'}(\mathbb{R}^n)} \|a_\lambda\|_{l^{q'}}.$$

We have used the fact that the supports of the functions $\phi_R(\xi - \lambda)$ have bounded overlap. This proves the first inequality.

For the second inequality, consider a finitely overlapping cover of $U$ containing $\sim R^{n-1}$ cubes of the form $\xi_i + [-10R^{-1}, 10R^{-1}]^{n-1}$, with $\min_{i \neq j} |\xi_i - \xi_j| > 1/R$ and $\min_i \text{dist}(\xi_i, \partial U) > 1/R$. By working with a partition of unity, we may in fact consider functions $f$ supported on a disjoint union of such cubes.

Note that for each $\tau$, the points $\lambda_{i,\tau} := (\xi_i + \tau, \psi(\xi_i + \tau))$ are $R^{-1}$-separated. Using Hölder's inequality twice and Fubini's theorem, we get for each $R \geq 1$

$$\|E^\psi f\|_{L^{p'}(B(0,R))}$$

$$= \left\| \int_{[-10R^{-1}, 10R^{-1}]^{n-1}} \sum_i f(\xi_i + \tau) e(\lambda_{i,\tau} \cdot x) \, d\tau \right\|_{L^{p'}(B(0,R),dx)}$$

$$\lesssim R^{-\frac{n-1}{p}} \left\| \left\| \sum_i f(\xi_i + \tau) e(\lambda_{i,\tau} \cdot x) \right\|_{L^{p'}([-10R^{-1}, 10R^{-1}]^{n-1}, d\tau)} \right\|_{L^{p'}(B(0,R),dx)}$$

$$= R^{-\frac{n-1}{p}} \left\| \left\| \sum_i f(\xi_i + \tau) e(\lambda_{i,\tau} \cdot x) \right\|_{L^{p'}(B(0,R),dx)} \right\|_{L^{p'}([-10R^{-1}, 10R^{-1}]^{n-1}, d\tau)}$$

$$\leq \mathrm{Discres}\,(\mathcal{M}^\psi, p, q) R^{\frac{n-1}{q} - \frac{n-1}{p}} \left\| \|f(\xi_i + \tau)\|_{l^{q'}} \right\|_{L^{p'}([-10R^{-1}, 10R^{-1}]^{n-1}, d\tau)}$$

$$\lesssim \mathrm{Discres}\,(\mathcal{M}^\psi, p, q) R^{(n-1)(\frac{1}{q} - \frac{1}{p} + \frac{1}{q'} - \frac{1}{p'})}$$

$$\left\| \|f(\xi_i + \tau)\|_{l^{q'}} \right\|_{L^{q'}([-10R^{-1}, 10R^{-1}]^{n-1}, d\tau)}$$

$$= \mathrm{Discres}\,(\mathcal{M}^\psi, p, q) \|f\|_{L^{q'}(U)}.$$

Letting $R \to \infty$ leads to the desired conclusion.                                    $\square$

When combined with Theorem 1.16, Proposition 1.29 leads to the following discrete restriction estimate. The sharpness of the exponent of $R$ is due to constructive interference.

**Corollary 1.30** *Let $\mathcal{M}$ be either $\mathbb{S}^{n-1}$ or $\mathbb{P}^{n-1}$ and let $p \geq 2(n+1)/(n-1)$. For each $R \geq 1$, each collection $\Lambda \subset \mathcal{M}$ consisting of $1/R$-separated points, each sequence $a_\lambda \in \mathbb{C}$ and each ball $B_R$ of radius $R$*

$$\left\| \sum_{\lambda \in \Lambda} a_\lambda e(\lambda \cdot x) \right\|_{L^p(B_R)} \lesssim R^{\frac{n-1}{2}} \|a_\lambda\|_{l^2}.$$

*Moreover, the exponent $(n-1)/2$ of $R$ cannot be lowered.*

**Exercise 1.31** (a) Show that for each collection of $1/R$-separated points $\Lambda \subset \mathbb{R}^n$ we have

$$\left\| \sum_{\lambda \in \Lambda} a_\lambda e(\lambda \cdot x) \right\|_{L^2(B_R)} \lesssim R^{\frac{n}{2}} \|a_\lambda\|_{l^2} \qquad (1.18)$$

and for each $\epsilon > 0$

$$R^{\frac{n}{2} - \epsilon} \|a_\lambda\|_{l^2} \lesssim_\epsilon \left\| \sum_{\lambda \in \Lambda} a_\lambda e(\lambda \cdot x) \right\|_{L^2(B_R)}. \qquad (1.19)$$

Conclude that if the points are on either $\mathbb{S}^{n-1}$ or $\mathbb{P}^{n-1}$ and $2 \leq p \leq 2(n+1)/(n-1)$, then

$$\left\| \sum_{\lambda \in \Lambda} a_\lambda e(\lambda \cdot x) \right\|_{L^p(B_R)} \lesssim R^{\frac{n-1}{4} + \frac{n+1}{2p}} \|a_\lambda\|_{l^2}. \qquad (1.20)$$

(b) Prove that the exponent of $R$ in (1.20) is sharp.

Hint for (b): Let $\Lambda$ be a collection of roughly $R^{(n-1)/2}$ points, within $O(R^{-1/2})$ from each other. Use constructive interference.

**Exercise 1.32** (Endpoint restriction estimates) Let $\mathcal{M}$ be either $\mathbb{S}^{n-1}$ or $\mathbb{P}^{n-1}$ and assume Conjecture 1.7 holds true for $\mathcal{M}$. Prove that the local restriction estimate $R^*_\mathcal{M}(2n/(n-1) \mapsto 2n/(n-1); \epsilon)$ holds for each $\epsilon > 0$.

**Exercise 1.33** Let $\Xi$ be a collection of $1/R$-separated points on $\mathbb{P}^1$ or $\mathbb{S}^1$. Prove that for each $a_\xi \in \mathbb{C}$ and each ball $B_R$ in $\mathbb{R}^2$

$$\left( \frac{1}{|B_R|} \int_{B_R} \left| \sum_{\xi \in \Xi} a_\xi e(\xi \cdot x) \right|^4 dx \right)^{\frac{1}{4}} \lesssim_\epsilon R^{\frac{1}{4}+\epsilon} \|a_\xi\|_{l^4}.$$

Prove that $R^{1/4}\|a_\xi\|_{l^4}$ cannot be replaced with $\|a_\xi\|_{l^2}$.

**Exercise 1.34** Assume that $R_{\mathcal{M}^\psi}(p \mapsto q; \alpha)$ for some $p' \geq q' \geq 2$. Prove that

$$\|\widehat{F}\|_{L^{p'}(\mathbb{R}^n)} \lesssim R^{\alpha - \frac{1}{q}} \|F\|_{L^{q'}(\mathcal{N}_\psi(R^{-1}))}$$

holds for each $F \in L^{q'}(\mathcal{N}_\psi(R^{-1}))$ supported on $\mathcal{N}_\psi(R^{-1})$.

## 1.7 Square Root Cancellation and the Role of Curvature

We consider the sequence $r_n$ of Rademacher functions defined on $[0, 1)$ as follows. The first function $r_1$ equals 1 on $[0, 1/2)$ and $-1$ on $[1/2, 1)$. If $\tilde{r}_1$ is its periodic extension to $\mathbb{R}$, then $r_n$ is defined to be the restriction to $[0, 1)$ of the function $t \mapsto \tilde{r}_1(2^{n-1}t)$. Considered as random variables, these are identically distributed and mutually independent.

Note that the sum $|\sum_{n=1}^N r_n(t)|$ can be as big as $N$ on some small subset of $[0, 1)$, e.g., on $[0, 2^{-N})$. But, as a result of the Central Limit Theorem, on a "large" part of $[0, 1)$ it will be "roughly" $N^{1/2}$. A variant of this phenomenon persists with arbitrary coefficients and manifests itself in an $L^p$ average sense.

The following well known result illustrates a widespread phenomenon in mathematics called *square root cancellation*. It is in some sense the opposite of constructive interference.

**Theorem 1.35** (Khinchin's inequality, [**71**]) *For each* $0 < p < \infty$ *and* $a_n \in \mathbb{C}$, *we have*

$$\left\| \sum_{n \geq 1} a_n r_n \right\|_{L^p([0,1))} \sim \|a_n\|_{l^2}.$$

We illustrate square root cancellation with another related example.

**Theorem 1.36** (Lacunary exponential sums, [**71**]) *Let* $\lambda_n$ *be a lacunary sequence of positive integers, that is,*

$$\frac{\lambda_{n+1}}{\lambda_n} > \lambda$$

*for all* $n \geq 1$ *and some* $\lambda > 1$. *Then for each* $1 \leq p < \infty$ *and* $a_n \in \mathbb{C}$

$$\left\| \sum_{n \geq 1} a_n e(\lambda_n t) \right\|_{L^p([0,1))} \sim \|a_n\|_{l^2}.$$

We will witness many instances of square root cancellation throughout this book. Most often they will amount to the $L^p$ norm of a function on a probability space being comparable to the $L^2$ norm, for some $p > 2$. We will alternatively refer to this as a *reverse Hölder's inequality*.

Let us next argue that square root cancellation is a key player in the Restriction Conjecture. Proposition 1.29 shows that Conjecture 1.10 admits a striking reformulation. It is equivalent with the following inequality being true for arbitrary $1/R$-separated sets of points $\Lambda$ on either $\mathbb{S}^{n-1}$ or $\mathbb{P}^{n-1}$, for each $p > 2n/(n-1)$

$$\left\| \sum_{\lambda \in \Lambda} a_\lambda e(\lambda \cdot x) \right\|_{L^p(B(0,R))} \lesssim R^{n-1} \|a_\lambda\|_{l^\infty}. \tag{1.21}$$

Let us state this in an even simpler form.

**Proposition 1.37** *Conjecture 1.10 is equivalent with the exponential sum estimate*

$$\sup_{\Lambda} \sup_{|a_\lambda| \sim 1} \left\| \sum_{\lambda \in \Lambda} a_\lambda e(\lambda \cdot x) \right\|_{L^{\frac{2n}{n-1}}(B(0,R))} \lesssim_\epsilon R^{n-1+\epsilon}, \tag{1.22}$$

*where* $\Lambda$ *ranges over all* $1/R$-*separated sets on either* $\mathbb{S}^{n-1}$ *or* $\mathbb{P}^{n-1}$.

*Proof* We need to prove that (1.22) implies (1.21). Invoking epsilon-removal technology (see for example [**110**]) and Hölder's inequality, it suffices to prove that

$$\left\| \sum_{\lambda \in \Lambda} a_\lambda e(\lambda \cdot x) \right\|_{L^{\frac{2n}{n-1}}(B(0,R))} \lesssim_\epsilon R^{n-1+\epsilon} \|a_\lambda\|_{l^\infty}.$$

Using the triangle inequality and a crude estimate, we may restrict the sum on the left to those $\lambda'$ for which

$$R^{-C_n} \|a_\lambda\|_\infty \leq |a_{\lambda'}| \leq \|a_\lambda\|_\infty.$$

There are $O(\log R)$ dyadic blocks in this range, and since we can afford an $R^\epsilon$-loss, we may restrict attention to one of them. The result follows by rescaling the coefficients and using (1.22). $\qquad\square$

Let us now rescale (1.22) and assume that $\Lambda$ consists of 1-separated points on $R\mathbb{S}^{n-1}$. In the extreme scenario (the saturated case) when $|\Lambda| \sim R^{n-1}$, (1.22) leads to

$$\left\| \sum_{\lambda \in \Lambda} a_\lambda e(\lambda \cdot x) \right\|_{L^{\frac{2n}{n-1}}(B(0,1))} \lesssim_\epsilon R^\epsilon \|a_\lambda\|_{l^2}, \qquad (1.23)$$

whenever $|a_\lambda|$ is roughly constant. This shows that, while $|\sum_{\lambda \in \Lambda} a_\lambda e(\lambda \cdot x)|$ may be comparable to $\|a_\lambda\|_{l^1}$ on a small subset of $B(0,1)$ (e.g., when $a_\lambda$ are all positive and $|x| \ll 1/R$), it is expected that for "most" $x$ the sum is of the much smaller order $O(R^\epsilon \|a_\lambda\|_{l^2})$. In other words, (almost) square root cancellation is supposed to govern the saturated case, at least in $L^{2n/(n-1)}$ average sense.

We point out that curvature plays a critical role in (1.23). Indeed, if $\Lambda$ is a subset of, say, a hyperplane, the exponentials $e(\lambda \cdot x)$ may conspire to make the sum $|\sum_{\lambda \in \Lambda} e(\lambda \cdot x)|$ comparable to $|\Lambda|$ on a large subset of $B(0,1)$. For example, assume $\Lambda$ is a subset of the ball $B_{n-1}(0,R)$ inside the hyperplane $x_n = 0$. Then for each $x = (\bar{x}, x_n)$ with $|\bar{x}| \ll R^{-1}$ we have constructive interference

$$\left| \sum_{\lambda \in \Lambda} e(\lambda \cdot x) \right| \gtrsim |\Lambda|.$$

In particular, if $|\Lambda| \sim R^{n-1}$ then for all $p \geq 1$

$$\left\| \sum_{\lambda \in \Lambda} e(\lambda \cdot x) \right\|_{L^p(B(0,1))} \gtrsim |\Lambda|^{1-\frac{1}{p}},$$

and the analog of (1.23) does not hold, even when $2n/(n-1)$ is replaced with another $p > 2$.

Let us close this chapter with an important observation. We have seen that the Restriction Conjecture is equivalent with estimating certain $L^p$ moments of exponential sums associated with $1/R$-separated frequency points, on spatial balls of radius $R$. A distinctly new situation arises when balls of larger radius $R^2$ are considered. This is part of a phenomenon called *decoupling*, which will be thoroughly investigated in the second part of the book.

# 2

# Wave Packets

Conjecture 1.10 allows us to measure our progress on the restriction problem for $S = \mathbb{S}^{n-1}, \mathbb{P}^{n-1}$ by means of how close we manage to get $p'$ to $2n/(n-1)$ in the estimate $R_S^*(\infty \mapsto p')$. The Stein–Tomas argument pushes $p'$ as low as $2(n+1)/(n-1)$, but has its limitations. It relies on a spatial Littlewood–Paley decomposition for the Fourier transform of the surface measure, which does not distinguish the contributions from different frequencies on $S$. To lower the range of $p'$, we need to employ new tools that have the ability to separate these contributions. In this chapter, we will seek to understand the quantity $Ef$ as a superposition of wave packets that are well localized in both space and frequency. The details are presented for the paraboloid, but the reader is invited to explore the general case in Exercise 2.8.

Consider a smooth function $\Upsilon : [-4, 4]^{n-1} \to [0, \infty)$ such that

$$\sum_{\mathbf{l} \in \mathbb{Z}^{n-1}} \Upsilon(\xi - \mathbf{l}) \equiv 1.$$

Let $f \in L^1(\mathbb{R}^{n-1})$ be a function that vanishes outside $[-1, 1]^{n-1}$ and let $R \gg 1$. Then

$$f(\xi) = \sum_{|\mathbf{l}| \lesssim R^{1/2}} f(\xi) \Upsilon(\xi R^{1/2} - \mathbf{l}).$$

The function $\Upsilon(\xi R^{1/2} - \mathbf{l})$ is supported inside a certain translate $\omega$ (centered at $c_\omega$) of the cube $[-4R^{-1/2}, 4R^{-1/2}]^{n-1}$, and equals $\Upsilon(R^{1/2}(\xi - c_\omega))$. Let us denote by $\Omega_R$ the collection of these $\omega$.

Note that the cubes $\omega \in \Omega_R$ overlap $O(1)$ many times, and only the cubes inside $[-2, 2]^{n-1}$ contribute to the summation. Expand $f$ into Fourier series on each $\omega \in \Omega_R$

26

$$f(\xi) = \sum_{q \in \mathcal{Q}_R} \left(\frac{R^{1/2}}{8}\right)^{n-1} \langle f 1_\omega, e(-c_q \cdot) \rangle e(-c_q \cdot \xi), \quad \xi \in \omega,$$

where $q$ ranges over a tiling $\mathcal{Q}_R$ of $\mathbb{R}^{n-1}$ consisting of cubes with side length $R^{1/2}/8$. Thus

$$f(\xi) = \sum_{q \in \mathcal{Q}_R} \sum_{\omega \in \Omega_R} 8^{1-n} e(-c_q \cdot c_\omega) \langle f 1_\omega, e(-c_q \cdot) \rangle \Upsilon_{q,\omega}(\xi), \quad \xi \in [-1,1]^{n-1},$$

where

$$\Upsilon_{q,\omega}(\xi) := R^{\frac{n-1}{2}} \Upsilon(R^{1/2}(\xi - c_\omega)) e(c_q \cdot (c_\omega - \xi)).$$

For future use, we record the following rather immediate almost orthogonality property, valid for each $w_{q,\omega} \in \mathbb{C}$

$$\left\| \sum_{q \in \mathcal{Q}_R} \sum_{\omega \in \Omega_R} w_{q,\omega} \Upsilon_{q,\omega} \right\|_2 \lesssim R^{\frac{n-1}{4}} \left( \sum_{q \in \mathcal{Q}_R} \sum_{\omega \in \Omega_R} |w_{q,\omega}|^2 \right)^{1/2}. \tag{2.1}$$

Let us now investigate the properties of $E\Upsilon_{q,\omega}(x)$. Changing variables $\eta = R^{1/2}(\xi - c_\omega)$, we find that

$$E\Upsilon_{q,\omega}(x) = e(x \cdot (c_\omega, |c_\omega|^2)) \int \Upsilon(\eta) e(\varphi_{x,q,\omega}(\eta)) \, d\eta,$$

with

$$\varphi_{x,q,\omega}(\eta) = \eta \cdot \frac{(\bar{x} - c_q) + 2c_\omega x_n}{R^{1/2}} + |\eta|^2 \frac{x_n}{R}.$$

**Definition 2.1** (Tubes and wave packets) We denote by $T_{q,\omega}$ the spatial tube in $\mathbb{R}^n$ given by

$$T_{q,\omega} = \{(\bar{x}, x_n) : |(\bar{x} - c_q) + 2c_\omega x_n| \leq R^{1/2}, \ |x_n| \leq R\}.$$

The collection of all such tubes will be denoted either by $\mathbb{T}_R$, $\mathbb{T}_{[-R,R]}$, or simply by $\mathbb{T}$, whenever there is no confusion about the parameter $R$.

For each tube $T = T_{q,\omega} \in \mathbb{T}_R$ we denote $\Upsilon_{q,\omega}$ by $\Upsilon_T$ and $E\Upsilon_T$ by $\phi_T$. We will call each $\phi_T$, and in fact any of its scalar multiples, a wave packet.

For $M \geq 1$, consider also the following dilation of $T_{q,\omega}$ around its central axis (see Figure 2.1):

$$MT_{q,\omega} = \{(\bar{x}, x_n) : |(\bar{x} - c_q) + 2c_\omega x_n| \leq MR^{1/2}, \ |x_n| \leq R\}.$$

The following theorem summarizes the main features of the wave packet decomposition.

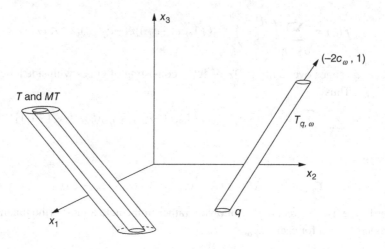

Figure 2.1 Tubes in $\mathbb{R}^3$.

**Theorem 2.2** (Wave packet decomposition of $Ef$ at scale $R$) *There is a decomposition*

$$f = \sum_{T \in \mathbb{T}_R} f_T$$

*with each $f_T$ supported on some cube $\omega_T \in \Omega_R$. We will write $Ef_T = a_T \phi_T$, $a_T \in \mathbb{C}$, so*

$$Ef = \sum_{T \in \mathbb{T}_R} a_T \phi_T. \tag{2.2}$$

*Then $\phi_T$, $a_T$ enjoy the following properties*

$$\|\phi_T\|_{L^\infty((\mathbb{R}^{n-1} \times [-R,R]) \setminus MT)} \lesssim_k M^{-k}, \quad \text{for each } M, k \geq 1, \tag{2.3}$$

$$\|\phi_T\|_{L^\infty(\mathbb{R}^n)} \lesssim 1, \tag{2.4}$$

$$R^{\frac{n-1}{2}} \sum_{\substack{T \in \mathbb{T}_R \\ \omega_T = \omega}} |a_T|^2 \sim \|f 1_\omega\|_2^2, \quad \text{for each } \omega \in \Omega_R, \tag{2.5}$$

$$R^{\frac{n-1}{2}} \sum_{T \in \mathbb{T}_R} |a_T|^2 \sim \|f\|_2^2, \tag{2.6}$$

$$\|f\|_2^2 \sim \sum_{T \in \mathbb{T}_R} \|f_T\|_2^2, \tag{2.7}$$

$$\text{supp}\,(\widehat{\phi_T}) \subset \{(\xi, |\xi|^2) \colon \xi \in \omega_T\}. \tag{2.8}$$

*Proof* For each $T = T_{q,\omega}$, let

$$a_T = e(-c_q \cdot c_\omega) \frac{\langle f1_\omega, e(-c_q \cdot) \rangle}{8^{n-1}}$$

and

$$f_T = a_T \Upsilon_T.$$

Inequality (2.3) follows from the principle of nonstationary phase (proposition 4, page 341 of [102]) since

$$\inf_{\substack{\eta \in \text{supp}(\Upsilon) \\ x \in (\mathbb{R}^{n-1} \times [-R, R]) \setminus MT}} |\nabla \varphi_{x,q,\omega}(\eta)| \gtrsim M.$$

Inequality (2.4) is immediate, while (2.5) follows from Parseval's identity

$$\sum_{q \in \mathcal{Q}_R} |\langle f1_\omega, e(-c_q \cdot) \rangle|^2 = 8^{n-1} R^{-\frac{n-1}{2}} \|f1_\omega\|_2^2.$$

To get (2.6), it suffices to sum this over all $\omega$ and to use the fact that these cubes overlap at most $O(1)$ many times. Note also that (2.7) follows from (2.8) and the fact that $\|\Upsilon_T\|_2^2 \sim R^{(n-1)/2}$.

Finally, (2.8) is a direct computation. Note that $\widehat{\phi_T}$ is a measure supported on a hypersurface. $\qquad\square$

When using wave packets, we will often localize the arguments to the spatial ball $|x| \leq R$. The previous theorem shows that

$$\phi_{T,loc}(x) := \phi_T(x) 1_{\{|x_n| \leq R\}}$$

has the following two main features. First, we can think of it as being essentially supported on $T$. Fully rigorous arguments would demand regarding $\phi_{T,loc}$ as being concentrated on the slight enlargement $R^\delta T$, for a small enough parameter $\delta$. Indeed, according to (2.3), outside this enlarged tube $\phi_{T,loc}$ will be $O(R^{-k})$ for all $k \geq 1$, and this will prove to be a negligible error term in all our applications. Also, the computation in (1.7) shows that we can think of $|\phi_{T,loc}|$ as being essentially constant on $T$.

The second main feature of $\phi_{T,loc}$ is its modulation $e(x \cdot (c_\omega, |c_\omega|^2))$. Thus, a slightly simplified but helpful way to represent $\phi_{T,loc}$ is as follows:

$$\phi_{T,loc}(x) \approx 1_T(x) e(x \cdot (c_\omega, |c_\omega|^2)). \tag{2.9}$$

According to (1.2), for each $T = T_{q,\omega}$ the wave packet $\phi_T$ is the solution of the free Schrödinger equation with initial data that is compactly supported in frequency on $\omega$, and highly concentrated near $q$ on the spatial side. The representation (2.9) indicates that at times $|x_n| \leq R$ the solution remains concentrated near the tube $T$. For later times $|x_n| \gtrsim R$, the principle of nonstationary phase shows that $\phi_T$ will be concentrated near wider and wider tubes that are part of an infinite cone. Thus, while the $L^2$ mass is conserved over time

$$\int_{\mathbb{R}^{n-1}} |\phi_T(\bar{x}, x_n)|^2 \, d\bar{x} = \int_{\mathbb{R}^{n-1}} |\phi_T(\bar{x}, 0)|^2 \, d\bar{x}, \quad x_n \in \mathbb{R}, \qquad (2.10)$$

this mass gets dispersed over larger spatial regions, as $|x_n| \to \infty$. This is quantified by the following general principle, an example of a so-called dispersive estimate.

**Theorem 2.3** (Dispersive estimate) *Let $u(\bar{x}, t)$ be the solution of the free Schrödinger equation (1.1) with initial data $\phi \in L^1$. Then*

$$\lim_{|t| \to \infty} \|u(\bar{x}, t)\|_{L_{\bar{x}}^{\infty}(\mathbb{R}^{n-1})} = 0.$$

*Proof* The proof is immediate using the following representation, valid for some $A_n \in \mathbb{C}$, $B_n \in \mathbb{R} \setminus 0$, whose exact values are irrelevant:

$$u(\bar{x}, t) = \frac{A_n}{|t|^{\frac{n-1}{2}}} \int_{\mathbb{R}^{n-1}} \phi(y) e^{i \frac{|\bar{x} - y|^2}{B_n t}} \, dy.$$

$\square$

We will also encounter a few slight variations of the wave packet decomposition from Theorem 2.2. For example, sometimes we will need to understand $Ef$ on a ball $B_R$ of radius $R$, which is not centered at the origin. If the ball lies in the horizontal strip $\{(\bar{x}, x_n) : x_n \in I\}$ for some interval $I = [c - R, c + R]$, then we may define a new collection $\mathbb{T}_I$ of tubes by translating the tubes in $\mathbb{T}_{[-R, R]}$ vertically by $c$. For each $T \in \mathbb{T}_I$, we define the corresponding wave packet

$$\phi_T(x) = \phi_{T-(0,c)}(x - (0, c))$$

and its local version

$$\phi_{T,loc}(x) = \phi_T(x) 1_{x_n \in I}.$$

Note that $\phi_T$ has the same frequency localization as $\phi_{T-(0,c)}$.

We apply Theorem 2.2 to the function $g(\xi) = f(\xi) e(c|\xi|^2)$

$$Eg(x) = \sum_{T \in \mathbb{T}_{[-R, R]}} a_T \phi_T(x).$$

Since $Eg(x) = Ef(x + (0, c))$, we get a wave packet decomposition involving tubes in $\mathbb{T}_I$:

$$Ef(x) = \sum_{T \in \mathbb{T}_I} a_T \phi_T(x).$$

We will refer to the approximate equality

$$Ef(x)1_{B_R}(x) \approx \sum_{\substack{T \in \mathbb{T}_I \\ T \cap B_R \neq \emptyset}} a_T \phi_{T, loc}(x) \qquad (2.11)$$

as the wave packet decomposition of $Ef$ on $B_R$.

**Example 2.4** We have seen that $Ef$ is one single wave packet if $f = \Upsilon_{q, \omega}$. This remains essentially true for any function $f$ that is supported on a cube of diameter $R^{-1/2}$, and whose Fourier transform is concentrated near a cube of diameter $R^{1/2}$.

Let us now narrow the support of $f$. More precisely, let

$$f(\xi) = \Upsilon(R(\xi - c_1))e(-\xi \cdot c_2)$$

with $c_1 \in [-1, 1]^{n-1}$ and $c_2 \in \mathbb{R}^{n-1}$. Since the support of $f$ fits inside one cube $\omega$ with diameter $R^{-1/2}$, all the wave packets of $Ef$ at scale $R$ will be spatially concentrated inside parallel tubes. But what is their magnitude? A similar computation to the one used to prove (1.7) shows that

$$Ef(x) \sim e(-c_1 \cdot c_2)R^{1-n}\|\Upsilon\|_1 e(x \cdot (c_1, |c_1|^2)),$$

if $|x - (c_2, 0)| \ll R$. Thus, a good way to think of the wave packet decomposition of $Ef$ on the ball $B((c_2, 0), R)$ is that it consists of roughly $R^{(n-1)/2}$ wave packets $a_T \phi_T$ with direction $(-2c_1, 1)$ and essentially identical coefficients

$$a_T \sim e(-c_1 \cdot c_2)R^{1-n}\|\Upsilon\|_1.$$

One may repeat the arguments from this section to produce wave packet decompositions associated with arbitrary hypersurfaces. This will be explored in Exercise 2.8

**Remark 2.5** If $\mathcal{M}^\psi$ has nonzero Gaussian curvature, the Inverse Function Theorem implies that $\nabla \psi$ is locally injective. Invoking compactness, such a manifold can be covered with finitely many patches, in such a way that on each patch the normal vector points in distinct directions. This implies that the wave packets for a given patch are spatially concentrated on distinct tubes.

On the other hand, it is easy to see that for the cone $\text{Cone}^{n-1}$ there are multiple wave packets corresponding to a given tube, only differentiated among themselves by their distinct oscillatory terms $e(x \cdot (c_\omega, |c_\omega|))$. Indeed, there is a collection $\mathcal{C}_T$ of roughly $R^{1/2}$ frequencies $c_\omega$, all sitting on a given line on the cone and giving rise to wave packets $\phi_{T,\omega}$ spatially concentrated near a given $T$. The wave packet decomposition for the cone then reads as follows:

$$E^{\text{Cone}^{n-1}} f = \sum_T \sum_{c_\omega \in \mathcal{C}_T} a_{T,\omega} \phi_{T,\omega} .$$

In some applications, it is more helpful to collect terms sharing the spatial localization into a single wave packet:

$$\Phi_T = \sum_{c_\omega \in \mathcal{C}_T} a_{T,\omega} \phi_{T,\omega}. \tag{2.12}$$

We will close this section with yet another type of wave packet decomposition. Let $\chi : \mathbb{R}^n \to \mathbb{R}$ be the function

$$\chi(x) = (1 + |x|)^{-100n}.$$

Given a rectangular box $B$ in $\mathbb{R}^n$, we define $\chi_B := \chi \circ A_B$, where $A_B$ is the affine function mapping $B$ to $[-1, 1]^n$.

**Definition 2.6** Two rectangular boxes $B_1, B_2 \subset \mathbb{R}^n$ with same orientation and with side lengths $(l_1^{(1)}, \dots, l_n^{(1)})$ and $(l_1^{(2)}, \dots, l_n^{(2)})$ are said to be dual if $l_i^{(1)} l_i^{(2)} = 1$, for each $1 \le i \le n$.

Exercise 2.7 provides an alternative route to the representation (2.11) for an arbitrary hypersurface $\mathcal{M}$. Indeed, let $\eta_{B_R}$ be a Schwartz function such that $1_{B_R} \approx \eta_{B_R}$ and supp $(\widehat{\eta_{B_R}}) \subset B(0, 1/R)$. Then $1_{B_R} E^{\mathcal{M}} f \approx \eta_{B_R} E^{\mathcal{M}} f$ and $\eta_{B_R} E^{\mathcal{M}} f$ has Fourier support in the $R^{-1}$-neighborhood of $\mathcal{M}$. This neighborhood can be covered with finitely overlapping rectangular boxes $B$ of dimensions $\sim (R^{-1/2}, R^{-1/2}, \dots, R^{-1})$ (see Figure 2.2). Using a partition of unity adapted to the family of boxes $B$ and applying Exercise 2.7 on each $B$ leads to a representation

$$\eta_{B_R} E f = \sum_B \sum_{T \in \mathcal{T}_B} w_T W_T \approx \sum_B \sum_{\substack{T \in \mathcal{T}_B \\ T \cap B_R \neq \emptyset}} w_T W_T$$

that is very similar to (2.11). Indeed, note that $\phi_{T,loc}$ and $|T|^{1/2} W_T$ have essentially the same normalization, as well as space-frequency localization.

**Exercise 2.7** Let $B$ be a rectangular box in $\mathbb{R}^n$ and let $\mathcal{T}_B$ be a tiling of $\mathbb{R}^n$ with rectangular boxes $T$ dual to $2B$. Prove that for each $T \in \mathcal{T}_B$ there is

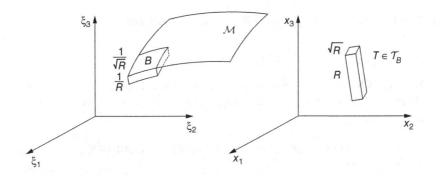

Figure 2.2 Dual boxes.

a Schwartz function $W_T$ (we will also refer to it as a wave packet) with the following properties:

(a) $\text{supp}(\widehat{W_T}) \subset 2B$ and $|W_T| \lesssim_M 1/|T|^{1/2}(\chi_T)^M$ for each $M \geq 1$.
(b) For each $w_T \in \mathbb{C}$,

$$\left\| \sum_{T \in \mathcal{T}_B} w_T W_T \right\|_2 \sim \|w_T\|_{l^2}.$$

(c) If $|w_T| \sim \lambda$ for $T \in \mathcal{T}_B' \subset \mathcal{T}_B$, then for each $2 \leq p < \infty$, we have

$$\left\| \sum_{T \in \mathcal{T}_B'} w_T W_T \right\|_p \sim \lambda \left( \sum_{T \in \mathcal{T}_B'} \|W_T\|_p^p \right)^{\frac{1}{p}}.$$

(d) For each $F$ with $\text{supp}(\widehat{F}) \subset B$, we have

$$F = \sum_{T \in \mathcal{T}_B} \langle F, W_T \rangle W_T \tag{2.13}$$

and

$$\left\| \sum_{\substack{T \in \mathcal{T}_B \\ |\langle F, W_T \rangle| \sim \lambda}} \langle F, W_T \rangle W_T \right\|_p \lesssim \|F\|_p$$

for each $\lambda > 0$ and $2 \leq p \leq \infty$.

Hint: Let $\eta_B$ be a Schwartz function satisfying $1_B \leq \eta_B \leq 1_{2B}$. Consider the Fourier expansion of $\widehat{F}$ on $2B$ and note that

$$\widehat{F} = |B|^{-1} \sum_{T \in \mathcal{T}_B} \langle \widehat{F} \eta_B, e(c_T \cdot) \rangle e(c_T \cdot) \eta_B.$$

To prove (c), let $f = \sum_{T \in T'_B} w_T W_T$ and use the fact that

$$\|f\|_2^{\frac{2(p-1)}{p}} \|f\|_1^{\frac{2-p}{p}} \leq \|f\|_p \leq \|f\|_1^{\frac{1}{p}} \|f\|_\infty^{\frac{1}{p'}}.$$

**Exercise 2.8** Let $\psi : U \to \mathbb{R}$ be a $C^2$ function. Repeat the analysis from this chapter to obtain a wave packet decomposition for $E^\psi$ as in Theorem 2.2. Note that this time

$$\phi_T(x) = e(x \cdot (c_\omega, \psi(c_\omega))) \int \Upsilon(\eta) e(\varphi_{x,q,\omega}(\eta)) \, d\eta,$$

with

$$\varphi_{x,q,\omega}(\eta) = \eta \cdot \frac{(\bar{x} - c_q)}{R^{1/2}} + \left( \psi \left( \frac{\eta}{R^{1/2}} + c_\omega \right) - \psi(c_\omega) \right) x_n$$

$$= \eta \cdot \frac{(\bar{x} - c_q) + \nabla \psi(c_\omega) x_n}{R^{1/2}} + \frac{x_n}{R} \varepsilon(\eta)$$

and $\|\nabla \varepsilon\|_{L^\infty(\operatorname{supp}(\Upsilon))} \lesssim 1$. Note also that $\phi_T$ is spatially concentrated near the tube

$$T_{q,\omega}^\psi = \{(\bar{x}, x_n) : |(\bar{x} - c_q) + \nabla \psi(c_\omega) x_n| \leq R^{1/2}, \ |x_n| \leq R\}.$$

# 3

# Bilinear Restriction Theory

In this chapter, we will explore the bilinear analog of the Stein–Tomas theorem. Most of our efforts will be devoted to the paraboloid, in particular to the proof of the following result. Recall that we denote by $E$ the extension operator for $\mathbb{P}^{n-1}$.

**Theorem 3.1** *For $n \geq 2$, let $\Omega_1, \Omega_2$ be two cubes in $[-1,1]^{n-1}$ with* $\operatorname{dist}(\Omega_1, \Omega_2) > 0$. *Then for each $p > 2(n+2)/n$ and $f : \Omega_1 \cup \Omega_2 \to \mathbb{C}$, we have*

$$\||E_{\Omega_1} f E_{\Omega_2} f|^{\frac{1}{2}}\|_{L^p(\mathbb{R}^n)} \lesssim (\|f\|_{L^2(\Omega_1)} \|f\|_{L^2(\Omega_2)})^{1/2}. \qquad (3.1)$$

*The implicit constant in* (3.1) *will depend on* $\operatorname{dist}(\Omega_1, \Omega_2)$.

Inequality (3.1) is an example of a *bilinear restriction estimate*. It was proved by Tao in [**108**]. His proof builds on an earlier argument of Wolff [**118**], which established a similar result for two angularly separated regions on the cone. A description of some fairly general conditions that guarantee a bilinear theorem in the same range as the preceding appears in [**4**] and [**82**].

The work of Tao, Vargas, and Vega [**114**] is perhaps the first systematic treatment of the bilinear phenomenon and its impact on the linear restriction problem. It also contains references to earlier papers, most notably by Bourgain, that have shaped the development of the topic since the early 1990s.

Simple examples demonstrate that (3.1) is false for $p < 2(n+2)/n$, even when $\mathbb{P}^{n-1}$ is replaced with arbitrary $C^2$ hypersurfaces. See Exercise 3.33. The validity of (3.1) at the endpoint remains open when $n \geq 3$, though the similar question for the cone has been settled in [**109**]. The full expected range of bilinear restriction estimates is presented in Conjecture 6.1.

Note that the case $p \geq 2(n+1)/(n-1)$ is covered by Theorem 1.16, while the result is genuinely bilinear in the range $2(n+2)/n < p \leq 2(n+1)/(n-1)$.

35

The main strength of the bilinear restriction inequality is not coming from the use of the $L^2$ norm on the right-hand side of (3.1) – this is merely the right space that provides us with enough tools to prove it – but from the fact that the range of $p$ breaks the $2(n + 1)/(n - 1)$ threshold. Theorem 3.1 will be used in Chapter 4 to extend the range in the linear restriction problem, beyond the Stein–Tomas exponent.

The case $n = 2$ of Theorem 3.1 is easier, in fact we will be able to prove the endpoint result for $p = 4$ in Sections 3.2 through 3.4. The key feature that simplifies the two-dimensional argument is a certain diagonal behavior of caps (arcs) on $\mathbb{P}^1$ which is captured by $L^4$, as proved in Proposition 3.2. This diagonal behavior is lost for higher-dimensional paraboloids, and the proof of Theorem 3.1 becomes significantly harder. In particular, while the proof in two dimensions requires only a frequency decomposition for the extension operator, the analysis for $n \geq 3$ will also dwell significantly on a spatial decomposition using tubes. The key innovation in this case is the fact that, after eliminating a certain small contribution that is estimated using induction on scales, one can in fact achieve a certain diagonal behavior at the level of tubes. The precise inequality that quantifies this is (3.34). To ease the understanding of this inequality for tubes, we will first prove a similar inequality for lines in Section 3.7. The reader may also benefit from the information contained in the otherwise independent Section 3.6, which compares some of the technical difficulties when dealing with tubes versus lines.

We will split the bilinear interactions between wave packets into two contributions. The so-called *local part* will be easily estimated using induction on scales. In Section 3.5, we prove a general $L^4$ estimate, which is sharp for arbitrary wave packets. Restricting to wave packets whose tubes exhibit diagonal behavior will lead us to the stronger $L^4$ inequality (3.29) for the *global part*. This will then be interpolated with a trivial $L^2$ estimate to get the expected estimate for the global part on the critical space $L^{2(n+2)/n}$.

In Section 3.10, we present a slightly easier version of this argument, originally due to Tao [110] and later streamlined by Bejenaru [4]. This approach relies on a different reshuffling of the wave packets that leads to an identical gain on the $L^4$ estimate; see Proposition 3.37.

The proof of the case $n = 3$ already encodes all difficulties present in dimensions $n \geq 3$, so we will explain both arguments in this case.

## 3.1 A Case Study

In this warm-up section, we will take a first look at the mechanism that allows bilinear estimates to be stronger than the linear ones. This is perhaps easiest to see in the case of $\mathbb{S}^1$. Let us revisit Example 1.9. For a closed interval $J \subset (-1,1)$, let $\eta_J$ be a smooth nonnegative function supported on $J$ and equal to 1 on its inner half $1/2J$. If $f_J(\xi) = \eta_J(\xi)/\sqrt{1-\xi^2}$ then

$$|E^{\mathbb{S}^1_+} f_J(x_1, x_2)| = \left| \int \eta_J(\xi) e\left( x_1\xi + x_2\sqrt{1-\xi^2} \right) \frac{d\xi}{\sqrt{1-\xi^2}} \right|.$$

Making the change of variables $\xi = \sin\theta$ ($|\theta| \leq \pi/2$) and writing $(x_1, x_2) = (r\cos\varphi, r\sin\varphi)$ with $\varphi \in [0, 2\pi)$, we find the preceding expression to be equal to

$$I_{J,\varphi}(r) := \left| \int_{-\frac{\pi}{2}+\varphi}^{\frac{\pi}{2}+\varphi} \eta_J(\sin(\theta - \varphi)) e(r\sin\theta) \, d\theta \right|.$$

Note that the derivative of $\sin\theta$ has only one zero $\theta_\varphi$ in $[-\pi/2 + \varphi, \pi/2 + \varphi]$, namely either $\pi/2$ or $3\pi/2$. A standard oscillatory integral estimate (proposition 3, page 334 of [**102**]) shows that

$$I_{J,\varphi}(r) = \eta_J(\sin(\theta_\varphi - \varphi)) \frac{1}{r^{1/2}} + O\left( \frac{1}{r} \right), \quad r \geq 1.$$

In particular, $I_{J,\varphi}(r) \sim 1/r^{1/2}$ for $\varphi$ in the union of two arcs, and thus

$$\| E^{\mathbb{S}^1_+} f_J \|_{L^4(\mathbb{R}^2)} = \infty$$

for each $J$.

Let now $J_1, J_2$ be two disjoint closed intervals in $(-1, 1)$. Since at least one of $\eta_{J_i}(\sin(\theta_\varphi - \varphi))$ must be zero for each $\varphi$, it follows that

$$\max\{I_{J_1,\varphi}(r), I_{J_2,\varphi}(r)\} \lesssim \frac{1}{(1+r)^{1/2}}, \quad \min\{I_{J_1,\varphi}(r), I_{J_2,\varphi}(r)\} \lesssim \frac{1}{(1+r)}.$$

Integration shows that we have the bilinear estimate

$$\| (E^{\mathbb{S}^1_+} f_{J_1} E^{\mathbb{S}^1_+} f_{J_2})^{\frac{1}{2}} \|_{L^4(\mathbb{R}^2)} < \infty.$$

In a nutshell, the separation of $J_1, J_2$ prevents $|E^{\mathbb{S}^1_+} f_{J_1}|$ and $|E^{\mathbb{S}^1_+} f_{J_2}|$ from being simultaneously big.

## 3.2 Biorthogonality: The Córdoba–Fefferman Argument

There are a few different proofs of Theorem 3.1 in the special case $n = 2$, some of which will be explored later. For example, Exercise 3.13(d) shows that the curvature of $\mathbb{P}^1$ is not essential; the only important feature is transversality.

As we will see in Chapter 6, this is just a particular instance of the more general phenomenon of multilinear restriction estimates.

We find it appropriate to present a proof of Theorem 3.1 for the parabola that will motivate the approach for the more complicated case $n \geq 3$. In this section, we introduce the first key idea behind this argument, which is entirely geometric and goes back to the work [47] of Fefferman. This exposes a certain diagonal behavior captured by the space $L^4$, according to which a pair of caps on $\mathbb{P}^1$ only interact with itself and its neighbors. We can view the following result as a strong manifestation of the fact that $\mathbb{P}^1$ cannot contain all four vertices of a nondegenerate parallelogram.

Given an interval $I$ and a scale $\sigma$, we let

$$\mathcal{N}_I(\sigma) = \{(\xi, \xi^2 + t) : \xi \in I, |t| \leq \sigma\}.$$

**Proposition 3.2** (Diagonal behavior for caps on $\mathbb{P}^1$) *Let $\delta \ll 1$.*

(i) *For each interval $I_1, I_2, I_2' \subset [-1, 1]$ with length $\delta$ and $\operatorname{dist}(I_2, I_2') \geq 100\delta$, we have*

$$(\mathcal{N}_{I_1}(\delta^2) - \mathcal{N}_{I_1}(\delta^2)) \cap (\mathcal{N}_{I_2}(\delta^2) - \mathcal{N}_{I_2'}(\delta^2)) = \emptyset.$$

(ii) *For each interval $I_1, I_1', I_2, I_2' \subset [-1, 1]$ with length $\delta$ and satisfying the separation condition*

$$\operatorname{dist}(I_1, I_1'), \ \operatorname{dist}(I_2, I_2'), \ \max(\operatorname{dist}(I_1, I_2), \operatorname{dist}(I_1', I_2')) \geq 100\delta,$$

*we have*

$$(\mathcal{N}_{I_1}(\delta^2) - \mathcal{N}_{I_1'}(\delta^2)) \cap (\mathcal{N}_{I_2}(\delta^2) - \mathcal{N}_{I_2'}(\delta^2)) = \emptyset. \tag{3.2}$$

*Proof* To see why (i) holds, note that each $\mathcal{N}_I(\delta^2)$ lives inside a certain ball with radius $2\delta$. Thus $\mathcal{N}_{I_1}(\delta^2) - \mathcal{N}_{I_1}(\delta^2)$ lives inside $B(0, 4\delta)$, while $\mathcal{N}_{I_2}(\delta^2) - \mathcal{N}_{I_2'}(\delta^2)$ lives inside $B(c, 4\delta)$, where $c = (c_2, c_2^2) - (c_2', (c_2')^2)$ and $c_2, c_2'$ are the centers of $I_2, I_2'$. Then (i) holds since $|c| \geq 100\delta$.

We next prove the contrapositive of (ii). We could use a more geometric approach, but choose instead to present an algebraic argument. Assume that the intersection in (3.2) is nonempty and also that

$$\operatorname{dist}(I_1, I_1'), \operatorname{dist}(I_2, I_2') \geq 100\delta.$$

We need to show that

$$\max(\operatorname{dist}(I_1, I_2), \operatorname{dist}(I_1', I_2')) \leq 100\delta. \tag{3.3}$$

Our assumption allows us to pick $\xi_1 \in I_1, \xi_1' \in I_1', \xi_2 \in I_2, \xi_2' \in I_2'$ such that

$$v := \xi_1 - \xi_1' = \xi_2 - \xi_2', \quad w_1 := \xi_1^2 - (\xi_1')^2, \quad w_2 := \xi_2^2 - (\xi_2')^2$$

satisfy

$$|w_1 - w_2| \le 4\delta^2 \text{ and } |v| \ge 100\delta. \tag{3.4}$$

The desired estimate (3.3) will follow if we prove that $|\xi_1 - \xi_2|, |\xi_1' - \xi_2'| \le 100\delta$. This in turn will immediately follow from the inequality

$$|\xi_1 + \xi_1' - (\xi_2 + \xi_2')| \le \delta,$$

which in turn is a consequence of (3.4) and

$$|\xi_1 + \xi_1' - (\xi_2 + \xi_2')| = \left| \frac{w_1 - w_2}{v} \right|. \qquad \square$$

Over the years, a few inequalities grew out of these geometric observations, some through the influential works [**41**], [**40**] of Córdoba. We next present one such variant, with a few more explored in the exercises.

Consider two intervals $J_1, J_2 \subset [-1, 1]$. We allow for the possibility that they overlap or even coincide. Let $\mathcal{P}_i$ be a partition of $J_i$ into intervals $I_i$ with length $\delta$. For each interval $H \subset \mathbb{R}$ and $F \colon \mathbb{R}^2 \to \mathbb{C}$, we let $F_H$ be the Fourier projection of $F$ onto $\mathcal{N}_H(\delta^2)$.

**Proposition 3.3** (Reverse square function estimate) *The following inequality holds with the implicit constant independent of $\delta$:*

$$\| |F_{J_1} F_{J_2}|^{1/2} \|_{L^4(\mathbb{R}^2)} \lesssim \left\| \left( \sum_{I_1 \in \mathcal{P}_1} |F_{I_1}|^2 \right)^{1/4} \left( \sum_{I_2 \in \mathcal{P}_2} |F_{I_2}|^2 \right)^{1/4} \right\|_{L^4(\mathbb{R}^2)}. \tag{3.5}$$

*In particular, for each partition $\mathcal{P}$ of $[-1, 1]$ into intervals of length $\delta$*

$$\| F_{[-1,1]} \|_{L^4(\mathbb{R}^2)} \lesssim \left\| \left( \sum_{I \in \mathcal{P}} |F_I|^2 \right)^{1/2} \right\|_{L^4(\mathbb{R}^2)}. \tag{3.6}$$

*Proof* We start by writing

$$F_{J_i} = \sum_{I_i \in \mathcal{P}_i} F_{I_i}.$$

Partition $\mathcal{P}_i = \cup_{j=1}^{101} \mathcal{P}_{i,j}$, with each $\mathcal{P}_{i,j}$ consisting of intervals separated by at least $100\delta$. Using elementary inequalities, it will suffice to prove that for each $1 \le j, j' \le 101$

$$\left\| \left| \sum_{I_1 \in \mathcal{P}_{1,j}} F_{I_1} \sum_{I_2 \in \mathcal{P}_{2,j'}} F_{I_2} \right|^{1/2} \right\|_{L^4(\mathbb{R}^2)}^4$$

$$\leq \left\| \left( \sum_{I_1 \in \mathcal{P}_{1,j}} |F_{I_1}|^2 \right)^{1/4} \left( \sum_{I_2 \in \mathcal{P}_{2,j'}} |F_{I_2}|^2 \right)^{1/4} \right\|_{L^4(\mathbb{R}^2)}^4. \tag{3.7}$$

A computation involving Plancherel's theorem shows that the left-hand side equals

$$\sum_{I_1, I_1' \in \mathcal{P}_{1,j}} \sum_{I_2, I_2' \in \mathcal{P}_{2,j'}} \int_{\mathbb{R}^2} F_{I_1} \bar{F}_{I_1'} F_{I_2} \bar{F}_{I_2'} = \sum_{I_1, I_1' \in \mathcal{P}_{1,j}} \sum_{I_2, I_2' \in \mathcal{P}_{2,j'}} \int_{\mathbb{R}^2} \widehat{F_{I_1} \bar{F}_{I_1'}} \overline{\widehat{F_{I_2} \bar{F}_{I_2'}}}.$$

Note that $\widehat{F_{I_1} \bar{F}_{I_1'}}$ is supported on $\mathcal{N}_{I_1}(\delta^2) - \mathcal{N}_{I_1'}(\delta^2)$, while $\widehat{F_{I_2} \bar{F}_{I_2'}}$ is supported on $\mathcal{N}_{I_2'}(\delta^2) - \mathcal{N}_{I_2}(\delta^2)$. Proposition 3.2 now shows that the contribution to the preceding sum only comes from two sources. The first one corresponds to the case when dist $(I_1, I_2') \leq 100\delta$ and dist $(I_1', I_2) \leq 100\delta$. Let us combine these two restrictions by writing $(I_2', I_2) \in \text{Nei}(I_1, I_1')$. The second contribution comes from the case $I_1 = I_1'$ and $I_2 = I_2'$.

We will analyze the first contribution, as the second one is similar but slightly easier. Note that each $\text{Nei}(I_1, I_1')$ contains at most $O(1)$ pairs. With this in mind, we write, using the Cauchy–Schwarz inequality twice along the way,

$$\int_{\mathbb{R}^2} \sum_{I_1, I_1' \in \mathcal{P}_{1,j}} \sum_{(I_2, I_2') \in \text{Nei}(I_1, I_1')} |F_{I_1} F_{I_1'} F_{I_2} F_{I_2'}|$$

$$\leq \int_{\mathbb{R}^2} \left( \sum_{I_1, I_1' \in \mathcal{P}_{1,j}} |F_{I_1} F_{I_1'}|^2 \right)^{1/2} \left( \sum_{I_1, I_1' \in \mathcal{P}_{1,j}} \left( \sum_{(I_2, I_2') \in \text{Nei}(I_1, I_1')} |F_{I_2} F_{I_2'}| \right)^2 \right)^{1/2}$$

$$\lesssim \int_{\mathbb{R}^2} \left( \sum_{I_1, I_1' \in \mathcal{P}_{1,j}} |F_{I_1} F_{I_1'}|^2 \right)^{1/2} \left( \sum_{I_1, I_1' \in \mathcal{P}_{1,j}} \sum_{(I_2, I_2') \in \text{Nei}(I_1, I_1')} |F_{I_2} F_{I_2'}|^2 \right)^{1/2}$$

$$\lesssim \int_{\mathbb{R}^2} \left( \sum_{I_1, I_1' \in \mathcal{P}_{1,j}} |F_{I_1} F_{I_1'}|^2 \right)^{1/2} \left( \sum_{I_2, I_2' \in \mathcal{P}_{2,j'}} |F_{I_2} F_{I_2'}|^2 \right)^{1/2}$$

$$= \int_{\mathbb{R}^2} \sum_{I_1 \in \mathcal{P}_{1,j}} |F_{I_1}|^2 \sum_{I_2 \in \mathcal{P}_{2,j'}} |F_{I_2}|^2. \qquad \square$$

**Exercise 3.4** (Reverse square function estimate for $\mathbb{S}^1$) Prove that the result in Proposition 3.2 continues to hold if the parabola $\mathbb{P}^1$ is replaced with a half circle

$$\mathbb{S}^1_\pm = \left\{ \left( \xi, \pm\sqrt{1-\xi^2} \right) : |\xi| \le 1 \right\}$$

and $\mathcal{N}_I(\delta^2)$ is replaced with the $\delta^2$-neighborhood of an arc $\tau$ of length $\delta$ on $\mathbb{S}^1_\pm$. Indeed, note that the half circle restriction is necessary, since

$$(\mathcal{N}_{\tau_1}(\delta^2) - \mathcal{N}_{\tau'_1}(\delta^2)) \cap (\mathcal{N}_{\tau_2}(\delta^2) - \mathcal{N}_{\tau'_2}(\delta^2)) \ne \emptyset$$

whenever both $\tau_1, \tau'_2$ and $\tau'_1, \tau_2$ are diametrically opposite arcs on $\mathbb{S}^1$. Prove, however, that the analog of Proposition 3.3 holds for $\mathbb{S}^1$, by exploiting the symmetry $\mathbb{S}^1_+ = -\mathbb{S}^1_-$.

Prove that the analog of (3.6) for neighborhoods of a line segment is false.

**Exercise 3.5** (Transverse reverse square function estimate) Assume now that dist $(J_1, J_2) \gtrsim 1$. Prove that the larger neighborhoods $\mathcal{N}_I(\delta)$ satisfy the following property. For each intervals $I_1, I'_1 \in \mathcal{P}_1$ and $I_2, I'_2 \in \mathcal{P}_2$ of length $\delta$, we have

$$(\mathcal{N}_{I_1}(\delta) - \mathcal{N}_{I'_1}(\delta)) \cap (\mathcal{N}_{I_2}(\delta) - \mathcal{N}_{I'_2}(\delta)) = \emptyset, \tag{3.8}$$

unless dist $(I_1, I'_1)$, dist $(I_2, I'_2) \le 100\delta$.

For $F \colon \mathbb{R}^2 \to \mathbb{C}$ and an interval $H \subset [-1, 1]$, let $F^*_H$ denote the Fourier projection of $F$ onto $\mathcal{N}_H(\delta)$. Use the previous observation to prove that the analog of (3.5) continues to hold

$$\left\| |F^*_{J_1} F^*_{J_2}|^{1/2} \right\|_{L^4(\mathbb{R}^2)} \lesssim \left\| \left( \sum_{I_1 \in \mathcal{P}_1} |F^*_{I_1}|^2 \right)^{1/4} \left( \sum_{I_2 \in \mathcal{P}_2} |F^*_{I_2}|^2 \right)^{1/4} \right\|_{L^4(\mathbb{R}^2)}.$$

**Exercise 3.6** Check that the results from Exercise 3.5 extend to the case when the regions $\mathcal{N}_{J_1}(\delta)$ and $\mathcal{N}_{J_2}(\delta)$ are replaced with $\delta$-neighborhoods of

(a) Closed pairwise disjoint arcs on $\mathbb{S}^1_+$.
(b) Two nonparallel line segments.

## 3.3 Bilinear Interaction of Transverse Wave Packets

The following lemma will be the second major ingredient in the proof of Theorem 3.1 when $n = 2$. This will handle the special case when the wave packet decompositions of both $E_{\Omega_1} f$ and $E_{\Omega_2} f$ consist only of wave packets spatially concentrated on parallel tubes. For later use, we present the result in arbitrary dimensions.

**Lemma 3.7** *Let $B_1, B_2$ be two rectangular boxes in $\mathbb{R}^n$, $n \ge 2$, with $n-1$ of the sides having length $\delta \le 1$ and the short side of length $\delta^2$. Assume that the angle $v$ between the axes corresponding to their short sides satisfies $v \gg \delta$.*

Assume also that $F_1, F_2 \colon \mathbb{R}^n \to \mathbb{C}$ have their Fourier transforms supported inside $B_1$ and $B_2$, respectively. Then

$$\| |F_1 F_2|^{1/2} \|_{L^4(\mathbb{R}^n)} \lesssim_\nu \delta^{\frac{n+2}{4}} (\|F_1\|_{L^2(\mathbb{R}^n)} \|F_2\|_{L^2(\mathbb{R}^n)})^{1/2}.$$

*Proof* We consider a wave packet decomposition as in Exercise 2.7:

$$F_i = \sum_{T \in \mathcal{T}_i} w_T W_T.$$

Here $T$ ranges over a tiling $\mathcal{T}_i$ of $\mathbb{R}^n$ with rectangular boxes of dimensions $\sim \delta^{-1} \times \cdots \times \delta^{-1} \times \delta^{-2}$, sharing the main axes with $B_i$. Also, $W_T$ has the Fourier transform supported inside $2B_i$, is $L^2$ normalized, and decays rapidly outside of $T$:

$$W_T(x) \lesssim \frac{1}{|T|^{1/2}} \chi_T(x).$$

Moreover,

$$\sum_{T \in \mathcal{T}_i} |w_T|^2 \sim \|F_i\|_2^2.$$

Using the triangle inequality, we write

$$\| |F_1 F_2|^{1/2} \|_{L^4(\mathbb{R}^n)}^4 \le \sum_{\substack{T_1, T_1' \in \mathcal{T}_1 \\ T_2, T_2' \in \mathcal{T}_2}} |w_{T_1} w_{T_1'} w_{T_2} w_{T_2'}| \int_{\mathbb{R}^n} |W_{T_1} W_{T_1'} W_{T_2} W_{T_2'}|.$$

Note that the angle separation condition forces the intersection of any two boxes in $\mathcal{T}_1$ and $\mathcal{T}_2$ to have volume $\lesssim_\nu \delta^{-n}$. If we combine this with the rapid decay of each $W_T$ outside of $T$, we arrive at the inequality

$$\int_{\mathbb{R}^n} |W_{T_1} W_{T_1'} W_{T_2} W_{T_2'}| \lesssim_\nu \delta^{n+2} c(T_1, T_1') c(T_2, T_2'), \quad T_i, T_i' \in \mathcal{T}_i,$$

where $c(T_i, T_i')$ are weights that decrease rapidly with the distance between $T_i, T_i'$.

We conclude that

$$\| |F_1 F_2|^{1/2} \|_{L^4(\mathbb{R}^n)}^4 \lesssim_\nu \delta^{n+2} \sum_{T_1, T_1' \in \mathcal{T}_1} |w_{T_1} w_{T_1'} c(T_1, T_1')| \sum_{T_2, T_2' \in \mathcal{T}_2} |w_{T_2} w_{T_2'} c(T_2, T_2')|.$$

By a few applications of the Cauchy–Schwarz inequality, this is further seen to be bounded by

$$\delta^{n+2} \sum_{T_1} |w_{T_1}|^2 \sum_{T_2} |w_{T_2}|^2.$$

This term is $O(\delta^{n+2} \|F_1\|_2^2 \|F_2\|_2^2)$, as desired. $\qquad \square$

## 3.4 Proof of Theorem 3.1 When $n = 2$

In this section, we assemble the ingredients from Sections 3.2 and 3.3 to provide the proof of Theorem 3.1 in two dimensions. We will in fact prove the endpoint result corresponding to $p = 4$.

Let $J_1, J_2$ be two closed, disjoint intervals in $[-1, 1]$ and let $f : \mathbb{R}^2 \to \mathbb{C}$. It suffices to prove the inequality

$$\int_{[-R, R]^2} |E_{J_1} f E_{J_2} f|^2 \lesssim \|f\|_{L^2(J_1)}^2 \|f\|_{L^2(J_2)}^2 \tag{3.9}$$

with an implicit constant independent of $R \gg 1$. To be more precise, we can assume $R \geq (\text{dist}\,(J_1, J_2))^{-2}$.

Fix such an $R$. Let $\eta_R : \mathbb{R}^2 \to [0, \infty]$ have the Fourier transform supported on $B(0, 1/R)$ and satisfy

$$1_{[-R, R]^2} \leq \eta_R,$$

$$\int_{\mathbb{R}} \sup_{x_1 \in \mathbb{R}} \eta_R(x_1, x_2)^2 \, dx_2 \lesssim R.$$

This function allows us to replace the rough spatial cut-off with a smooth one, which will secure frequency localization

$$\int_{[-R, R]^2} |E_{J_1} f E_{J_2} f|^2 \leq \int_{\mathbb{R}^2} |\eta_R(E_{J_1} f) \eta_R(E_{J_2} f)|^2.$$

Indeed, using the notation from Section 3.2, $E_{J_i} f \eta_R$ has the Fourier transform supported inside $\mathcal{N}_{J_i'}(R^{-1})$. Here $J_i'$ is a slight enlargement of $J_i$.

Let $\mathcal{P}_i$ be a partition of $J_i$ into intervals $I_i$ with length $R^{-1/2}$, so that

$$\eta_R E_{J_i} f = \sum_{I_i \in \mathcal{P}_i} \eta_R E_{I_i} f.$$

A small modification of (3.5) leads to

$$\int_{\mathbb{R}^2} |\eta_R(E_{J_1} f) \eta_R(E_{J_2} f)|^2 \lesssim \int_{\mathbb{R}^2} \sum_{I_1 \in \mathcal{P}_1} |\eta_R E_{I_1} f|^2 \sum_{I_2 \in \mathcal{P}_2} |\eta_R E_{I_2} f|^2.$$

Next, note that the frequency support of $\eta_R E_{I_i} f$ is inside a rectangle $B_i$ with dimensions $\sim R^{-1/2} \times R^{-1}$. The disjointness of $J_1$ and $J_2$ forces the angle between the short sides of any such $B_1, B_2$ to be at least $\nu$, for some $\nu > 0$ depending on $\text{dist}\,(J_1, J_2)$. We can thus apply Lemma 3.7 with $\delta \sim R^{-1/2}$ to estimate

$$\int_{\mathbb{R}^2} |\eta_R(E_{I_1} f) \eta_R(E_{I_2} f)|^2 \lesssim R^{-2} \|\eta_R E_{I_1} f\|_2^2 \|\eta_R E_{I_2} f\|_2^2.$$

Finally, we invoke the conservation of mass (1.17) and Fubini's theorem to write for each $I_i$

$$\|\eta_R E_{I_i} f\|_2^2 \leq \left( \sup_{x_2 \in \mathbb{R}} \int_{\mathbb{R}} |E_{I_i} f(x_1, x_2)|^2 \, dx_1 \right) \int_{\mathbb{R}} \sup_{x_1 \in \mathbb{R}} |\eta_R(x_1, x_2)|^2 \, dx_2$$

$$\lesssim R \|f\|_{L^2(I_i)}^2.$$

Combining the last three inequalities leads to the desired estimate (3.9).

## 3.5 A General Bilinear Restriction Estimate in $L^4(\mathbb{R}^3)$

The space $L^4$ lies at the heart of the proof of Theorem 3.1 in every dimension. To simplify notation, we restrict our discussion to $\mathbb{R}^3$, but all arguments can be extended to higher dimensions relatively easily. The material in this section is a warm-up meant to introduce the reader to some of the $L^4$ manipulations needed in three and higher dimensions. It will be superseded by stronger arguments in the following sections. Along the way, we present two different arguments that explain why transversality alone guarantees a bilinear inequality in the smaller range $p \geq 4$. We also briefly explore the insufficiency of transversality in order to cover the larger range $p > 10/3$.

To emphasize the universality of the $L^4$ argument, we will describe it for two fairly general $C^2$ surfaces $\mathcal{M}_1, \mathcal{M}_2$ in $\mathbb{R}^3$ given by

$$\mathcal{M}_i = \{(\xi_1, \xi_2, \psi_i(\xi_1, \xi_2)) : (\xi_1, \xi_2) \in R_i\}, \ 1 \leq i \leq 2.$$

Here $R_1, R_2$ are two rectangles inside $[-1, 1]^2$ that are not necessarily disjoint. We do not impose any curvature condition on $\mathcal{M}_1, \mathcal{M}_2$, but we will require that their normals point in separated directions.

**Definition 3.8** (Transversality) Let $0 < \nu \leq \pi/2$. We will say that $\mathcal{M}_1$ and $\mathcal{M}_2$ are $\nu$-transverse if for each unit normal vectors $\mathbf{n}_1, \mathbf{n}_2$ to $\mathcal{M}_1$ and $\mathcal{M}_2$ we have

$$|\mathbf{n}_1 \wedge \mathbf{n}_2| > \nu. \tag{3.10}$$

If we do not want to emphasize the value of $\nu$, we will simply call $\mathcal{M}_1, \mathcal{M}_2$ transverse.

Note that transversality does not force any type of curvature. It is satisfied when $\mathcal{M}_1, \mathcal{M}_2$ are subsets of nonparallel planes, disjoint regions on the paraboloid $\mathbb{P}^2$, angularly separated regions on the cone, and even for subsets of a cylinder of the form

$$\mathcal{M}_i = \{(\xi_1, \xi_2, \xi_1^2) : (\xi_1, \xi_2) \in I_i \times [-1, 1]\}, \tag{3.11}$$

with $I_1, I_2$ disjoint intervals.

Let us now explore one useful consequence of transversality. Write for each $\omega \subset R_i$

$$\mathcal{M}_i(\omega) = \{(\xi_1, \xi_2, \psi_i(\xi_1, \xi_2)) : (\xi_1, \xi_2) \in \omega\}, \ 1 \le i \le 2.$$

Given $0 < \delta \ll 1$, let $\mathcal{P}_i = \mathcal{P}_i(\delta)$ be a partition of $R_i$ into squares $\omega_i$ with side length $\sim \delta$.

For $r > 0$, we will write

$$S_r(\omega_1, \omega_2') := \{(\omega_1', \omega_2) \in \mathcal{P}_1 \times \mathcal{P}_2 : \text{dist}\,((\mathcal{M}_1(\omega_1)$$
$$+ \mathcal{M}_2(\omega_2)), (\mathcal{M}_1(\omega_1') + \mathcal{M}_2(\omega_2')) \le r\}. \tag{3.12}$$

The most important value for us is $r = \delta$, and we will sometimes simply write $S(\omega_1, \omega_2')$ to denote $S_\delta(\omega_1, \omega_2')$.

Exercise 3.13 shows that if $\mathcal{M}_1, \mathcal{M}_2$ are $\nu$-transverse, then for each $\omega_1 \in \mathcal{P}_1$ and $\omega_2' \in \mathcal{P}_2$ we have

$$|S(\omega_1, \omega_2')| \le M\delta^{-1}, \tag{3.13}$$

for some $M$ that depends on $\nu$. Note that $M\delta^{-1}$ is much smaller than the cardinality of $\mathcal{P}_1 \times \mathcal{P}_2$, which is $\sim \delta^{-4}$. This should be compared with (3.8). Inequality (3.13) is weaker than what we previously called diagonal behavior; it allows for multiple interactions between pairs of caps on $\mathcal{M}_1 \times \mathcal{M}_2$.

Before we move on, let us explain the upper bound (3.13) in two concrete cases.

**Example 3.9** In our first example, $\mathcal{M}_1$ and $\mathcal{M}_2$ will be two separated regions on $\mathbb{P}^2$. Let us assume that $P_1 \ne P_1' \in \mathcal{M}_1$ and $P_2 \ne P_2' \in \mathcal{M}_2$ satisfy

$$P_1 + P_2 = P_1' + P_2' . \tag{3.14}$$

Writing $P_1 = (a_1, b_1, a_1^2 + b_1^2)$, etc., we find that

$$\begin{cases} a_1 + a_2 = a_1' + a_2', \\ b_1 + b_2 = b_1' + b_2', \\ a_1^2 + b_1^2 + a_2^2 + b_2^2 = (a_1')^2 + (b_1')^2 + (a_2')^2 + (b_2')^2. \end{cases}$$

If we denote by $X, Y, Z$, the values of the preceding three expressions, a simple computation reveals that the points $(a_1, b_1)$, $(a_2, b_2)$, $(a_1', b_1')$, and $(a_2', b_2')$ belong to the circle of radius $\sqrt{Z/2 - X^2 + Y^2/4}$ centered at $(X/2, Y/2)$. Moreover, they determine a rectangle.

This observation extends to the case when points are replaced with caps as follows. Given two squares $\omega_1 \in \mathcal{P}_1(\delta)$ and $\omega_2' \in \mathcal{P}_2(\delta)$ with centers $c_1$ and $c_2'$, let $l(c_1, c_2')$ be the line in the $(\xi_1, \xi_2)$-plane containing $c_1$ and perpendicular to $c_1 - c_2'$. Then for any pair $(\omega_1', \omega_2) \in S(\omega_1, \omega_2')$ centered at $(c_1', c_2)$, a computation similar to the preceding one reveals that

$$\text{dist}(c_1', l(c_1, c_2')) \lesssim \delta.$$

The implicit constant in $\lesssim$ only depends on the distance between $\mathcal{M}_1$ and $\mathcal{M}_2$. Moreover, $c_1'$ determines $c_2$ up to $O(1)$ choices. For future reference, we write this as

$$c_2 \in \mathcal{F}(c_1, c_1', c_2'),$$

with $|\mathcal{F}(c_1, c_1', c_2')| \lesssim 1$. These properties restrict the cardinality of $S(\omega_1, \omega_2')$ to $O(\delta^{-1})$. See Figure 3.1.

**Example 3.10** Let us now assume that $\mathcal{M}_1$ and $\mathcal{M}_2$ are two nonparallel planes intersecting along a line $L$. For each $P_1 \neq P_1' \in \mathcal{M}_1$ and $P_2 \neq P_2' \in \mathcal{M}_2$ satisfying (3.14), elementary geometry shows that the line segments $P_1 P_1'$ and $P_2 P_2'$ must be parallel to $L$.

This observation extends to the case when points are replaced with caps, as follows. For each $(\omega_1', \omega_2) \in S(\omega_1, \omega_2')$, the cap $\mathcal{M}_1(\omega_1')$ $(\mathcal{M}_2(\omega_2))$ is confined to the $O(\delta)$-wide strip on $\mathcal{M}_1$ $(\mathcal{M}_2)$ centered at $\mathcal{M}_1(\omega_1)$ $(\mathcal{M}_2(\omega_2'))$ and parallel to $L$. We conclude as in the previous example that $|S(\omega_1, \omega_2')| \lesssim \delta^{-1}$.

The next result shows that (3.10) is enough to guarantee a bilinear restriction inequality in the range $p \geq 4$. Note that since we do not assume any curvature, this result does not follow from Stein–Tomas-type arguments. There is a slightly easier proof that is explored in Exercise 3.13. However, the argument we present here will serve as better preparation for the material in the next sections. The reader will notice that this argument is very similar to the one we have introduced in the previous sections.

**Proposition 3.11** *Assume that $\mathcal{M}_1, \mathcal{M}_2$ satisfy (3.10), and denote by $E^1, E^2$ their extension operators. Then for each $f_i \colon R_i \to \mathbb{C}$, we have the following $L^4$ bilinear estimate:*

$$\| |E^1 f_1 E^2 f_2|^{\frac{1}{2}} \|_{L^4(\mathbb{R}^3)} \lesssim_\nu (\|f_1\|_{L^2} \|f_2\|_{L^2})^{1/2}.$$

*Proof* It suffices to prove the inequality

$$\int_{[-R, R]^3} |E^1 f_1 E^2 f_2|^2 \lesssim_\nu \|f_1\|_2^2 \|f_2\|_2^2$$

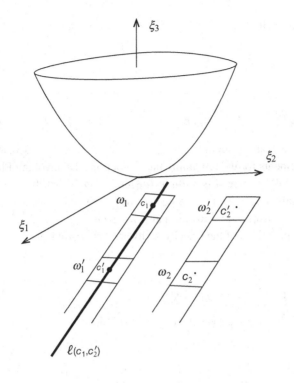

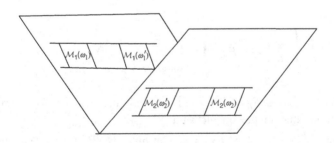

Figure 3.1 Collections $S(\omega_1, \omega_2')$ for the paraboloid and for planes.

uniformly over $R \gg 1$. Fix $R$ and let $\mathcal{P}_i$ be a partition of $R_i$ into squares with side length $R^{-1/2}$. Let $\eta_R \colon \mathbb{R}^3 \to [0, \infty]$ have the Fourier transform supported inside $B(0, 1/R)$ and satisfy

$$1_{[-R,R]^3} \leq \eta_R,$$

$$\int_{\mathbb{R}} \sup_{(x_1, x_2) \in \mathbb{R}^2} |\eta_R(x_1, x_2, x_3)|^2 \, dx_3 \lesssim R.$$

Thus we may write

$$\int_{[-R,R]^3} |E^1 f_1 E^2 f_2|^2 \leq \sum_{\omega_i, \omega_i' \in \mathcal{P}_i} \int_{\mathbb{R}^3} F_{\omega_1} \bar{F}_{\omega_1'} F_{\omega_2} \bar{F}_{\omega_2'}. \qquad (3.15)$$

Here

$$F_{\omega_1} = \eta_R E^1_{\omega_1} f_1, \quad F_{\omega_1'} = \eta_R E^1_{\omega_1'} f_1, \quad F_{\omega_2} = \eta_R E^2_{\omega_2} f_2, \quad F_{\omega_2'} = \eta_R E^2_{\omega_2'} f_2.$$

Note that the Fourier support of $F_{\omega_1}$ is inside $\mathcal{M}_1(\omega_1) + B(0, R^{-1})$, with similar statements for the remaining three functions. By invoking Plancherel's theorem, it follows that a quadruple $(\omega_1, \omega_1', \omega_2, \omega_2')$ contributes to the sum in (3.15) only if $(\omega_1', \omega_2) \in S_{R^{-1/2}}(\omega_1, \omega_2')$. We have used the fact that $4B(0, R^{-1}) \subset B(0, R^{-1/2})$. Recall that $|S_{R^{-1/2}}(\omega_1, \omega_2')| \leq M R^{1/2}$.

A few applications of the Cauchy–Schwarz inequality lead to

$$\int_{[-R,R]^3} |E^1 f_1 E^2 f_2|^2$$

$$\leq \sum_{\substack{\omega_1 \in \mathcal{P}_1 \\ \omega_2' \in \mathcal{P}_2}} \int_{\mathbb{R}^3} |F_{\omega_1} F_{\omega_2'}| \sum_{(\omega_1', \omega_2) \in S_{R^{-1/2}}(\omega_1, \omega_2')} |F_{\omega_1'} F_{\omega_2}|$$

$$\lesssim_M R^{1/4} \int_{\mathbb{R}^3} \sum_{\substack{\omega_1 \in \mathcal{P}_1 \\ \omega_2' \in \mathcal{P}_2}} |F_{\omega_1} F_{\omega_2'}| \left( \sum_{(\omega_1', \omega_2) \in S_{R^{-1/2}}(\omega_1, \omega_2')} |F_{\omega_1'} F_{\omega_2}|^2 \right)^{1/2}$$

$$\leq R^{1/4} \int_{\mathbb{R}^3} \left( \sum_{\substack{\omega_1 \in \mathcal{P}_1 \\ \omega_2' \in \mathcal{P}_2}} |F_{\omega_1} F_{\omega_2'}|^2 \right)^{1/2} \left( \sum_{\substack{\omega_1 \in \mathcal{P}_1 \\ \omega_2' \in \mathcal{P}_2}} \sum_{(\omega_1', \omega_2) \in S_{R^{-1/2}}(\omega_1, \omega_2')} |F_{\omega_1'} F_{\omega_2}|^2 \right)^{1/2}.$$

Note that $(\omega_1', \omega_2) \in S_{R^{-1/2}}(\omega_1, \omega_2')$ if and only if $(\omega_1, \omega_2') \in S_{R^{-1/2}}(\omega_1', \omega_2)$. This implies that in the double sum $\sum_{\omega_1 \in \mathcal{P}_1, \omega_2' \in \mathcal{P}_2} \sum_{(\omega_1', \omega_2) \in S_{R^{-1/2}}(\omega_1, \omega_2')}$, each pair $(\omega_1', \omega_2) \in \mathcal{P}_1 \times \mathcal{P}_2$ appears at most $M R^{1/2}$ times. Thus, the last expression is in fact dominated by

$$\lesssim_M R^{1/2} \int_{\mathbb{R}^3} \sum_{\omega_1 \in \mathcal{P}_1, \omega_2 \in \mathcal{P}_2} |F_{\omega_1} F_{\omega_2}|^2. \qquad (3.16)$$

Next, note that $\mathcal{M}_i(\omega_i) + B(0, R^{-1})$ lies inside a rectangular box $B_i$ with dimensions $\sim R^{-1/2} \times R^{-1/2} \times R^{-1}$. Our transversality requirement forces the angle between the short sides of any such $B_1, B_2$ to be at least $\nu$. We can thus apply Lemma 3.7 with $\delta \sim R^{-1/2}$ to estimate

$$\int_{\mathbb{R}^3} |F_{\omega_1} F_{\omega_2}|^2 \lesssim_\nu R^{-5/2} \|F_{\omega_1}\|_2^2 \|F_{\omega_2}\|_2^2. \qquad (3.17)$$

Finally, we invoke the conservation of mass (1.17) and Fubini's theorem to write

$$\|\eta_R E^i_{\omega_i} f_i\|_2^2$$

$$\leq \left( \sup_{x_3 \in \mathbb{R}} \int_{\mathbb{R}^2} |E^i_{\omega_i} f_i(x_1, x_2, x_3)|^2 \, dx_1 dx_2 \right) \int_{\mathbb{R}} \sup_{(x_1, x_2) \in \mathbb{R}^2} |\eta_R(x_1, x_2, x_3)|^2 \, dx_3$$

$$\lesssim R \|f_i\|_{L^2(\omega_i)}^2.$$

Combining this with (3.16) and (3.17) leads to the desired estimate.  □

To recap, there are two major ingredients in this proof: the upper bound (3.13) and inequality (3.17). It is worth noting that transversality plays a key role in both of them.

Proposition 3.11 does not in general hold true with 4 replaced with any $p < 4$, if $\mathcal{M}_1, \mathcal{M}_2$ are only subject to requirement (3.10). The simplest example to check is that of two planar surfaces. For future purposes, it will help to draw a first comparison between the cone and the cylinder. The two surfaces are locally isometric, having one zero and one nonzero principal curvature. As we have mentioned earlier (see also Exercise 3.32), the cone admits a full range of bilinear restriction estimates, $p \geq 10/3$. However, we will next show that the sharp range for the cylinder is $p \geq 4$. The feature responsible for this contrast between the two manifolds is explained in Remark 3.21.

**Example 3.12** Consider two cylinders $\mathcal{M}_1, \mathcal{M}_2$ as in (3.11), with $I_1 = [c_1, c_1 + \delta]$ and $I_2 = [c_2, c_2 + \delta]$ disjoint intervals of length $\delta \ll 1$ inside $[-1, 1]$. Let $E^i$ be the extension operator associated with $\mathcal{M}_i$. Let

$$f_i(\xi_1, \xi_2) = 1_{I_i}(\xi_1)\eta(\xi_2),$$

where $\eta$ is supported on $[-1, 1]$ and satisfies $|\widehat{\eta}| \geq 1_{[-1,1]}$. Then a simple computation involving a change of variables shows that

$$|E^i f_i(x_1, x_2, x_3)| = \delta \left| \widehat{\eta}(-x_2) \int_0^1 e(t\delta(x_1 + 2c_i x_3) + t^2 \delta^2 x_3) \, dt \right|.$$

Note that

$$|E^i f_i(x_1, x_2, x_3)| \gtrsim \delta$$

in the region $U_i$ defined by $|x_1 + 2c_i x_3| \ll \delta^{-1}$, $|x_2| \ll 1$ and $|x_3| \ll \delta^{-2}$. Even under the separation condition $|c_1 - c_2| \sim 1$, which guarantees (3.10), we still have $|U_1 \cap U_2| \gtrsim \delta^{-2}$. It thus follows that for each $p \geq 1$

$$\||E^1 f_1 E^2 f_2|^{1/2}\|_{L^p(\mathbb{R}^3)} \gtrsim \delta^{1 - \frac{2}{p}}.$$

Combining this with the fact that $\|f_i\|_2 \lesssim \delta^{1/2}$ proves that the inequality in Proposition 3.11 fails to hold in the range $p < 4$ for this choice of $\mathcal{M}_1, \mathcal{M}_2$.

We remark that in this example each $U_i$ is an almost rectangular box with dimensions $\sim \delta^{-1} \times 1 \times \delta^{-2}$. Also, $U_1, U_2$ can be thought of as being coplanar, since the axes corresponding to their sides of length 1 point in the same direction. This is precisely what makes $|U_1 \cap U_2|$ big. The reader is invited to redo these computations for angularly separated subsets of the cone. In this case, the angular separation will force $|U_1 \cap U_2| \lesssim \delta^{-1}$.

**Exercise 3.13** Let $\mathcal{M}_i = \{(\xi, \psi_i(\xi)) : \xi \in R_i\}$ be two $\nu$-transverse $C^2$ hypersurfaces in $\mathbb{R}^n$, equipped with the pullbacks $d\sigma^{\psi_i}$ of the Lebesgue measure from $\mathbb{R}^{n-1}$.

(a) Prove that if $\psi \in C^1(R)$, $R$ is a rectangular box in $\mathbb{R}^{n-1}$, and

$$\inf_{\xi \in R} |\nabla \psi(\xi)| \geq c > 0$$

then

$$|\{\xi \in R : |\psi(\xi)| \leq \delta\}| \lesssim_c \delta, \quad \delta \ll 1.$$

(b) Prove that $d\sigma^{\psi_1} * d\sigma^{\psi_2}$ is absolutely continuous with respect to the Lebesgue measure, with density satisfying $\|F\|_{L^\infty(\mathbb{R}^n)} \lesssim_\nu 1$.

(c) Use (b) to prove (3.13).

(d) Use (b) to give a slightly different proof of Proposition 3.11 for arbitrary $n \geq 2$.

Hint for (a): Prove it first for $n = 2$. When $n > 2$, partition $R$ into rectangular boxes, on each of which a given partial derivative is $\gtrsim c$, and reduce it to $n = 2$.

Hint for (b): Reduce the problem to proving that

$$|(\mathcal{M}_1 + B(0, \delta)) \cap (\mathcal{M}_2 + B(\eta, \delta))| \lesssim_\nu \delta^2$$

holds for $\delta \ll 1$ and $\eta \in \mathbb{R}^n$. Deduce this from (a) and Fubini's theorem.

Hint for (c): Fix $\eta_1 \in \mathcal{M}_1(\omega_1)$ and $\eta_2' \in \mathcal{M}_2(\omega_2')$. Find an upper bound for

$$d\sigma^{\psi_1} \times d\sigma^{\psi_2}(\{(\eta_1', \eta_2) : \eta_2 - \eta_1' \in B(\eta_2' - \eta_1, 10\delta)\}).$$

**Exercise 3.14** (a) Given a hypersurface $\mathcal{M} = \{(\xi, \psi(\xi)) : \xi \in U\}$ in $\mathbb{R}^n$ and $F : \mathcal{N}_\psi(\delta) \to \mathbb{C}$, let $f_t(\xi) = F(\xi, \psi(\xi) + t)$ for $|t| < \delta$. Let also $E^{\psi_t}$ be the extension operator associated with the vertical translate $te_n + \mathcal{M}$. Prove that

$$\widehat{F}(x) = \int_{-\delta}^{\delta} E^{\psi_t} f_t(-x)\, dt.$$

(b) Consider two hypersurfaces $\mathcal{M}_i = \{(\xi, \psi_i(\xi)) : \xi \in U_i\}$ in $\mathbb{R}^n$ with extension operators $E^i$. Let $C_{1,p}$ be the smallest constant so that

$$\||E^1 f_1 E^2 f_2|^{\frac{1}{2}}\|_{L^p(\mathbb{R}^n)} \leq C_{1,p}(\|f_1\|_{L^2(U_1)}\|f_2\|_{L^2(U_2)})^{\frac{1}{2}}$$

holds for each $f_i \in L^2(U_i)$. Let also $C_{2,p}$ be the smallest constant so that

$$\||\widehat{F}_1\widehat{F}_2|^{\frac{1}{2}}\|_{L^p(\mathbb{R}^n)} \leq C_{2,p}R^{-\frac{1}{2}}(\|F_1\|_{L^2(\mathcal{N}_{\psi_1}(R^{-1}))}\|F_2\|_{L^2(\mathcal{N}_{\psi_2}(R^{-1}))})^{\frac{1}{2}} \tag{3.18}$$

holds for each $R \geq 1$ and each $F_i \in L^2(\mathcal{N}_{\psi_i}(R^{-1}))$.
Prove that $C_{1,p} \lesssim C_{2,p}$, and moreover, if $2 \leq p \leq 4$, then $C_{2,p} \lesssim C_{1,p}$.
Conclude using Exercise 3.13 (d) that (3.18) holds with $C_2 = O(1)$ for $p = 4$, whenever $\mathcal{M}_1, \mathcal{M}_2$ are transverse.

**Exercise 3.15** (Improved $L^4$ estimate) Let $l_1, l_2$ be two nonparallel line segments in $[-1, 1]^2$. Assume that $f_i$ is supported on the $R^{-1/2}$-neighborhood on $l_i$. Prove that

$$\||Ef_1 Ef_2|^{1/2}\|_{L^4(B(0,R))} \lesssim R^{-1/8}(\|f_1\|_2\|f_2\|_2)^{1/2}.$$

If $l_1, l_2$ are parallel, construct $f_1, f_2$ supported as before such that

$$\||Ef_1 Ef_2|^{1/2}\|_{L^4(B(0,R))} \sim (\|f_1\|_2\|f_2\|_2)^{1/2}.$$

**Exercise 3.16** Rework the computations in Example 3.9 to identify the structure of $S(\omega_1, \omega_2')$ in the case of $\text{Cone}^2$ and $\mathbb{S}^2$.

## 3.6 From Point-Line to Cube-Tube Incidences

In this section, we illustrate some of the differences and similarities between the way lines intersect and the way tubes intersect. This discussion will pave the way toward the more intricate analysis in the following sections.

Let $\mathcal{L}$ and $\mathcal{P}$ be finite families of lines and points in $\mathbb{R}^n$. The collection of their incidences is defined as follows:

$$\mathcal{I}(\mathcal{L}, \mathcal{P}) = \{(L, P) \in \mathcal{L} \times \mathcal{P} : P \in L\}.$$

The next result presents a basic estimate on the size of $\mathcal{I}(\mathcal{L}, \mathcal{P})$.

**Proposition 3.17** *The following inequality holds:*

$$|\mathcal{I}(\mathcal{L},\mathcal{P})| \leq |\mathcal{L}| + |\mathcal{P}||\mathcal{L}|^{1/2}. \tag{3.19}$$

*Proof* Using the Cauchy–Schwarz inequality and the fact that there is at most one line passing through two distinct points, we get

$$|\mathcal{I}(\mathcal{L},\mathcal{P})|^2 = \left(\sum_L \sum_P 1_L(P)\right)^2$$

$$\leq |\mathcal{L}| \sum_L \left(\sum_P 1_L(P) + \sum_{P_1 \neq P_2} 1_L(P_1)1_L(P_2)\right)$$

$$< |\mathcal{L}|(|\mathcal{I}(\mathcal{L},\mathcal{P})| + |\mathcal{P}|^2).$$

The desired inequality is now immediate. $\qquad\square$

The term $|\mathcal{L}|$ in the upper bound for $|\mathcal{I}(\mathcal{L},\mathcal{P})|$ cannot in general be neglected, as can be seen in the case when all lines intersect in one point $P$ and $\mathcal{P} = \{P\}$. However, we will show that if we ignore (at most) one pair $(L, P)$ for each line $L$, the size of the remaining incidences is $O(|\mathcal{P}||\mathcal{L}|^{1/2})$. Such an upper bound is more convenient for applications.

More precisely, let us consider the subset of $\mathcal{L} \times \mathcal{P}$ consisting of all pairs $(L, P)$ such that $P$ is the only point in $\mathcal{P}$ that belongs to $L$. For each such pair, we write $P \sim L$, and we will otherwise write $P \not\sim L$.

We define

$$\mathcal{I}^{\not\sim}(\mathcal{L},\mathcal{P}) = \{(L,P) \in \mathcal{L} \times \mathcal{P} : P \in L \text{ and } P \not\sim L\}.$$

**Proposition 3.18** *The following two inequalities hold:*

$$|\{P \in \mathcal{P} : P \sim L\}| \leq 1, \quad \text{for each } L \in \mathcal{L}$$

*and*

$$|\mathcal{I}^{\not\sim}(\mathcal{L},\mathcal{P})| \leq \sqrt{3}|\mathcal{P}||\mathcal{L}|^{1/2}.$$

*Proof* The first inequality is obvious. To prove the second one, we start by noting that $\mathcal{I}^{\not\sim}(\mathcal{L},\mathcal{P}) = \mathcal{I}(\mathcal{L}',\mathcal{P})$, where $\mathcal{L}'$ consists of those lines in $\mathcal{L}$ that contain at least two points from $\mathcal{P}$. Since $|\mathcal{L}'| \leq |\mathcal{P}|^2$, (3.19) gives the estimate

$$|\mathcal{I}(\mathcal{L}',\mathcal{P})| \leq 2|\mathcal{P}|^2.$$

We combine this with the inequality

$$|\mathcal{I}(\mathcal{L}',\mathcal{P})|^2 \leq |\mathcal{L}'|(|\mathcal{I}(\mathcal{L}',\mathcal{P})| + |\mathcal{P}|^2)$$

from the proof of Proposition 3.17 to conclude that

$$|\mathcal{I}^{\not\sim}(\mathcal{L},\mathcal{P})|^2 \leq |\mathcal{I}(\mathcal{L}',\mathcal{P})|^2 \leq |\mathcal{L}|(2|\mathcal{P}|^2 + |\mathcal{P}|^2). \qquad\square$$

We will next show how to construct a similar relation $\sim$ for tubes. Let $\mathcal{T}$ be a finite collection consisting of cylinders with radius $\delta$ and length 1 inside $[-1, 1]^n$. Assume the separation condition

$$T \not\subset 100T', \text{ for each } T, T' \in \mathcal{T}. \tag{3.20}$$

Let also $\mathcal{Q}$ be a collection of pairwise disjoint cubes $q \subset [-1, 1]^n$ with side length $\delta$. We introduce the following quantity

$$\mathcal{I}(\mathcal{T}, \mathcal{Q}) = \delta^{-n} \sum_{(T,q) \in \mathcal{T} \times \mathcal{Q}} |T \cap q|.$$

Roughly speaking, this measures the number of pairs $(T, q) \in \mathcal{T} \times \mathcal{Q}$ with $T \cap q \neq \emptyset$.

It is not difficult to arrange for $\mathcal{I}(\mathcal{T}, \mathcal{Q})$ to be much larger than $|\mathcal{T}| + |\mathcal{Q}||\mathcal{T}|^{1/2}$. Indeed, let us consider a collection $\mathcal{T}$ of tubes containing the origin, and with directions separated by $\gg \delta$. Note that in particular (3.20) is satisfied. If we require that the direction of each tube is within $\alpha$ from a fixed direction $\mathbf{v}$, $|\mathcal{T}|$ may be as large as $\sim (\alpha/\delta)^{n-1}$. Let $\mathcal{Q}$ be a collection of roughly $\alpha^{-1}$ pairwise disjoint cubes along the line segment $[-\delta/\alpha, \delta/\alpha]\mathbf{v}$. It is easy to see that each $T \in \mathcal{T}$ intersects significantly each $q \in \mathcal{Q}$, and thus

$$\mathcal{I}(\mathcal{T}, \mathcal{Q}) \sim |\mathcal{T}||\mathcal{Q}|.$$

To obtain a result similar to Proposition 3.18, we will define a relation $\sim$ between the tubes in $\mathcal{T}$ and the cubes with scale much larger than $\delta$. More precisely, we will partition $[-1, 1]^n$ into $M^n$ cubes $Q$ of side length $2M^{-1}$.

For each given $T$, we write $Q \sim T$ if $Q$ either maximizes, or if it is one of the (at most) $3^n - 1$ neighbors of the cube that maximizes the quantity

$$I(T, Q) = \sum_{q \in \mathcal{Q}} |T \cap q \cap Q|.$$

We define the restricted cube-tube incidences:

$$\mathcal{I}^{\not\sim}(\mathcal{T}, \mathcal{Q}) = \delta^{-n} \sum_{T} \sum_{Q \not\sim T} \sum_{q \in \mathcal{Q}} |T \cap q \cap Q|.$$

**Proposition 3.19** *The following two inequalities hold*

$$|\{Q : Q \sim T\}| \leq 3^n, \text{ for each } T \in \mathcal{T} \tag{3.21}$$

*and*

$$\mathcal{I}^{\not\sim}(\mathcal{T}, \mathcal{Q}) \lesssim M^{\frac{3n+1}{2}} |\mathcal{Q}||\mathcal{T}|^{1/2}, \tag{3.22}$$

*where $\lesssim$ denotes the logarithmic loss $\log(1/\delta)^{O(1)}$.*

*Proof* Let $\mathcal{T}_k$ be the collection of those tubes $T \in \mathcal{T}$ satisfying

$$2^k \leq \delta^{-n} \sum_{Q \not\sim T} \sum_{q \in Q} |T \cap q \cap Q| < 2^{k+1}.$$

There are $\lesssim 1$ values of $k$ that need to be considered. It thus suffices to show that for each $k$

$$|\mathcal{T}_k|2^k \lesssim M^{\frac{3n+1}{2}}|Q||\mathcal{T}_k|^{1/2}. \tag{3.23}$$

Since at most $O(M)$ cubes $Q$ intersect each tube, it follows that for each $T \in \mathcal{T}_k$, there is $Q'_T \not\sim T$ such that

$$I(T, Q'_T) \gtrsim 2^k M^{-1} \delta^n.$$

Our definition of the relation $\sim$ guarantees that we also have

$$I(T, Q_T) \gtrsim 2^k M^{-1} \delta^n$$

for some $Q_T \sim T$ with $\text{dist}(Q_T, Q'_T) \gtrsim M^{-1}$.

Since there are $O(M^{2n})$ pairs of $Q, Q'$, it follows that there are two cubes $Q, Q'$ that are not neighbors, and a subset $\mathcal{T}'_k \subset \mathcal{T}_k$ with size $\gtrsim |\mathcal{T}_k|M^{-2n}$ such that $Q_T = Q$ and $Q'_T = Q'$ for each $T \in \mathcal{T}'_k$.

On the one hand, we have

$$\sum_{T \in \mathcal{T}'_k} I(T, Q)I(T, Q') \gtrsim |\mathcal{T}_k|M^{-2n}(2^k M^{-1} \delta^n)^2.$$

On the other hand, writing $S = \cup_{q \in Q} q$ we also have

$$\sum_{T \in \mathcal{T}'_k} I(T, Q)I(T, Q') = \int_{S \cap Q} \int_{S \cap Q'} \sum_{T \in \mathcal{T}'_k} 1_T(x)1_T(x')\, dx dx'$$

$$\lesssim M^{n-1}|S \cap Q||S \cap Q'|$$

$$\lesssim M^{n-1}|Q|^2 \delta^{2n}.$$

We have used that $\text{dist}(Q, Q') \gtrsim M^{-1}$ together with (3.20) to argue that there are $O(M^{n-1})$ tubes in $\mathcal{T}$ that contain both $x \in Q$ and $x' \in Q'$. The combination of the last two inequalities proves (3.23). $\qquad\square$

Let us briefly comment on the usefulness of Proposition 3.19 and its close relatives. There are many interesting problems that amount to estimating various quantities involving tubes and cubes. It is often helpful to have a control like (3.22) over the size of the incidences. Typically, Proposition 3.19 is applied with $M = \delta^{-\epsilon}$, for a small enough $\epsilon$, so that the loss $M^{(3n+1)/2}$ in (3.22) is tiny. On the other hand (3.21) shows that for each $T$ only the tiny fraction $O((3/M)^n)$ of the cubes $Q$ are discarded from the restricted incidences. The contribution from $Q \sim T$ is typically estimated rather easily by rescaling $Q$ into $[-1, 1]^n$ and invoking induction on scales.

## 3.7 A Bilinear Incidence Result for Lines

The main results in this section model the interactions between wave packets appearing in the proof of Theorem 3.1 using lines instead of tubes. This model is technically simpler, but otherwise a rather accurate replica of the "real-life" scenario involving tubes.

We consider two finite families of lines $\mathcal{L}_1, \mathcal{L}_2$ in $\mathbb{R}^3$. We allow multiple lines to be parallel to each other within each family. For each $L_1 \in \mathcal{L}_1, L_2 \in \mathcal{L}_2$, we also consider subfamilies $\mathcal{L}(L_1, L_2) \subset \mathcal{L}_1$ with the following property:

(P1)  Given $L_1 \in \mathcal{L}_1, L_2, L_2' \in \mathcal{L}_2$ and two nonparallel lines $L_1', L_1''$ in $\mathcal{L}(L_1, L_2')$, the lines $L_1', L_1'', L_2$ cannot lie in the same plane.

Let $F$ be a map from $\mathcal{L}_1 \times \mathcal{L}_1 \times \mathcal{L}_2$ to $\mathcal{L}_2$. Given subsets $\mathcal{L}_1' \subset \mathcal{L}_1$ and $\mathcal{L}_2' \subset \mathcal{L}_2$, we introduce the associated bilinear incidences, defined as the following collection of 4-tuples

$$\mathcal{BI}(\mathcal{L}_1', \mathcal{L}_2') := \{(L_1, L_1', L_2, L_2') \in (\mathcal{L}_1')^2 \times (\mathcal{L}_2')^2 :$$
$$L_1' \in \mathcal{L}(L_1, L_2'), \; L_1 \in \mathcal{L}(L_1', L_2),$$
$$L_2 = F(L_1, L_1', L_2'), \; L_2' = F(L_1', L_1, L_2),$$
$$L_1 \cap L_1' \cap L_2 \cap L_2' \neq \emptyset\}.$$

We call any 4-tuple such as the preceding ones *admissible*. We summarize the requirements $L_2 = F(L_1, L_1', L_2')$ and $L_2' = F(L_1', L_1, L_2)$ as follows:

(P2)  In each admissible 4-tuple $(L_1, L_1', L_2, L_2')$, the entries $L_1, L_1', L_2'$ uniquely determine the entry $L_2$ and the entries $L_1, L_1', L_2$ uniquely determine the entry $L_2'$.

**Remark 3.20** The motivation for considering such families of lines comes from Example 3.9, and we borrow the notation from there to explain it. Consider two families $\mathcal{L}_i$ of lines $L_i$ having directions parallel to normal vectors $(-2c_i, 1)$ to $\mathbb{P}^2$ at points $(c_i, |c_i|^2)$, $c_i \in R_i$. We will assume $R_1, R_2$ to be planar neighborhoods of $(1, 0, 0)$ and $(0, 1, 0)$, respectively. If we choose them small enough, an immediate inspection shows that for each collection of distinct line $L_1', L_1'', L_1 \in \mathcal{L}_1$ and $L_2, L_2' \in \mathcal{L}_2$ corresponding to points $c_1', c_1'', c_1, c_2, c_2'$ that satisfy

$$c_1' \neq c_1'' \text{ and } c_1', c_1'' \in l(c_1, c_2'),$$

the lines $L_1', L_1'', L_2$ cannot lie in the same plane. Thus (P1) will be satisfied if we let $\mathcal{L}(L_1, L_2)$ consist of the lines $L_1' \in \mathcal{L}_1$ corresponding to points $c_1'$ such that $c_1' \in l(c_1, c_2)$.

**Remark 3.21** It is rather immediate that (P1) is satisfied for each choice of $\mathcal{L}(L_1, L_2)$, if the lines $L_i \in \mathcal{L}_i$ are parallel to normal vectors at points from two angularly separated regions on the cone. Note, however, that this observation does not extend to the cylinder, as all normal vectors are now parallel to a plane.

We are interested in the situation where the number $|\mathcal{BI}(\mathcal{L}_1, \mathcal{L}_2)|$ of bilinear incidences is not much larger than $|\mathcal{L}_1||\mathcal{L}_2|$. This is similar to what we have previously called diagonal behavior. The next example shows that requirements (P1) and (P2) are not enough to guarantee such an upper bound.

**Example 3.22** Fix a point $P \in \mathbb{R}^3$ and two distinct planes $\pi_1, \pi_2$ through $P$. Let $\mathcal{L}_i$ be an arbitrary collection of lines in $\pi_i$, all containing $P$. We assume that none of these lines is $\pi_1 \cap \pi_2$ and we let each $\mathcal{L}(L_1, L_2)$ consist of the whole $\mathcal{L}_1$. Note that for each choice of $F$, the requirements (P1) and (P2) will be satisfied. Moreover, $|\mathcal{BI}(\mathcal{L}_1, \mathcal{L}_2)|$ is as large as $|\mathcal{L}_1|^2|\mathcal{L}_2|$.

We next aim at strengthening the definition of an admissible 4-tuple, with the intent of getting the better upper bound on $|\mathcal{BI}(\mathcal{L}_1, \mathcal{L}_2)|$. We will tolerate multiple logarithmic losses of the order $(\log N)^{O(1)}$, where $N = \max (|\mathcal{L}_1|, |\mathcal{L}_2|)$. We will hide these losses in the notation $\lesssim$.

Let $\mathcal{P}$ be the collection of points in $\mathbb{R}^3$ lying at the intersection of at least one line from $\mathcal{L}_1$ and at least one line from $\mathcal{L}_2$. There are at most $N^2$ such points. For $1 \le \mu \le N^2$ of the form $\mu \in 2^{\mathbb{N}}$, we let $\mathcal{P}_\mu$ be all the points in $\mathcal{P}$ through which pass at least $\mu$ and at most $2\mu - 1$ lines from $\mathcal{L}_2$. There are $\lesssim 1$ such sets $\mathcal{P}_\mu$ and they partition $\mathcal{P}$.

For $P \in \mathcal{P}_\mu$ and $L_1 \in \mathcal{L}_1$, we write $P \sim L_1$ if $P \in L_1$ and $L_1$ contains no other point from $\mathcal{P}_\mu$. For $\mathcal{L}_1' \subset \mathcal{L}_1$ and $\mathcal{L}_2' \subset \mathcal{L}_2$, we refine the collection of bilinear incidences as follows:

$$\mathcal{BI}^{\nsim}(\mathcal{L}_1', \mathcal{L}_2') := \{(L_1, L_1', L_2, L_2') \in \mathcal{BI}(\mathcal{L}_1', \mathcal{L}_2'): P \nsim L_1, \ P \nsim L_1',$$

$$\text{where } P = L_1 \cap L_1' \cap L_2 \cap L_2'\}.$$

The lines $L_1, L_1'$ in this new definition will necessarily contain at least one point in $\mathcal{P}_\mu \setminus \{P\}$, where $\mu$ is the unique number such that $P \in \mathcal{P}_\mu$. Note that in the previous example where $|\mathcal{BI}(\mathcal{L}_1, \mathcal{L}_2)|$ was very big, we now have $\mathcal{BI}^{\nsim}(\mathcal{L}_1, \mathcal{L}_2) = \emptyset$. The next theorem shows that we always have a favorable estimate on the size of $\mathcal{BI}^{\nsim}(\mathcal{L}_1, \mathcal{L}_2)$.

**Theorem 3.23** (Diagonal behavior for line incidences) *For each $\mathcal{L}_1, \mathcal{L}_2$ such as the preceding, we have*

$$|\mathcal{BI}^{\not\sim}(\mathcal{L}_1, \mathcal{L}_2)| \lesssim |\mathcal{L}_1||\mathcal{L}_2|.$$

It is natural to ask whether the extra requirement $P \not\sim L_1, L_1'$ does not trivialize the 4-tuples in $\mathcal{BI}^{\not\sim}(\mathcal{L}_1, \mathcal{L}_2)$, to the extent that the inequality in Theorem 3.23 is too weak and potentially not useful in applications. Fortunately, this is not the case. At this point, we can only give a somewhat vague justification of the strength of this estimate. Note that

$$|\{P \in \mathcal{P} : P \sim L_1\}| \lesssim 1 \text{ for each } L_1 \in \mathcal{L}_1. \tag{3.24}$$

Thus, for each line in $\mathcal{L}_1$, we only discard bilinear incidences corresponding to at most $\lesssim 1$ points in $\mathcal{P}$. The precise justification of why this is acceptable will come in the next sections, where we apply these ideas to the setting of tubes. In short, the tube incidences that we will discard will be easily estimated using induction on scales.

For subsets $\mathcal{P}' \subset \mathcal{P}$, $\mathcal{L}_1', \mathcal{L}_1'' \subset \mathcal{L}_1$ and $\mathcal{L}_2' \subset \mathcal{L}_2$, we let

$$\begin{aligned}\mathcal{BI}(\mathcal{L}_1', \mathcal{L}_1'', \mathcal{L}_2', \mathcal{P}') := \{&(L_1, L_1', L_2, L_2') \in (\mathcal{L}_1' \times \mathcal{L}_1'' \times \mathcal{L}_2' \times \mathcal{L}_2')\\ &\cap \mathcal{BI}(\mathcal{L}_1, \mathcal{L}_2) : L_1 \cap L_1' \cap L_2 \cap L_2' \in \mathcal{P}'\}.\end{aligned}$$

In order to prove Theorem 3.23, we need another dyadic partitioning. For each $1 \leq \lambda \leq N$ of the form $\lambda \in 2^{\mathbb{N}}$ and each $\mu$ as shown previously, we let $\mathcal{L}_1[\lambda, \mu]$ be the collection of all lines from $\mathcal{L}_1$ containing between $\lambda + 1$ and $2\lambda$ points from $\mathcal{P}_\mu$.

Note that since $\lambda + 1 \geq 2$, we have the following partition

$$\mathcal{BI}^{\not\sim}(\mathcal{L}_1, \mathcal{L}_2) = \cup_\lambda \cup_\lambda' \cup_\mu \mathcal{BI}(\mathcal{L}_1[\lambda, \mu], \mathcal{L}_1[\lambda', \mu], \mathcal{L}_2, \mathcal{P}_\mu).$$

Thus, Theorem 3.23 will follow once we prove the next proposition.

**Proposition 3.24** *For each $\lambda, \lambda', \mu$ as shown previously, we have the estimate*

$$|\mathcal{BI}(\mathcal{L}_1[\lambda, \mu], \mathcal{L}_1[\lambda', \mu], \mathcal{L}_2, \mathcal{P}_\mu)| \leq 4|\mathcal{L}_1||\mathcal{L}_2|.$$

Let us fix $\lambda, \lambda', \mu$ throughout the rest of the argument. Because of the symmetry between the role of the first and second entries in the definition of admissible 4-tuples, we may assume that $\lambda' \geq \lambda$. The proof of the proposition will follow from a sequence of simple lemmas.

For each $P \in \mathcal{P}_\mu$, we introduce the quantity

$$\nu(P) = \max_{\substack{L_1 \in \mathcal{L}_1[\lambda, \mu] \\ L_2' \in \mathcal{L}_2}} |\{L_1' \in \mathcal{L}_1[\lambda', \mu] \cap \mathcal{L}(L_1, L_2') : P \in L_1'\}|.$$

In the following, the addition of $P$ as a variable to various quantities has the effect of restricting lines to those going through $P$.

**Lemma 3.25**

$$|\mathcal{BI}(\mathcal{L}_1[\lambda,\mu], \mathcal{L}_1[\lambda',\mu], \mathcal{L}_2, \mathcal{P}_\mu)| \leq 2 \sum_{P \in \mathcal{P}_\mu} |\mathcal{L}_1[\lambda,\mu,P]|\mu\nu(P).$$

*Proof* First we observe that

$$\mathcal{BI}(\mathcal{L}_1[\lambda,\mu], \mathcal{L}_1[\lambda',\mu], \mathcal{L}_2, \mathcal{P}_\mu) = \bigcup_{P \in \mathcal{P}_\mu} \mathcal{BI}(\mathcal{L}_1[\lambda,\mu,P], \mathcal{L}_1[\lambda',\mu,P], \mathcal{L}_2(P), \{P\}).$$

Second, we invoke (P2) and the fact that $|\mathcal{L}_2(P)| \leq 2\mu$ to write

$$|\mathcal{BI}(\mathcal{L}_1[\lambda,\mu,P], \mathcal{L}_1[\lambda',\mu,P], \mathcal{L}_2(P), \{P\})|$$

$$= \sum_{L_1 \in \mathcal{L}_1[\lambda,\mu,P]} \sum_{L_2' \in \mathcal{L}_2(P)} \sum_{L_1' \in \mathcal{L}_1[\lambda',\mu,P] \cap \mathcal{L}(L_1, L_2')} 1$$

$$\leq 2|\mathcal{L}_1[\lambda,\mu,P]|\mu\nu(P). \qquad \square$$

**Lemma 3.26**

$$\sum_{P \in \mathcal{P}_\mu} |\mathcal{L}_1[\lambda,\mu,P]| \leq 2\lambda|\mathcal{L}_1|.$$

*Proof* It suffices to note that each line in $\mathcal{L}_1$ belongs to at most $2\lambda$ sets $\mathcal{L}_1[\lambda,\mu,P]$, as $P$ ranges over all points in $\mathcal{P}_\mu$. $\qquad \square$

**Lemma 3.27** *For each* $P \in \mathcal{P}_\mu$, *we have*

$$\nu(P) \leq \frac{|\mathcal{L}_2|}{\mu\lambda'}.$$

*Proof* Fix $L_1 \in \mathcal{L}_1[\lambda,\mu]$ and $L_2' \in \mathcal{L}_2$. We need to prove that

$$|\mathcal{L}_1'| \leq \frac{|\mathcal{L}_2|}{\mu\lambda'},$$

where

$$\mathcal{L}_1' := \mathcal{L}_1[\lambda',\mu,P] \cap \mathcal{L}(L_1, L_2').$$

Figure 3.2 shows three lines $L_1', L_1'', L_1''' \in \mathcal{L}_1'$ and a few clusters. Each of these lines contains at least $\lambda'$ points different from $P$, each with a cluster of at least $\mu$ lines from $\mathcal{L}_2$ through it. One should keep in mind that all lines in the figure live in $\mathbb{R}^3$.

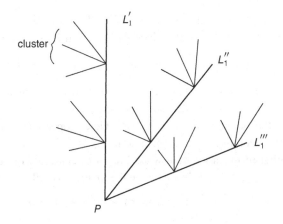

Figure 3.2 The clusters.

Property (P1) forces the lines from various clusters to be distinct. Indeed, any line appearing in two distinct clusters, say the cluster through $P' \in L_1'$ and the cluster through $P'' \in L_1''$, would necessarily be coplanar with $L_1'$ and $L_1''$. We remark that it is in this part of the argument where the fact that $\lambda' > 0$ plays a critical role. The inequality $\lambda' > 0$ is a consequence of the fact that we only consider lines satisfying $P \not\subset L_1'$. It is precisely this restriction that allows us to work with $P', P'' \neq P$.

As a result of all these observations, the union of all clusters will contain at least $\lambda' \mu |\mathcal{L}_1'|$ lines from $\mathcal{L}_2$. This proves the inequality from our lemma. □

The proof of Proposition 3.24 follows by combining Lemmas 3.25 through 3.27, recalling also that $\lambda \leq \lambda'$.

## 3.8 Achieving Diagonal Behavior for Tubes

Fix $R \gg 1$. Consider two finite sets $\Xi_1, \Xi_2 \subset \mathbb{R}^2$, lying inside small enough neighborhoods of $(1,0)$ and $(0,1)$, respectively. The relevance of this smallness requirement is explained in Remark 3.20.

We assume $\operatorname{dist}(\xi, \xi') \geq R^{-1/2}$ whenever $\xi, \xi' \in \Xi_1$ or $\xi, \xi' \in \Xi_2$. Consider also a finite set $\Theta \subset [-10R, 10R]^2$ with $\operatorname{dist}(\theta, \theta') \geq R^{1/2}$ for each $\theta \neq \theta' \in \Theta$.

Let $\mathcal{L}_1, \mathcal{L}_2$ be two finite collections of lines on $\mathbb{R}^3$, with the following properties. Each line in $\mathcal{L}_i$ has direction $(-2\xi, 1)$, for some $\xi \in \Xi_i$ and passes through a point $(\theta, 0)$, for some $\theta \in \Theta$. Given any such line $L$, we denote by $(\xi_L, \theta_L)$ the data associated with it this way.

For $\xi \neq \xi' \in \mathbb{R}^2$, we will denote by $l(\xi, \xi')$ the line in $\mathbb{R}^2$ through $\xi$ and perpendicular to the vector $\xi - \xi'$. Given $L_1 \in \mathcal{L}_1$ and $L_2 \in \mathcal{L}_2$, we let

$$\mathcal{L}(L_1, L_2) := \{L_1' \in \mathcal{L}_1 : \text{dist}\,(\xi_{L_1'}, l(\xi_{L_1}, \xi_{L_2})) \lesssim R^{-1/2}\}.$$

The definition of $\mathcal{L}$ is motivated by Remark 3.20, with an eye on the application from the next section. The reader can check that this setup is consistent with the one from the previous section; in particular, property (P1) is satisfied. Consequently, Theorem 3.23 is applicable to $\mathcal{L}_1$ and $\mathcal{L}_2$. However, we want to have a similar result in the case when lines are replaced with tubes. More precisely, for each $L \in \mathcal{L}_1 \cup \mathcal{L}_2$ we define the tube $T_L$ as follows:

$$T_L := \{x = (x_1, x_2, x_3) : \text{dist}\,(x, L) \le R^{1/2}, \ |x_3| \le R\}. \tag{3.25}$$

The separation conditions imposed on $\Xi_1$, $\Xi_2$, and $\Theta$ have the effect of making these tubes sufficiently distinct. By that we mean that there are only $O(1)$ tubes $T_{L'}$ inside the fatter tube $CT_L$, whenever $C = O(1)$.

We fix a small parameter $\delta \ll 1$ and also define the slight enlargement of $T_L$:

$$R^\delta T_L := \{x = (x_1, x_2, x_3) : \text{dist}\,(x, L) \le R^{\frac{1}{2}+\delta}, \ |x_3| \le R\}.$$

We call $\mathbb{T}_i$ the collection of all tubes $T_{L_i}$ with $L_i \in \mathcal{L}_i$. Recall the definition of $\mathcal{F}$ from Example 3.9. For each $L_1, L_1' \in \mathcal{L}_1$ and $L_2' \in \mathcal{L}_2$, we abuse notation and let $\mathcal{F}(L_1, L_1', L_2')$ be the subset of $\mathcal{L}_2$ given by

$$\mathcal{F}(L_1, L_1', L_2') = \{L_2 \in \mathcal{L}_2 : \xi_{L_2} \in \mathcal{F}(\xi_{L_1}, \xi_{L_1'}, \xi_{L_2'})\}.$$

For each $\mathbb{T}_1' \subset \mathbb{T}_1$ and $\mathbb{T}_2' \subset \mathbb{T}_2$, we define the collection of their bilinear incidences as follows:

$$\mathcal{BI}(\mathbb{T}_1', \mathbb{T}_2') := \{(T_{L_1}, T_{L_1'}, T_{L_2}, T_{L_2'}) \in (\mathbb{T}_1')^2 \tag{3.26}$$

$$\times (\mathbb{T}_2')^2 : L_1' \in \mathcal{L}(L_1, L_2'), \ L_1 \in \mathcal{L}(L_1', L_2),$$

$$L_2 \in \mathcal{F}(L_1, L_1', L_2'), \ L_2' \in \mathcal{F}(L_1', L_1, L_2),$$

$$R^\delta T_{L_1} \cap R^\delta T_{L_1'} \cap R^\delta T_{L_2} \cap R^\delta T_{L_2'} \neq \emptyset\}.$$

Note that in order for four tubes to contribute as a bilinear incidence, we do not require that their central axes intersect, but rather that there is a point at distance $\lesssim R^{\delta+1/2}$ from each of these lines. We will again be interested in the situation where, after a small calibration, the size of $\mathcal{BI}(\mathbb{T}_1, \mathbb{T}_2)$ is not much larger than $|\mathbb{T}_1||\mathbb{T}_2|$; in fact, at most it is $R^{O(\delta)}|\mathbb{T}_1||\mathbb{T}_2|$.

Let $\mathcal{B}$ be the partition of $[-R, R]^3$ into cubes $B$ with side length $2R^{1-\delta}$. There are $\sim R^{3\delta}$ such cubes. Given any relation $B \sim T_1$ with $T_1 \in \mathbb{T}_1$ and

$B \in \mathcal{B}$ – in other words, a subset of $\mathcal{B} \times \mathbb{T}_1$ – we define the set of restricted bilinear incidences:

$$\mathcal{BI}^{\not\sim}(\mathbb{T}_1', \mathbb{T}_2') := \{(T_1, T_1', T_2, T_2') \in \mathcal{BI}(\mathbb{T}_1', \mathbb{T}_2') \colon B \not\sim T_1, T_1'$$
$$\text{for each } B \in \mathcal{B} \text{ such that}$$
$$R^\delta T_1 \cap R^\delta T_1' \cap R^\delta T_2 \cap R^\delta T_2' \subset 2B\}.$$

For each 4-tuple of tubes, there are only $O(1)$ cubes $B$ as before. Note that compared to the case of lines from the previous section, we are pruning more bilinear incidences from the original collection $\mathcal{BI}(\mathbb{T}_1', \mathbb{T}_2')$, as the scale $R^{1-\delta}$ of each $B$ is much larger than $R^{1/2+\delta}$, the typical diameter of the intersection of two transverse tubes.

We have the following analog of Theorem 3.23 and inequality (3.24).

**Theorem 3.28** (Diagonal behavior for tube incidences) *Given two families of tubes* $\mathbb{T}_1, \mathbb{T}_2$ *as before, there is a relation* $\sim$ *between the cubes in* $\mathcal{B}$ *and the tubes in* $\mathbb{T}_1$ *such that*

$$|\mathcal{BI}^{\not\sim}(\mathbb{T}_1, \mathbb{T}_2)| \lesssim R^{O(\delta)} |\mathbb{T}_1| |\mathbb{T}_2| \tag{3.27}$$

*and*

$$|\{B \in \mathcal{B} \colon B \sim T_1\}| \lesssim (\log R)^{O(1)}, \text{ for each } T_1 \in \mathbb{T}_1.$$

The proof is an adaptation of the one of Theorem 3.23 from the previous section. We will omit it here, but the interested reader can either consider it as an exercise or check the details in [**108**]. The extra room provided by working with the big scale cubes $B$ is needed to accommodate the fact that, unlike lines, two nonparallel tubes may have large intersection. The reader is referred to Section 3.6, which analyzes the technical differences between tubes and lines in a related context.

The following simple example gives a wave packet perspective on the tightness of the $L^4$ estimate from Section 3.5.

**Example 3.29** (Multiple clustering) Let $\mathcal{M}_1, \mathcal{M}_2$ be two surfaces in $\mathbb{R}^3$. Consider $f_1, f_2$. Let us consider the wave packet decompositions at scale $R$ (see Exercise 2.8):

$$E^{\mathcal{M}_i} f_i = \sum_{T_i \in \mathbb{T}_i} z_{T_i} \phi_{T_i},$$

with

$$\sum_{T_i \in \mathbb{T}_i} |z_{T_i}|^2 \sim R^{-1} \|f_i\|_2^2.$$

We choose $z_{T_i}$ to be complex numbers of modulus roughly one. If we assume that all tubes in $\mathbb{T}_1, \mathbb{T}_2$ contain the origin, by choosing $z_{T_i}$ appropriately, it can easily be arranged that constructive interference holds

$$|E^{\mathcal{M}_i} f_i(x)| \gtrsim |\mathbb{T}_i|$$

for $|x| \ll 1$. Note that if $|\mathbb{T}_1|, |\mathbb{T}_2| \sim R$ then

$$\||E^{\mathcal{M}_1} f_1 E^{\mathcal{M}_2} f_2|^{\frac{1}{2}}\|_{L^4(B(0,1))} \sim (\|f_1\|_2 \|f_2\|_2)^{\frac{1}{2}}. \tag{3.28}$$

In this example, most of the mass of $|E^{\mathcal{M}_1} f_1 E^{\mathcal{M}_2} f_2|^{1/2}$ is concentrated in a tiny spatial region inside just one cube $B \in \mathcal{B}$. It turns out that (3.28) can be improved dramatically if, loosely speaking, this type of multiple clustering scenario is avoided. In more precise terms, the forbidden clusters will be specified by the relation $\sim$ introduced in Theorem 3.28.

**Theorem 3.30** (Improved $L^4$ estimate) *Let $E = E^{\mathbb{P}^2}$. Let $\mathbb{T}_1, \mathbb{T}_2$ and $\sim$ be as in Theorem 3.28. Given $B \in \mathcal{B}$, consider two functions $f_1, f_2$ such that*

$$Ef_1 = \sum_{\substack{T_1 \in \mathbb{T}_1 \\ B \not\sim T_1}} \phi_{T_1}$$

*and*

$$Ef_2 = \sum_{T_2 \in \mathbb{T}_2} \phi_{T_2}.$$

*Then*

$$\|Ef_1 Ef_2|^{\frac{1}{2}}\|_{L^4(B)} \lesssim_\delta R^{-\frac{1}{8}} R^{C\delta} (\|f_1\|_2 \|f_2\|_2)^{1/2}. \tag{3.29}$$

*Proof* Note that the inequality we have to prove is equivalent to

$$\left\| \sum_{\substack{T_1 \in \mathbb{T}_1 \\ B \not\sim T_1}} \phi_{T_1} \sum_{T_2 \in \mathbb{T}_2} \phi_{T_2} \right\|_{L^2(B)} \lesssim_\delta R^{\frac{3}{4}} R^{C\delta} (|\mathbb{T}_1||\mathbb{T}_2|)^{1/2}.$$

To prove this, we will take advantage of both transversality and biorthogonality. First, note that (2.3) allows us to restrict both $T_1$ and $T_2$ to smaller families $\mathbb{T}'_1, \mathbb{T}'_2$ consisting of tubes $T$ such that $R^\delta T \cap 2B \neq \emptyset$. At this point, we can replace the smaller domain of integration $B$ with $[-R, R]^3$. Let $\eta_R$ be as in the proof of Proposition 3.11. We may then write

$$\left\| \sum_{\substack{T_1 \in \mathbb{T}'_1 \\ B \not\sim T_1}} \phi_{T_1} \sum_{T_2 \in \mathbb{T}'_2} \phi_{T_2} \right\|_{L^2(B)}^2 \leq \sum_{\substack{T_1, T'_1 \in \mathbb{T}'_1 \\ B \not\sim T_1, T'_1}} \sum_{T_2, T'_2 \in \mathbb{T}'_2} \int_{\mathbb{R}^3} \phi_{T_1} \eta_R \bar{\phi}_{T'_1} \eta_R \phi_{T_2} \eta_R \bar{\phi}_{T'_2} \eta_R.$$

Example 3.9 shows that a 4-tuple $(T_1, T_1', T_2, T_2')$ contributes to this sum only if it belongs to $\mathcal{BI}^{\neq}(\mathbb{T}_1', \mathbb{T}_2')$. We can thus refine our estimate as follows:

$$\left\| \sum_{\substack{T_1 \in \mathbb{T}_1' \\ B \not\sim T_1}} \phi_{T_1} \sum_{T_2 \in \mathbb{T}_2'} \phi_{T_2} \right\|_{L^2(B)}^2 \leq \sum_{(T_1, T_2) \in \mathcal{BI}^{\neq}(\mathbb{T}_1', \mathbb{T}_2')} \int_{\mathbb{R}^3} |\phi_{T_1} \eta_R \bar{\phi}_{T_1'} \eta_R \phi_{T_2} \eta_R \bar{\phi}_{T_2'} \eta_R|.$$

Invoking (2.3), (2.4), and transversality shows that each integral in the preceding summation is $O(R^{3/2})$. If we combine this with (3.27), the desired estimate will follow. $\qquad\square$

## 3.9 Induction on Scales and the Proof of Theorem 3.1

We are now in position to prove Theorem 3.1 when $n = 3$. The argument will be structured around an induction on scales. The quantity *scale* will refer to the side length $R$ of the spatial cubes.

Let us fix two small enough squares $S_1, S_2$ centered at $(1, 0)$ and $(0, 1)$, respectively. The reader is referred to Remark 3.20 for the relevance of this assumption. The mechanism that allows for the reduction to such choices for $S_1$ and $S_2$ will be explored in Exercise 4.10.

For $R \gg 1$, we let $C_R$ be the smallest number such that the inequality

$$\left\| |E_{S_1} f_1 E_{S_2} f_2|^{\frac{1}{2}} \right\|_{L^{\frac{10}{3}}(Q_R)} \leq C_R (\|f_1\|_{L^2(S_1)} \|f_2\|_{L^2(S_2)})^{1/2}$$

holds true for each cube $Q_R$ in $\mathbb{R}^3$ with side length $2R$ and each $f_i : S_i \to \mathbb{C}$. It suffices to test this inequality with $Q_R = [-R, R]^3$. This fact is due to the following identity, valid for each $S \subset \mathbb{R}^2$ and $c_1, c_2, c_3 \in \mathbb{R}$:

$$E_S f(x_1, x_2, x_3) = E_S g(x_1 - c_1, x_2 - c_2, x_3 - c_3), \tag{3.30}$$

where

$$g(\xi_1, \xi_2) = f(\xi_1, \xi_2) e(\xi_1 c_1 + \xi_2 c_2 + (\xi_1^2 + \xi_2^2) c_3).$$

Theorem 3.1 will follow by employing the $\epsilon$-removal machinery (see, for example, lemma 2.4 in [112]), once we prove the estimate

$$C_R \lesssim_\epsilon R^\epsilon \tag{3.31}$$

for arbitrarily small $\epsilon > 0$. To achieve this, we will prove the following result.

**Theorem 3.31** *Assume that*

$$C_R \lesssim R^\alpha \tag{3.32}$$

*for some $\alpha > 0$ and all $R \geq 1$. Then for each $0 < \epsilon, \delta \ll 1$, we have the stronger inequality*

$$C_R \lesssim_{\epsilon,\delta} R^{\max((1-\delta)\alpha, C\delta)+\epsilon},$$

*for some constant $C$ independent of $\epsilon, \delta, \alpha, R$.*

This provides a self-improving mechanism (also known as bootstrapping) for the exponent $\alpha$ in (3.32). An elementary analysis shows that iterating Theorem 3.31 drives $\alpha$ arbitrarily close to zero, leading to (3.31). We could rephrase the main result as follows:

$$C_R \lesssim_{\epsilon,\delta} R^\epsilon \max(C_{R^{1-\delta}}, R^{C\delta}).$$

This is an example of *induction on scales*. We estimate the quantity $C_R$ at scale $R$ assuming some information about the same quantity at the smaller scale $R^{1-\delta}$.

Let us now start the proof of Theorem 3.31. Fix $0 < \delta \ll 1$. It suffices to prove the estimate

$$\| |E_{S_1} f_1 E_{S_2} f_2|^{\frac{1}{2}} \|_{L^{\frac{10}{3}}([-R,R]^3)} \lesssim_{\epsilon,\delta} R^{\max((1-\delta)\alpha, C\delta)+\epsilon}, \tag{3.33}$$

assuming $\|f_i\|_2 = 1$. We expand each $Ef_i$ into wave packets at scale $R$ as in Theorem 2.2:

$$E_{S_i} f_i = \sum_{T_i \in \tilde{\mathbb{T}}_i} a_{T_i} \phi_{T_i},$$

with

$$\sum_{T_i \in \tilde{\mathbb{T}}_i} |a_{T_i}|^2 \sim R^{-1}.$$

Thus $|a_{T_i}| = O(R^{-1/2})$ and recall that $\|\phi_{T_i}\|_\infty \lesssim 1$. Invoking (2.3) and the triangle inequality allows us to restrict attention to the tubes $T_i$ that lie inside $[-10R, 10R]^2 \times [-R, R]$. Indeed, the contribution to the left-hand side of (3.33) from the other tubes is $O(1)$. Similarly, the contribution coming from the wave packets with small coefficients $|a_{T_i}| \leq R^{-100}$ is also $O(1)$. Thus we are left with analyzing the contribution from the range $R^{-100} \leq |a_{T_i}| \lesssim R^{-1/2}$. Note that there are $O(\log R)$ dyadic blocks in this range. Since we tolerate logarithmic losses in (3.33), the triangle inequality will allow us to restrict attention to just one such block for each $i$.

Call $\mathbb{T}_i$ the collection of tubes with this property. It will suffice to prove that

$$\left\| \left| \sum_{T_1 \in \mathbb{T}_1} a_{T_1} \phi_{T_1} \sum_{T_2 \in \mathbb{T}_2} a_{T_2} \phi_{T_2} \right|^{\frac{1}{2}} \right\|_{L^{\frac{10}{3}}([-R,R]^3)} \lesssim_{\epsilon,\delta} R^{\max((1-\delta)\alpha, C\delta)+\epsilon}.$$

Assume that $A_i \leq |a_{T_i}| \leq 2A_i$ for each $T_i \in \mathbb{T}_i$ and call $\Phi_{T_i} = a_{T_i}/A_i\phi_{T_i}$. Note that $\Phi_{T_i}$ has the same essential properties as $\phi_{T_i}$. Since $A_i|\mathbb{T}_i|^{1/2} \leq R^{-1/2}$, it will be enough to prove the inequality

$$\left\| \left| \sum_{T_1 \in \mathbb{T}_1} \Phi_{T_1} \sum_{T_2 \in \mathbb{T}_2} \Phi_{T_2} \right|^{\frac{1}{2}} \right\|_{L^{\frac{10}{3}}([-R,R]^3)} \lesssim_{\epsilon,\delta} R^{\frac{1}{2}+\epsilon} R^{\max((1-\delta)\alpha, C\delta)} (|\mathbb{T}_1||\mathbb{T}_2|)^{1/4}.$$

$$(3.34)$$

Let $\mathcal{B}$ be the partition of $[-R, R]^3$ into cubes $B$ with side length $2R^{1-\delta}$. The families $\mathbb{T}_1$ and $\mathbb{T}_2$ are of the type considered in the previous section. But the same remains true if we reverse their order. Recall the definitions of $\mathcal{L}$ and $\mathcal{F}$ from the previous section. These led to the definition of $\mathcal{BI}(\mathbb{T}_1, \mathbb{T}_2)$ from (3.26). A similar definition exists for $\mathcal{BI}(\mathbb{T}_2, \mathbb{T}_1)$. A double application of Theorem 3.28 guarantees the existence of a relation $\sim$ on $\mathcal{B} \times (\mathbb{T}_1 \cup \mathbb{T}_2)$ such that the collections of restricted bilinear incidences

$$\mathcal{BI}^{\nsim}(\mathbb{T}_1, \mathbb{T}_2) := \{(T_1, T_1', T_2, T_2') \in \mathcal{BI}(\mathbb{T}_1, \mathbb{T}_2) : B \nsim T_1, T_1'$$

$$\text{for each } B \in \mathcal{B} \text{ such that } R^\delta T_1 \cap R^\delta T_1' \cap R^\delta T_2 \cap R^\delta T_2' \subset 2B\}$$

and

$$\mathcal{BI}^{\nsim}(\mathbb{T}_2, \mathbb{T}_1) := \{(T_2, T_2', T_1, T_1') \in \mathcal{BI}(\mathbb{T}_2, \mathbb{T}_1) : B \nsim T_2, T_2'$$

$$\text{for each } B \in \mathcal{B} \text{ such that } R^\delta T_1 \cap R^\delta T_1' \cap R^\delta T_2 \cap R^\delta T_2' \subset 2B\}$$

satisfy

$$|\mathcal{BI}^{\nsim}(\mathbb{T}_1, \mathbb{T}_2)|, |\mathcal{BI}^{\nsim}(\mathbb{T}_2, \mathbb{T}_1)| \lesssim R^{O(\delta)}|\mathbb{T}_1||\mathbb{T}_2|, \qquad (3.35)$$

and moreover

$$|\{B \in \mathcal{B} : B \sim T\}| \lesssim (\log R)^{O(1)}, \text{ for each } T \in \mathbb{T}_1 \cup \mathbb{T}_2. \qquad (3.36)$$

We separate the term we need to estimate into four parts, using the triangle inequality

$$\left\| \sum_{T_1 \in \mathbb{T}_1} \Phi_{T_1} \sum_{T_2 \in \mathbb{T}_2} \Phi_{T_2} \right\|_{L^{\frac{5}{3}}([-R,R]^3)}$$

$$\leq \sum_{B \in \mathcal{B}} \left\| \sum_{B \sim T_1} \Phi_{T_1} \sum_{B \sim T_2} \Phi_{T_2} \right\|_{L^{\frac{5}{3}}(B)} \qquad \text{(local part)}$$

$$+ \sum_{B \in \mathcal{B}} \left\| \sum_{B \nsim T_1} \Phi_{T_1} \sum_{T_2 \in \mathbb{T}_2} \Phi_{T_2} \right\|_{L^{\frac{5}{3}}(B)} \qquad \text{(global part)}$$

$$+ \sum_{B \in \mathcal{B}} \left\| \sum_{T_1 \in \mathbb{T}_1} \Phi_{T_1} \sum_{B \nsim T_2} \Phi_{T_2} \right\|_{L^{\frac{5}{3}}(B)} \quad \text{(global part)}$$

$$+ \sum_{B \in \mathcal{B}} \left\| \sum_{B \nsim T_1} \Phi_{T_1} \sum_{B \nsim T_2} \Phi_{T_2} \right\|_{L^{\frac{5}{3}}(B)} \quad \text{(global part)}.$$

We first estimate the local contribution by writing for each $B \in \mathcal{B}$

$$\left\| \sum_{B \sim T_1} \Phi_{T_1} \sum_{B \sim T_2} \Phi_{T_2} \right\|_{L^{\frac{5}{3}}(B)} \lesssim R(C_{R^{1-\delta}})^2 |\{T_1 : B \sim T_1\}|^{1/2} |\{T_2 : B \sim T_2\}|^{1/2}.$$

(3.37)

This can be seen by recalling that $\Phi_{T_i} = E(Ra_{T_i}/A_i \psi_{T_i})$, together with (2.1). It is at this point only where we use the induction hypothesis at smaller scales, and we apply it with $f_i' = \sum_{T_i \in \mathbb{T}_i} Ra_{T_i}/A_i \psi_{T_i}$. We sum (3.37) taking advantage of (3.36) and the Cauchy–Schwarz inequality. The local part can be dominated by

$$R(C_{R^{1-\delta}})^2 \sum_{B \in \mathcal{B}} |\{T_1 : B \sim T_1\}|^{1/2} |\{T_2 : B \sim T_2\}|^{1/2}$$

$$\leq R(C_{R^{1-\delta}})^2 |\{(B, T_1) \in \mathcal{B} \times \mathbb{T}_1 : B \sim T_1\}|^{\frac{1}{2}} |\{(B, T_2) \in \mathcal{B} \times \mathbb{T}_2 : B \sim T_2\}|^{\frac{1}{2}}$$

$$= R(C_{R^{1-\delta}})^2 \left( \sum_{T_1 \in \mathbb{T}_1} |\{B \in \mathcal{B} : B \sim T_1\}| \right)^{\frac{1}{2}} \left( \sum_{T_2 \in \mathbb{T}_2} |\{B \in \mathcal{B} : B \sim T_2\}| \right)^{\frac{1}{2}}$$

$$\lesssim (\log R)^{O(1)} R^{1+2(1-\delta)\alpha} (|\mathbb{T}_1||\mathbb{T}_2|)^{1/2}.$$

Thus, the contribution to (3.34) of the local part is acceptable.

It remains to deal with the global contributions. We will estimate the first term, the other two are very similar. It suffices to prove that for each $B \in \mathcal{B}$

$$\left\| \sum_{B \nsim T_1} \Phi_{T_1} \sum_{T_2 \in \mathbb{T}_2} \Phi_{T_2} \right\|_{L^{\frac{5}{3}}(B)} \lesssim_{\epsilon, \delta} R^{1+\epsilon} R^{C\delta} (|\mathbb{T}_1||\mathbb{T}_2|)^{1/2}. \quad (3.38)$$

This will immediately follow by interpolating two estimates. The first one is rather trivial; it follows from Hölder's inequality and the conservation of mass (2.10)

$$\left\| \sum_{B \nsim T_1} \Phi_{T_1} \sum_{T_2 \in \mathbb{T}_2} \Phi_{T_2} \right\|_{L^1(B)} \leq \left\| \sum_{B \nsim T_1} \Phi_{T_1} \right\|_{L^2(\mathbb{R}^2 \times [-R, R])} \left\| \sum_{T_2 \in \mathbb{T}_2} \Phi_{T_2} \right\|_{L^2(\mathbb{R}^2 \times [-R, R])}$$

$$\lesssim R^2 (|\mathbb{T}_1||\mathbb{T}_2|)^{1/2}.$$

Note that neither transversality nor the relation $\sim$ play any role in this estimate. However, the second estimate makes use of both of these features, and is a consequence of Theorem 3.30:

$$\left\| \sum_{B \not\sim T_1} \Phi_{T_1} \sum_{T_2 \in \mathbb{T}_2} \Phi_{T_2} \right\|_{L^2(B)} \lesssim_{\epsilon, \delta} R^{\frac{3}{4}+\epsilon} R^{C\delta} (|\mathbb{T}_1||\mathbb{T}_2|)^{1/2}.$$

The proof of Theorem 3.31 is now complete. Let us summarize our findings. The local part is "small" because the contribution of each tube is localized inside a small number of cubes $B$. On the other hand, the smallness of the global part is captured by $L^4$ and reflects the diagonal behavior of the restricted bilinear incidences.

**Exercise 3.32** Modify the arguments presented here to prove the analog of Theorem 3.1 for the cone.

Hint: Start with a wave packet decomposition (Exercise 2.8). Use Remark 3.21 to count bilinear incidences. Keep in mind that there are many wave packets spatially concentrated inside a given tube. Check how the earlier results about lines and tubes extend to the case when repetitions are allowed.

**Exercise 3.33** Let $\psi : [-1, 1]^2 \to \mathbb{R}$ be $C^2$ and let $S_1, S_2$ be arbitrary squares inside $[-1, 1]^2$. Prove that if $p < 10/3$, the bilinear inequality

$$\| E_{S_1}^{\psi} f \, E_{S_2}^{\psi} f |^{\frac{1}{2}} \|_{L^p(\mathbb{R}^3)} \lesssim (\|f\|_{L^2(S_1)} \|f\|_{L^2(S_2)})^{1/2}$$

fails for some $f$.

Hint: Use Exercise 2.8. For each $R \gg 1$, choose $f$ so that

$$E_{S_i}^{\psi} f = \sum_{T \in \mathcal{T}_i} \phi_T$$

with $\mathcal{T}_i$ a family of roughly $R^{1/2}$ parallel tubes sitting inside a rectangular box $B$ with dimensions $\sim R \times R \times R^{1/2}$.

**Exercise 3.34** Let $\Omega_1, \Omega_2$ be two cubes in $[-1, 1]^{n-1}$ with $\text{dist}(\Omega_1, \Omega_2) > 0$. Let also $p > 2(n+2)/n$. Prove the following bilinear discrete restriction: for each collections $\Lambda_1, \Lambda_2 \subset \mathbb{P}^{n-1}$ consisting of $1/R$-separated points $\lambda :=$ $(\xi_\lambda, |\xi_\lambda|^2)$ with $\xi_\lambda \in \Omega_i$ if $\lambda \in \Lambda_i$, each sequence $a_\lambda \in \mathbb{C}$ and each ball $B_R$

$$\left\| \prod_{i=1}^{2} \sum_{\lambda \in \Lambda_i} a_\lambda e(\lambda \cdot x) \right|^{\frac{1}{2}} \right\|_{L^p(B_R)} \lesssim R^{\frac{n-1}{2}} \left( \prod_{i=1}^{2} \|a_\lambda\|_{l^2(\Lambda_i)} \right)^{\frac{1}{2}}.$$

Use constructive interference to prove that the exponent of $R$ is sharp.

**Exercise 3.35** (Additive energy of $1/R$-separated $R^{-1}$-balls) Let $\Omega_1, \Omega_2$ be two squares in $[-1,1]^2$ with $\text{dist}(\Omega_1, \Omega_2) > 0$. Let $\Lambda_1, \Lambda_2 \subset \mathbb{P}^2$ consist of $1/R$-separated points $\lambda := (\xi_\lambda, |\xi_\lambda|^2)$ with $\xi_\lambda \in \Omega_i$ if $\lambda \in \Lambda_i$.

$$\mathbb{E}_2(\Lambda_1, \Lambda_2, R^{-1}) =$$

$$|\{(\lambda_1, \lambda_1', \lambda_2, \lambda_2') \in \Lambda_1 \times \Lambda_1 \times \Lambda_2 \times \Lambda_2 : |\lambda_1 + \lambda_2' - (\lambda_1' + \lambda_2)| \lesssim R^{-1}\}|.$$

(a) Use (3.13) to prove that

$$\mathbb{E}_2(\Lambda_1, \Lambda_2, R^{-1}) \lesssim |\Lambda_1||\Lambda_2| \min(R, |\Lambda_1|, |\Lambda_2|).$$

Also, prove that this upper bound is sharp when $\Lambda_1$ and $\Lambda_2$ consist of roughly $R^{1/2}$ points whose projections to $\mathbb{R}^2$ lie on parallel line segments of length $\sim R^{-1/2}$.

(b) Use Tao's bilinear restriction theorem to prove the stronger estimate

$$\mathbb{E}_2(\Lambda_1, \Lambda_2, R^{-1}) \lesssim_\epsilon (|\Lambda_1||\Lambda_2|)^{\frac{7}{6}} R^{\frac{1}{3}+\epsilon}.$$

(c) Let $1 \ll N \lesssim R$. Assume that $\Lambda_i$ is a maximal $1/R$-separated set whose projection to $\mathbb{R}^2$ lives inside a square $\Omega_i' \subset \Omega_i$ with side length $NR^{-1}$. In particular, $|\Lambda_i| \sim N^2$. Prove that

$$\mathbb{E}_2(\Lambda_1, \Lambda_2, R^{-1}) \sim (|\Lambda_1||\Lambda_2|)^{\frac{5}{4}}.$$

This exercise should be compared with Exercise 13.34, which implies the sharp upper bound

$$\mathbb{E}_2(\Lambda_1, \Lambda_2, R^{-1}) \lesssim_\epsilon R^\epsilon |\Lambda_1||\Lambda_2|$$

under the stronger $1/\sqrt{R}$-separation assumption.

Hint for (b): Interpolate the inequality from Exercise 3.34 (use $p$ arbitrarily close to $10/3$) with an $L^\infty$ estimate.

Hint for (c): See the computations in Example 3.9.

## 3.10 Weighted Wave Packets: Another Proof of Theorem 3.1

We will now explore a slightly different way to achieve an improved bilinear restriction estimate in $L^4$, as the main step toward the proof of Theorem 3.1. This type of argument was introduced in [109] in order to prove the sharp global endpoint bilinear estimate for the cone, and was later used in [4] for more general hypersurfaces. As we will soon see, this argument relies very little on incidence geometry. We will again restrict attention to $n = 3$, which is entirely representative for the general case.

Fix two small enough neighborhoods of $(1,0,0)$ and $(0,1,0)$, $S_1, S_2 \subset [-1,1]^2$, as in the previous section. For $R \gg 1$ and $p \geq 10/3$ we let $C_{R,p}$ denote the smallest constant such that the inequality

$$\||E_{S_1} f_1 E_{S_2} f_2|^{\frac{1}{2}}\|_{L^p(Q_R)} \leq C_{R,p}(\|f_1\|_{L^2(S_1)}\|f_2\|_{L^2(S_2)})^{1/2}$$

holds true for each cube $Q_R$ in $\mathbb{R}^3$ with side length $2R$ and each $f_i \in L^2(S_i)$.

We aim to prove the following result.

**Theorem 3.36** *Let $p \geq 10/3$. For each integer $l \geq 1$, there is $A_l$, and there is $A$ independent of $l$ such that for each large enough $R$ (depending on $l$) we have*

$$C_{R,p}^2 \leq A C_{\frac{R}{2^l},p}^2 + A_l. \tag{3.39}$$

It is rather immediate that iterating (3.39) leads to the inequality $C_{R,p} \lesssim_l R^{O(\log A/l)}$. Since we may choose $l$ as large as we wish, we conclude that $C_{R,p} \lesssim_\epsilon R^\epsilon$.

As explained earlier in the book, we may restrict attention to $Q_R = [-R,R]^3$. Let $\mathcal{Q}_l$ be the partition of $[-R,R]^3$ into cubes $Q$ with side length $2^{1-l}R$. For $Q, Q' \in \mathcal{Q}_l$, we will write $Q \sim Q'$ if $Q$ and $Q'$ are neighbors, and $Q \nsim Q'$ otherwise. Each $Q$ has at most 27 neighbors in $\mathcal{Q}_l$.

Here is the key result.

**Proposition 3.37** (Improved $L^4$ estimate) *Given $f_i \in L^2(S_i)$ for $i \in \{1,2\}$, there are functions $f_{i,Q}: S_i \to \mathbb{C}$ for each $Q \in \mathcal{Q}_l$, such that*

$$f_i = \sum_{Q \in \mathcal{Q}_l} f_{i,Q}, \tag{3.40}$$

$$\sum_{Q \in \mathcal{Q}_l} \|f_{i,Q}\|_2^2 \leq B\|f_i\|_2^2 \tag{3.41}$$

*and*

$$\||Ef_{1,Q_1} Ef_{2,Q_2}|^{\frac{1}{2}}\|_{L^4(Q)} \lesssim_l R^{-\frac{1}{8}}(\|f_1\|_2\|f_2\|_2)^{\frac{1}{2}}, \tag{3.42}$$

*unless $Q \sim Q_1$ and $Q \sim Q_2$. The constant $B$ is universal.*

Let us use this proposition to prove Theorem 3.36. We write

$$\|Ef_1 Ef_2\|_{L^{\frac{p}{2}}(Q_R)} = \left\|\sum_{Q_1 \in \mathcal{Q}_l} \sum_{Q_2 \in \mathcal{Q}_l} \sum_{Q \in \mathcal{Q}_l} 1_Q Ef_{1,Q_1} Ef_{2,Q_2}\right\|_{L^{\frac{p}{2}}(Q_R)}$$

$$\leq \sum_{Q \in \mathcal{Q}_l} \sum_{\substack{Q_1 \sim Q \\ Q_2 \sim Q}} \|Ef_{1,Q_1} Ef_{2,Q_2}\|_{L^{\frac{p}{2}}(Q)}$$

$$+ \sum_{Q \in \mathcal{Q}_l} \sum_{\substack{Q_1 \nsim Q \\ \text{or } Q_2 \nsim Q}} \|Ef_{1,Q_1} Ef_{2,Q_2}\|_{L^{\frac{p}{2}}(Q)}.$$

The first term is dominated by

$$C^2_{\frac{R}{2^l},p} \sum_{Q \in \mathcal{Q}_l} \sum_{Q_1 \sim Q} \sum_{Q_2 \sim Q} \|f_{1,Q_1}\|_2 \|f_{2,Q_2}\|_2.$$

Since each $Q \in \mathcal{Q}_l$ has at most 27 neighbors, the preceding expression is further dominated via the Cauchy–Schwarz inequality and (3.41) by $3^6 C^2_{R/2^l,p} B \|f_1\|_2 \|f_2\|_2$.

To estimate the second term, assume that either $Q_1 \not\sim Q$ or $Q_2 \not\sim Q$. Then, combining (3.42) with the easy estimate

$$\||Ef_{1,Q_1} Ef_{2,Q_2}|^{\frac{1}{2}}\|_{L^2(Q)} \lesssim R^{1/2} (\|f_1\|_2 \|f_2\|_2)^{\frac{1}{2}}$$

implies that

$$\|Ef_{1,Q_1} Ef_{2,Q_2}\|_{L^{\frac{p}{2}}(Q)} \leq C_l \|f_1\|_2 \|f_2\|_2,$$

for some $C_l$ independent of $R$. It is here that the restriction $p \geq 10/3$ plays a critical role.

Inequality (3.39) will now follow with $A = 3^6 B$ and $A_l = 2^{27l} C_l$.

Let us next construct the functions $f_{i,Q}$ from Proposition 3.37. Consider the wave packet decomposition at scale $R$, as in Theorem 2.2:

$$Ef_1 = \sum_{T_1 \in \mathbb{T}_1} \Phi_{T_1}.$$

We have that $\Phi_{T_1} = Ef_{T_1}$, for some $f_{T_1}$ supported on a square $\omega_{T_1} \in \Omega_R^{(1)}$ with side length $\sim R^{-1/2}$.

For $Q \in \mathcal{Q}_l$ and $T_1 \in \mathbb{T}_1$, we introduce the weights

$$w_{Q,T_1} = \|1_{4T_1} Ef_2\|^2_{L^2(Q)}$$

and

$$w_{T_1} = \sum_{Q \in \mathcal{Q}_l} w_{Q,T_1} = \|1_{4T_1} Ef_2\|^2_{L^2([-R,R]^3)}.$$

To avoid minor technicalities, we will assume that all $w_{T_1}$ are nonzero. We now define

$$f_{1,Q} = \sum_{T_1 \in \mathbb{T}_1} \frac{w_{Q,T_1}}{w_{T_1}} f_{T_1}.$$

Consider also the following wave packet decomposition at scale $R$

$$Ef_2 = \sum_{T_2 \in \mathbb{T}_2} \Psi_{T_2}.$$

with $\Psi_{T_2} = E f_{T_2}$, for some $f_{T_2}$ supported on a square $\omega_{T_2} \in \Omega_R^{(2)}$ with side length $\sim R^{-1/2}$. The functions $f_{2,Q}$ are defined in a similar fashion, by switching the roles of $f_1$ and $f_2$.

Let us observe that (3.40) is immediate, while (3.41) follows from the almost orthogonality (2.1) and (2.7) of the families of functions $f_{T_1}$ and $f_{T_2}$. The remaining inequality (3.42) will be a straightforward consequence of the following result and its symmetric version.

**Proposition 3.38** *For each $T_2 \in \mathbb{T}_2$, let $\alpha_{T_2} \in [0, 1]$ be an arbitrary weight. Define*

$$g = \sum_{T_2 \in \mathbb{T}_2} \alpha_{T_2} f_{T_2}. \tag{3.43}$$

*Then*

$$\|E f_{1,Q} E g\|_{L^2(Q')} \lesssim_l R^{-\frac{1}{4}} \|f_1\|_2 \|f_2\|_2$$

*whenever $Q \not\sim Q'$.*

Fix $Q \in \mathcal{Q}_l$. Let $\mathcal{Q}_Q$ be a finitely overlapping cover of $Q_R \setminus 2Q$ with cubes $q$ of side length $R^{1/2}$ satisfying $\mathrm{dist}\,(q, Q) \gtrsim_l R$. We will prove that

$$\sum_{q \in \mathcal{Q}_Q} \|E f_{1,Q} E g\|_{L^2(q)}^2 \lesssim_l R^{-\frac{1}{2}} \|f_1\|_2^2 \|f_2\|_2^2. \tag{3.44}$$

The proof of (3.44) will follow from a sequence of lemmas. Our discussion will be somewhat informal, and we will sometimes use the symbol $\approx$ to denote the fact that two quantities are comparable in size, apart from small error terms that will be neglected. For example, we will think of the wave packets $\Phi_{T_1}, \Psi_{T_2}$ as being compactly supported on (rather than concentrated near) the corresponding tubes. Also, we will assume that the functions $\Psi_{T_2}$ are almost orthogonal on each $q$, that is,

$$\left\| \sum_{T_2} \Psi_{T_2} 1_q \right\|_2^2 \approx \sum_{T_2} \|\Psi_{T_2} 1_q\|_2^2. \tag{3.45}$$

This can be made precise if $1_q$ is replaced with a smooth approximation $\eta_q$ whose Fourier transform is supported on $B(0, R^{-1/2})$. In a similar spirit, recall that

$$\int \Phi_{T_1} \overline{\Psi_{T_2} \Phi_{T_1'}} \Psi_{T_2'} |\eta_q|^4 = 0$$

unless, using the notation from (3.12), we have $(\omega_{T_1}, \omega_{T_2}) \in S_{R^{-1/2}}(\omega_{T_1'}, \omega_{T_2'})$. Because of this, we will ignore the terms $\int_q \Phi_{T_1} \overline{\Psi_{T_2} \Phi_{T_1'}} \Psi_{T_2'}$ corresponding to $(\omega_{T_1}, \omega_{T_2}) \notin S_{R^{-1/2}}(\omega_{T_1'}, \omega_{T_2'})$.

The argument will rely on the following decomposition of $\mathbb{T}_1 \times \mathbb{T}_2$, which was also implicitly used in the previous sections. We write, in line with our earlier notation

$$\mathcal{M}(\omega) = \{(\xi_1, \xi_2, \xi_1^2 + \xi_2^2) : (\xi_1, \xi_2) \in \omega\}.$$

**Proposition 3.39** *There is a partition of $\mathbb{T}_1 \times \mathbb{T}_2$ into collections $\mathcal{P}_j$ consisting of pairs of tubes satisfying the following properties:*

*(T1) The sets*

$$\{\mathcal{M}(\omega_{T_2}) - \mathcal{M}(\omega_{T_1}) + B(0, R^{-1/2}) : (T_1, T_2) \in \mathcal{P}_j\}$$

*have $O_l(1)$ overlap.*

*(T2) If $(T_1, T_2)$ and $(T_1, T_2')$ are in the same $\mathcal{P}_j$ then $\omega_{T_2} = \omega_{T_2'}$.*

*(T3) If $(T_1, T_2)$ and $(T_1', T_2')$ are in the same $\mathcal{P}_j$ and $T_1, T_1'$ are distinct tubes intersecting outside $2Q$, then for each $T_2'' \in \mathbb{T}_2$ either $4T_1 \cap 4T_2'' \cap Q = \emptyset$ or $4T_1' \cap 4T_2'' \cap Q = \emptyset$.*

*Proof* Let $\mathcal{B}$ be a cover of $\mathbb{R}^3$ with balls $B_j$ of radius $\sim R^{-1/2}$, such that the collection of balls $100B_j$ is finitely overlapping. We partition the pairs of squares $(\omega_1, \omega_2) \in \Omega_R^{(1)} \times \Omega_R^{(2)}$ into collections $\Omega_j$, each of which satisfies the following two properties. First, $\mathcal{M}(\omega_2) - \mathcal{M}(\omega_1)$ intersects $B_j$ for each $(\omega_1, \omega_2) \in \Omega_j$. Second, by placing neighboring pairs in distinct collections, we can make sure that $\omega_2 = \omega_2'$ whenever $(\omega_1, \omega_2) \in \Omega_j$ and $(\omega_1, \omega_2') \in \Omega_j$.

We now define

$$\mathcal{P}_j = \{(T_1, T_2) \in \mathbb{T}_1 \times \mathbb{T}_2 : (\omega_{T_1}, \omega_{T_2}) \in \Omega_j\}.$$

Properties (T1) and (T2) are immediate. Property (T3) is a quantitative version of (P1) from Section 3.7, and can be easily verified using the fact that $S_1, S_2$ are small enough neighborhoods of $(1, 0, 0)$ and $(0, 1, 0)$. See also Remark 3.20. $\qquad\square$

The following lemma exploits biorthogonality. The main gain here is that the contributions from distinct $\mathcal{P}_j$ are almost orthogonal.

**Lemma 3.40** (Biorthogonality) *For each $q$ with side length $R^{1/2}$ and $g$ as in (3.43), we have*

$$\left\| \sum_{T_1 \in \mathbb{T}_1} \frac{w_{Q, T_1}}{w_{T_1}} \Phi_{T_1} \overline{Eg} \right\|_{L^2(q)}^2 \approx \sum_j \left\| \sum_{(T_1, T_2) \in \mathcal{P}_j} \frac{w_{Q, T_1}}{w_{T_1}} \alpha_{T_2} \Phi_{T_1} \overline{\Psi_{T_2}} \right\|_{L^2(q)}^2. \tag{3.46}$$

*Proof* Use (T1) and the fact that $\Phi_{T_1}\overline{\Psi_{T_2}}$ has the Fourier transform supported on the set $\mathcal{M}(\omega_{T_2}) - \mathcal{M}(\omega_{T_1})$.    □

For each fixed $j$, we will restrict the sum in (3.46) to those $T_1$ that, in addition to serving as first entries for pairs in $\mathcal{P}_j$, also intersect $q$. Denote this collection by $\mathbb{T}_{1,j}(q)$. We will also restrict attention to those $T_2$ intersecting $q$. Thus, property (T2) determines essentially uniquely the second entry $T_2$ as a function of $T_1, q,$ and $j$, and we will denote it by $T_2(T_1, q, j)$. Note that $\mathbb{T}_{1,j}(q)$ and $\mathbb{T}'_{1,j}(q)$ may share tubes, but this will not cause any problem.

We have

$$\left\| \sum_{(T_1, T_2) \in \mathcal{P}_j} \frac{w_{Q,T_1}}{w_{T_1}} \alpha_{T_2} \Phi_{T_1} \overline{\Psi_{T_2}} \right\|^2_{L^2(q)}$$

$$\approx \left\| \sum_{T_1 \in \mathbb{T}_{1,j}(q)} \frac{w_{Q,T_1}}{w_{T_1}} \alpha_{T_2(T_1,q,j)} \Phi_{T_1} \overline{\Psi_{T_2(T_1,q,j)}} \right\|^2_{L^2(q)}.$$

Since $w_{Q,T_1}/w_{T_1} \alpha_{T_2(T_1,q,j)} \leq (w_{Q,T_1}/w_{T_1})^{1/2}$, we may dominate this term using the Cauchy–Schwarz inequality by

$$\left( \sum_{T_1 \in \mathbb{T}_{1,j}(q)} \frac{\|\Phi_{T_1} \Psi_{T_2(T_1,q,j)}\|^2_{L^2(q)}}{w_{T_1}} \right) \left( \sum_{T_1 \in \mathbb{T}_{1,j}(q)} w_{Q,T_1} \right). \qquad (3.47)$$

Let us first examine the second sum from (3.47). We will use the following lemma, which is a rather direct consequence of the conservation of mass (1.17). It is ultimately responsible for the gain of the factor $R^{-1/8}$ in (3.42), since the thickness of the wall $W$ is $O(R^{1/2})$, much smaller than the diameter $\sim R$ of $Q$.

**Lemma 3.41** (Transverse wall estimate) *Let $W$ be a subset of $Q_R$ with the property that*

$$\sup_{T_2 \in \mathbb{T}_2} \operatorname{diam}(4T_2 \cap W) \lesssim R^{1/2}. \qquad (3.48)$$

*Then*

$$\|Ef_2\|^2_{L^2(W)} \lesssim R^{1/2} \|f_2\|^2_2.$$

*Proof* Cover $W$ with a family $\mathcal{Q}_W$ of pairwise disjoint cubes $q$ with side length $R^{1/2}$, such that each $T_2 \in \mathbb{T}_2$ intersects only $O(1)$ of them. Note that (1.17) implies that

$$\|\Psi_{T_2}\|^2_{L^2(q)} \lesssim R^{1/2} \|f_{T_2}\|^2_2.$$

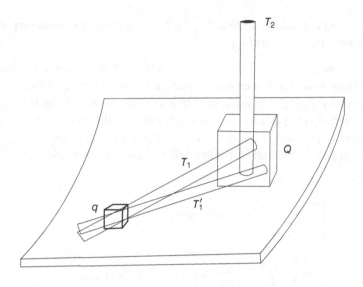

Figure 3.3 Transverse tube-wall interaction.

Following our earlier convention, we only consider the terms $\|\Psi_{T_2}\|^2_{L^2(q)}$ with $q$ intersecting $T_2$. Combining these observations with (3.45), we conclude that

$$
\|Ef_2\|^2_{L^2(W)} \le \sum_{q \in \mathcal{Q}_W} \left\| \sum_{T_2 \in \mathbb{T}_2} \Psi_{T_2} \right\|^2_{L^2(q)}
$$

$$
\approx \sum_{T_2 \in \mathbb{T}_2} \sum_{\substack{q \in \mathcal{Q}_W \\ q \cap T_2 \neq \emptyset}} \|\Psi_{T_2}\|^2_{L^2(q)}
$$

$$
\lesssim R^{1/2} \sum_{T_2 \in \mathbb{T}_2} \|f_{T_2}\|^2_2
$$

$$
\lesssim R^{1/2} \|f_2\|^2_2. \qquad \square
$$

The wave packets $\Phi_{T_1}$ and $\Psi_{T_2}$ have a transverse interaction, in the sense that diam $(T_1 \cap T_2) \lesssim R^{1/2}$. The next lemma proves that, when restricted to pairs from a fixed $\mathcal{P}_j$, each $\Psi_{T_2}$ has a globally transverse interaction with the family of wave packets $\Phi_{T_1}$. See Figure 3.3.

**Lemma 3.42** *For each $j$ and $q \in \mathcal{Q}_Q$, the "wall"*

$$
W := \left( \bigcup_{T_1 \in \mathbb{T}_{1,j}(q)} 4T_1 \right) \cap Q \tag{3.49}
$$

*satisfies (3.48).*

*Proof* First, due to (T3), each $4T_2$ can only intersect one set $4T_1 \cap Q$ with $T_1 \in \mathbb{T}_{1,j}(q)$. Also, transversality forces diam $(4T_2 \cap 4T_1) \lesssim R^{1/2}$.     $\square$

Note also that due to the separation between $q$ and $Q$, the tubes $4T_1$ with $T_1 \in \mathbb{T}_{1,j}(q)$ have a small overlap inside $Q$:

$$\left\| \sum_{T_1 \in \mathbb{T}_{1,j}(q)} 1_{4T_1} \right\|_{L^\infty(Q)} \lesssim_l 1.$$

Combining this observation with the previous two lemmas leads to the following corollary.

**Corollary 3.43** *For each $j$ and $q \in \mathcal{Q}_Q$,*

$$\sum_{T_1 \in \mathbb{T}_{1,j}(q)} w_{Q,T_1} \lesssim R^{1/2} \|f_2\|_2^2.$$

*Proof* Write, with $W$ as in (3.49)

$$\sum_{T_1 \in \mathbb{T}_{1,j}(q)} w_{Q,T_1} = \sum_{T_1 \in \mathbb{T}_{1,j}(q)} \|1_{4T_1} Ef_2\|_{L^2(Q)}^2$$

$$\lesssim_l \|Ef_2\|_{L^2(W)}^2$$

$$\lesssim R^{1/2} \|f_2\|_2^2.$$     $\square$

Let us next estimate the first term in (3.47).

**Lemma 3.44** *We have*

$$\sum_q \sum_j \sum_{T_1 \in \mathbb{T}_{1,j}(q)} \frac{\|\Phi_{T_1} \Psi_{T_2(T_1,q,j)}\|_{L^2(q)}^2}{w_{T_1}} \lesssim R^{-1} \|f_1\|_2^2.$$

*The first sum is considered over cubes $q$ in a finitely overlapping cover of $Q_R$, with no further restriction relative to $Q$.*

*Proof* The left-hand side is certainly smaller than

$$\sum_q \sum_{T_1 \in \mathbb{T}_1(q)} \sum_{T_2 \in \mathbb{T}_2} \frac{\|\Phi_{T_1} \Psi_{T_2}\|_{L^2(q)}^2}{w_{T_1}}, \tag{3.50}$$

with $\mathbb{T}_1(q)$ denoting the tubes $T_1 \in \mathbb{T}_1$ that intersect $q$. Since $\Phi_{T_1} \Psi_{T_2} 1_q$ has the Fourier transform essentially supported on a ball of radius $\sim R^{-1/2}$, we may write

$$\|\Phi_{T_1} \Psi_{T_2}\|_{L^2(q)}^2 \lesssim R^{-3/2} \|\Phi_{T_1} \Psi_{T_2}\|_{L^1(q)}^2.$$

Combining this with Hölder's inequality, we can dominate (3.50) by

$$R^{-3/2} \sum_q \sum_{T_1 \in \mathbb{T}_1(q)} \sum_{T_2 \in \mathbb{T}_2} \frac{\|\Phi_{T_1}\|_{L^2(q)}^2 \|\Psi_{T_2}\|_{L^2(q)}^2}{w_{T_1}}.$$

Conservation of mass forces

$$\sup_q \|\Phi_{T_1}\|_{L^2(q)}^2 \lesssim R^{1/2} \|f_{T_1}\|_2^2.$$

Finally, we rearrange the multiple sum to get

$$\sum_q \sum_{T_1 \in \mathbb{T}_1(q)} \sum_{T_2 \in \mathbb{T}_2} \frac{\|\Phi_{T_1}\|_{L^2(q)}^2 \|\Psi_{T_2}\|_{L^2(q)}^2}{w_{T_1}}$$

$$\lesssim \sum_{T_1 \in \mathbb{T}_1} \frac{R^{1/2} \|f_{T_1}\|_2^2}{w_{T_1}} \sum_{q : T_1 \in \mathbb{T}_1(q)} \sum_{T_2 \in \mathbb{T}_2} \|\Psi_{T_2}\|_{L^2(q)}^2$$

$$\approx R^{1/2} \sum_{T_1 \in \mathbb{T}_1} \frac{\|f_{T_1}\|_2^2}{w_{T_1}} \sum_{q : T_1 \in \mathbb{T}_1(q)} \left\| \sum_{T_2 \in \mathbb{T}_2} \Psi_{T_2} \right\|_{L^2(q)}^2$$

$$\leq R^{1/2} \sum_{T_1 \in \mathbb{T}_1} \frac{\|f_{T_1}\|_2^2}{w_{T_1}} \|1_{4T_1} E f_2\|_{L^2(Q_R)}^2$$

$$\sim R^{1/2} \|f_1\|_2^2.$$

We have used that the wave packets $\Psi_{T_2}$ are almost orthogonal on each $q$, and the fact that $T_1 \in \mathbb{T}_1(q)$ forces $q \subset 4T_1$. $\qquad\square$

Combining Lemma 3.40 with Corollary 3.43 and Lemma 3.44 leads to the proof of inequality (3.44) as follows:

$$\sum_{q \in \mathcal{Q}_Q} \|E f_{1,Q} E g\|_{L^2(q)}^2 \lesssim \sum_j \sum_{q \in \mathcal{Q}_Q} \left( \sum_{T_1 \in \mathbb{T}_{1,j}(q)} \frac{\|\Phi_{T_1} \Psi_{T_2(T_1,q,j)}\|_{L^2(q)}^2}{w_{T_1}} \right)$$

$$\times \left( \sum_{T_1 \in \mathbb{T}_{1,j}(q)} w_{Q,T_1} \right)$$

$$\lesssim R^{1/2} \|f_2\|_2^2 \sum_j \sum_{q \in \mathcal{Q}_Q} \sum_{T_1 \in \mathbb{T}_{1,j}(q)} \frac{\|\Phi_{T_1} \Psi_{T_2(T_1,q,j)}\|_{L^2(q)}^2}{w_{T_1}}$$

$$\lesssim R^{-1/2} \|f_1\|_2^2 \|f_2\|_2^2.$$

The arguments presented in this chapter beg the question as to whether there is an approach to Theorem 3.1 that does not rely on biorthogonality. A positive answer would be of independent interest, but may also provide a clue on how to prove multilinear restriction estimates of higher order.

Exercise 3.45 provides an affirmative answer in the case of the cone, by slightly reshuffling the arguments from this section. It relies on three observations, recorded in parts (a), (b), and (c) of the exercise. First, there is a gain in the bilinear estimate consisting of a factor of $R^{-1/4}$, if one of the functions is supported inside a thin, $R^{-1/2}$-wide sector. Moreover, this can be proved without using biorthogonality, cf. Exercise 6.35. Second, the tubes normal to the cone automatically satisfy the transversality requirement (3.48). Third, the contributions from different sectors can be separated by simply using the triangle inequality. The reader may check that this line of reasoning fails to produce a biorthogonality-free argument for $\mathbb{P}^2$, essentially because of the inefficiency of the triangle inequality in the last step.

Interestingly, however, this type of argument can be used to prove sharp trilinear restriction estimates for certain conical hypersurfaces with two zero principal curvatures in four and higher dimensions, as well as higher-order multilinear restriction estimates (cf. Conjecture 6.1) in the presence of sufficiently many zero principal curvatures. See [5], [7].

**Exercise 3.45** Let $S_1$, $S_2$ be two angularly separated sectors of the annulus $B_2(0,2) \setminus B_2(0,1)$ in $\mathbb{R}^2$, and let $f_i : S_i \to \mathbb{C}$. Write as in (2.12)

$$E^{\mathbb{C}\mathrm{one}^2} f_1 = \sum_{T_1 \in \mathbb{T}_1} \Phi_{T_1}, \quad \Phi_{T_1} = E^{\mathbb{C}\mathrm{one}^2} f_{T_1},$$

$$E^{\mathbb{C}\mathrm{one}^2} f_2 = \sum_{T_2 \in \mathbb{T}_2} \Psi_{T_2}, \quad \Psi_{T_2} = E^{\mathbb{C}\mathrm{one}^2} f_{T_2}.$$

Each $f_{T_i}$ is supported on some $R^{-1/2}$-wide sector in the annulus, and $\Phi_{T_1}, \Psi_{T_2}$ are sums of roughly $R^{1/2}$ wave packets at scale $R$, spatially concentrated near $T_1$ and $T_2$, respectively. We will use the earlier notation from this section.

(a) Prove that for each cube $q$ with side length $R^{1/2}$

$$\|\Phi_{T_1} E^{\mathbb{C}\mathrm{one}^2} g\|_{L^2(q)} \lesssim_\epsilon R^{\epsilon - \frac{3}{4}} \|\Phi_{T_1}\|_{L^2(\chi_q)} \|E^{\mathbb{C}\mathrm{one}^2} g\|_{L^2(\chi_q)}.$$

(b) Prove that

$$W := \left( \bigcup_{T_1 \in \mathbb{T}_1(q)} 4T_1 \right) \cap Q$$

satisfies

$$\mathrm{diam}\,(4T_2 \cap W) \lesssim R^{1/2},$$

for each cube $q \in \mathcal{Q}_Q$ with side length $R^{1/2}$ and each $T_2 \in \mathbb{T}_2$.

(c) Prove that

$$\sum_{q \in \mathcal{Q}_Q} \left( \sum_{T_1 \in \mathbb{T}_1} \frac{w_{Q,T_1}}{w_{T_1}} \|\Phi_{T_1}\|_{L^2(\chi_q)} \|E^{\mathrm{Cone}^2} g\|_{L^2(\chi_q)} \right)^2 \lesssim_{\epsilon} R^{1+\epsilon} \|f_1\|_2^2 \|f_2\|_2^2.$$

Conclude that Proposition 3.38 holds for the cone, via a biorthogonality-free argument.

Hint for (a): This is a local version of the inequality in Exercise 6.35.

# 4

# Parabolic Rescaling and a
# Bilinear-to-Linear Reduction

In this chapter, we will use Theorem 3.1 to improve the range for the linear restriction problem associated with the paraboloid, breaking the Stein–Tomas barrier.

We will introduce a few new tools. One of them is *parabolic rescaling,* which relies on the following two observations. First, the affine functions ($\xi_0 \in \mathbb{R}^{n-1}, \delta > 0$)

$$\mathbb{R}^{n-1} \times \mathbb{R} \ni (\xi, \xi_n) \mapsto \left( \frac{\xi - \xi_0}{\delta}, \frac{\xi_n - 2\xi_0 \cdot \xi + |\xi_0|^2}{\delta^2} \right)$$

map the (infinite) paraboloid to itself. Second, each nonsingular affine map $T(\eta) = A\eta + \mathbf{v}$ interacts nicely with the Fourier transform, in the sense that

$$\widehat{G} = \widehat{F} \circ T \implies G(x) = \frac{1}{\det(A)} F((A^{-1})^t x) e(-\mathbf{v} \cdot (A^{-1})^t x). \quad (4.1)$$

A simple change of variables will allow us to convert inequalities involving functions with spectrum localized near a small cap on $\mathbb{P}^{n-1}$ into similar inequalities involving functions with spectrum spread over neighborhoods of the whole $\mathbb{P}^{n-1}$. This principle will appear in various contexts throughout this book. The first application will appear in the proof of Proposition 4.2.

**Definition 4.1** (Bilinear restriction constants) Let $1 \le p, q \le \infty$ and let $0 < D \le 1$. We denote by $BR^*_{\mathbb{P}^{n-1}}(q \times q \mapsto p, D)$ the smallest constant $C$ such that for each set of cube $\Omega_1, \Omega_2 \subset [-1, 1]^{n-1}$ with dist $(\Omega_1, \Omega_2) \ge D$ and each $f : \Omega_1 \cup \Omega_2 \to \mathbb{C}$, we have

$$\| |E_{\Omega_1} f E_{\Omega_2} f|^{\frac{1}{2}} \|_{L^p(\mathbb{R}^n)} \le C (\|f\|_{L^q(\Omega_1)} \|f\|_{L^q(\Omega_2)})^{1/2}.$$

If we assume the validity of the (linear) Restriction Conjecture 1.7, Hölder's inequality forces $BR^*_{\mathbb{P}^{n-1}}(\infty \times \infty \mapsto p, D)$ to be uniformly bounded in $D$ when $p > 2(n + 2)/n$. We will be able to extract a slightly weaker version of this from Theorem 3.1, by simply invoking parabolic rescaling. We will then combine this with a Whitney decomposition to derive new linear restriction estimates.

**Proposition 4.2** *Let $\Omega_1, \Omega_2$ be two cubes in $[-1, 1]^{n-1}$ with side length $\delta$. Assume that the distance $D$ between their centers satisfies $D \geq 4\delta$. Then for each $1 \leq p, q \leq \infty$ and each $f : \Omega_1 \cup \Omega_2 \to \mathbb{C}$, we have*

$$\left\| |E_{\Omega_1} f E_{\Omega_2} f|^{\frac{1}{2}} \right\|_{L^p(\mathbb{R}^n)}$$
$$\leq D^{\frac{n-1}{q'} - \frac{n+1}{p}} BR^*_{\mathbb{P}^{n-1}}\left( q \times q \mapsto p, \frac{1}{2} \right) \left( \|f\|_{L^q(\Omega_1)} \|f\|_{L^q(\Omega_2)} \right)^{1/2}.$$

Note that the exponent of $D$ in this inequality is nonnegative when $p, q$ are in the linear restriction range (cf. (1.5)), so we have an extra gain as $D$ gets smaller.

*Proof* Let $\xi_0$ be the midpoint of the line segment joining the centers of $\Omega_1$ and $\Omega_2$. Define the affine transformation on $\mathbb{R}^{n-1}$:

$$L(\xi) = L_{\xi_0, D}(\xi) = \frac{\xi - \xi_0}{D}.$$

A simple computation shows that

$$|E_{\Omega_i} f(\bar{x}, x_n)| = D^{n-1} |E_{L(\Omega_i)} f_L(D(\bar{x} + 2x_n \xi_0), D^2 x_n)|,$$

where $f_L = f \circ L^{-1}$. Note that $L(\Omega_1)$ and $L(\Omega_2)$ are cubes in $[-1, 1]^{n-1}$, separated by at least $1/2$. After a change of variables on the spatial side, we get

$$\left\| |E_{\Omega_1} f E_{\Omega_2} f|^{1/2} \right\|_{L^p(\mathbb{R}^n)}$$
$$= D^{n-1-\frac{n+1}{p}} \left\| |E_{L(\Omega_1)} f_L E_{L(\Omega_2)} f_L|^{1/2} \right\|_{L^p(\mathbb{R}^n)}$$
$$\leq D^{n-1-\frac{n+1}{p}} BR^*_{\mathbb{P}^{n-1}}\left( q \times q \mapsto p, \frac{1}{2} \right) \left( \|f_L\|_{L^q(L(\Omega_1))} \|f_L\|_{L^q(L(\Omega_2))} \right)^{1/2}$$
$$= D^{\frac{n-1}{q'} - \frac{n+1}{p}} BR^*_{\mathbb{P}^{n-1}}\left( q \times q \mapsto p, \frac{1}{2} \right) \left( \|f\|_{L^q(\Omega_1)} \|f\|_{L^q(\Omega_2)} \right)^{1/2}. \quad \square$$

Recall that a dyadic interval is one of the form $[l2^k, (l+1)2^k]$, with $l, k \in \mathbb{Z}$. A dyadic cube is the Cartesian product of dyadic intervals of equal length. If two dyadic cubes intersect, one must be a subset of the other.

**Proposition 4.3** *Let S be a closed set in $\mathbb{R}^m$. There is a collection $\mathcal{C}$ of closed dyadic cubes $\Omega$ with pairwise disjoint interiors such that*

$$\mathbb{R}^m \setminus S = \bigcup_{\Omega \in \mathcal{C}} \Omega$$

*and*

$$4l(\Omega) \leq \operatorname{dist}(\Omega, S) \leq 50l(\Omega), \tag{4.2}$$

*where $l(\Omega)$ is the side length of $\Omega$.*

*Proof* Let $\mathcal{C}'$ be the collection of all dyadic cubes $\Omega$ in $\mathbb{R}^m$ such that $8\Omega \cap S = \emptyset$. Let $\mathcal{C}$ consist of the cubes in $\mathcal{C}'$ that are maximal with respect to inclusion. It is easy to check that this collection satisfies the desired properties. $\square$

We will need this in the following particular case. See Figure 4.1.

**Corollary 4.4** *Let $n \geq 2$. There is a collection $\mathcal{C}$ of closed cubes $\Omega = \Omega_1 \times \Omega_2 \subset [-1, 1]^{n-1} \times [-1, 1]^{n-1}$ with pairwise disjoint interiors such that*

$$[-1, 1]^{2n-2} \setminus \{(\xi, \xi) \colon \xi \in [-1, 1]^{n-1}\} = \bigcup_{\Omega \in \mathcal{C}} \Omega$$

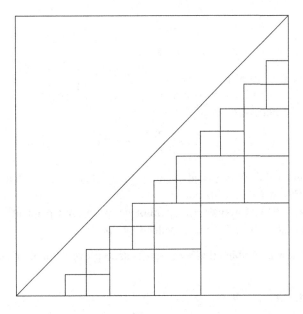

Figure 4.1 A Whitney decomposition of $[-1, 1]^2$ in the lower triangle.

*and*

$$4l(\Omega) \leq \text{dist}(\Omega_1, \Omega_2) \leq 100l(\Omega). \tag{4.3}$$

*Proof* It suffices to achieve a similar decomposition with $[-1, 1]$ replaced with $[0, 2]$ and then translate the cubes by $(-1, \ldots, -1)$. The advantage of working with $[0, 2]$ is that it is a dyadic interval.

Use the family from Proposition 4.3 with $m = 2n - 2$ and $S = \{(\xi, \xi) \colon \xi \in \mathbb{R}^{n-1}\}$, and only keep the cubes that are inside $[0, 2]^{2n-2}$. They obviously cover $[0, 2]^{2n-2}$. Also, (4.3) follows from (4.2) using elementary inequalities. $\square$

The following inequality is a hybrid between $L^2$ orthogonality and the triangle inequality in $L^1$ and $L^\infty$.

**Lemma 4.5** *Let $\mathcal{R}$ be a finite collection of rectangular boxes in $\mathbb{R}^n$ with $2R \cap 2R' = \emptyset$ whenever $R \neq R' \in \mathcal{R}$. For $R \in \mathcal{R}$, let $F_R \colon \mathbb{R}^n \to \mathbb{C}$ be an $L^s$ function for some $1 \leq s \leq \infty$. Assume that $\text{supp}(\widehat{F_R}) \subset R$. Then*

$$\left\| \sum_R F_R \right\|_s \lesssim \left( \sum_R \|F_R\|_s^s \right)^{1/s}, \text{ if } 1 \leq s \leq 2$$

*and*

$$\left\| \sum_R F_R \right\|_s \lesssim \left( \sum_R \|F_R\|_s^{s'} \right)^{1/s'}, \text{ if } s \geq 2.$$

*The implicit constants do not depend on $\mathcal{R}$.*

*Proof* Let $\phi_R \in \mathcal{S}(\mathbb{R}^n)$ with $1_R \leq \widehat{\phi_R} \leq 1_{2R}$ and $\|\phi_R\|_1 = 1$. Note that $F_R = F_R * \phi_R$. Consider the operator $T$ acting on an arbitrary family $G_{\mathcal{R}} = (G_R)_{R \in \mathcal{R}}$ of functions $G_R$ via

$$T(G_{\mathcal{R}}) = \sum_R G_R * \phi_R.$$

By invoking orthogonality and Young's inequality we see that $T$ maps $L^r(\mathbb{R}^n, l^r(\mathcal{R}))$ to $L^r(\mathbb{R}^n)$ for both $r = 1$ and $r = 2$. It also maps $L^\infty(\mathbb{R}^n, l^1(\mathcal{R}))$ to $L^\infty(\mathbb{R}^n)$ Moreover, the operator norms are independent of $\mathcal{R}$. The conclusion follows by using vector-valued interpolation. $\square$

We will now assemble all these ingredients to prove the main result of this chapter.

**Theorem 4.6** *Assume that*

$$BR^*_{\mathbb{P}^{n-1}}\left(\infty \times \infty \mapsto p, \frac{1}{2}\right) < \infty$$

*for some $2n/(n-1) < p \leq 4$ when $n \geq 3$, or for some $p > 4$ when $n = 2$. Then the linear restriction estimate $R^*_{\mathbb{P}^{n-1}}(\infty \mapsto p)$ holds.*

*Proof* For $f \colon [-1,1]^{n-1} \to \mathbb{C}$ and $C$ as in Corollary 4.4, we may write

$$
\begin{aligned}
Ef(x)^2 &= \int_{[-1,1]^{n-1} \times [-1,1]^{n-1}} f(\xi_1) f(\xi_2) e((\xi_1 + \xi_2) \cdot \bar{x} \\
&\quad + (|\xi_1|^2 + |\xi_2|^2)x_n)\, d\xi_1 d\xi_2 \\
&= \sum_{\Omega = \Omega_1 \times \Omega_2 \in C} \int_{\Omega_1 \times \Omega_2} f(\xi_1) f(\xi_2) e((\xi_1 + \xi_2) \cdot \bar{x} \\
&\quad + (|\xi_1|^2 + |\xi_2|^2)x_n)\, d\xi_1 d\xi_2 \\
&= \sum_{\Omega \in C} E_{\Omega_1} f(x) E_{\Omega_2} f(x).
\end{aligned}
$$

For $k \geq 1$ define $C_k$ to consist of those cubes in $C$ whose side length is $2^{-k}$. We use the triangle inequality to separate scales:

$$
\|Ef\|_p^2 = \|(Ef)^2\|_{\frac{p}{2}} \leq \sum_{k \geq 1} \left\| \sum_{\Omega \in C_k} E_{\Omega_1} f E_{\Omega_2} f \right\|_{\frac{p}{2}}. \tag{4.4}
$$

Next, note that as $\Omega$ ranges through $C_k$, the collection of cubes $4(\Omega_1 + \Omega_2)$ overlap at most $C$ times, for some $C$ independent of $k$. This follows by combining two observations. One is the fact that the upper bound in (4.3) forces $\Omega_1 + \Omega_2 \subset \Omega_1 + 1{,}000\Omega_1$. The other one is the fact that each $\Omega_1$ appears at most $O(1)$ times as the first component of some $\Omega \in C_k$. This observation will allow us to exploit orthogonality within each family $C_k$.

The Fourier transform of $E_{\Omega_1} f E_{\Omega_2} f$ is supported inside a rectangular box $R_\Omega \subset \mathbb{R}^{n-1} \times \mathbb{R}$ whose projection to $\mathbb{R}^{n-1}$ lies inside $2(\Omega_1 + \Omega_2)$. It follows that we can split $C_k$ into $C = O(1)$ families, such that the boxes $2R_\Omega$ are pairwise disjoint for $\Omega$ in each family. By applying Lemma 4.5 with $s = p/2$ to each family, we get

$$
\left\| \sum_{\Omega \in C_k} E_{\Omega_1} f E_{\Omega_2} f \right\|_{\frac{p}{2}} \lesssim \left( \sum_{\Omega \in C_k} \|E_{\Omega_1} f E_{\Omega_2} f\|_{\frac{p}{2}}^{\frac{p}{2}} \right)^{\frac{2}{p}}, \quad \text{if } n \geq 3
$$

and

$$
\left\| \sum_{\Omega \in C_k} E_{\Omega_1} f E_{\Omega_2} f \right\|_{\frac{p}{2}} \lesssim \left( \sum_{\Omega \in C_k} \|E_{\Omega_1} f E_{\Omega_2} f\|_{\frac{p}{2}}^{\frac{p}{p-2}} \right)^{\frac{p-2}{p}}, \quad \text{if } n = 2,
$$

with implicit constants independent of $k$.

The lower bound in inequality (4.3) allows us to apply Proposition 4.2 to each term in the sum:

$$\|E_{\Omega_1} f E_{\Omega_2} f\|_{\frac{p}{2}}^{\frac{1}{2}} \lesssim 2^{-k\left(n-1-\frac{n+1}{p}\right)} \|f\|_{L^\infty([-1,1]^{n-1})}.$$

Note that there are $O(2^{k(n-1)})$ cubes in $\mathcal{C}_k$, so

$$\left\| \sum_{\Omega \in \mathcal{C}_k} E_{\Omega_1} f E_{\Omega_2} f \right\|_{\frac{p}{2}} \lesssim \begin{cases} \left( 2^{k(n-1)} 2^{-kp\left(n-1-\frac{n+1}{p}\right)} \right)^{\frac{2}{p}} \|f\|_{L^\infty([-1,1]^{n-1})}^2, & \text{if } n \geq 3, \\ \left( 2^{k} 2^{-k\frac{2p}{p-2}\left(1-\frac{3}{p}\right)} \right)^{\frac{p-2}{p}} \|f\|_{L^\infty([-1,1]^{n-1})}^2, & \text{if } n = 2. \end{cases}$$

In both cases, the upper bound is $O\left( 2^{-k\epsilon_p} \|f\|_{L^\infty([-1,1]^{n-1})}^2 \right)$, for some $\epsilon_p > 0$ (since $p > 2n/(n-1)$). Combining this with (4.4) finishes the proof of the theorem. $\qquad\square$

Theorems 3.1 and 4.6 lead to the following conclusion.

**Corollary 4.7** *The restriction estimate $R^*_{\mathbb{P}^{n-1}}(\infty \mapsto p)$ holds for all $p > 2(n+2)/n$. In particular, the full Restriction Conjecture 1.10 is verified when $n = 2$.*

The arguments in this chapter can be adapted to prove the full expected range of linear restriction estimates ($R^*_{\text{Cone}^{n-1}}(\infty \mapsto p)$ with $p > 2(n-1)/(n-2)$, cf. Conjecture 1.11), for the cone in three and four dimensions. The fact that the linear restriction theory of $\text{Cone}^3$ has been resolved may come as a great surprise, given the fact that this manifold contains (and in fact can be foliated into rescaled copies of) the much less understood paraboloid $\mathbb{P}^2$.

The short explanation for the sharp contrast with the paraboloid is that the range for bilinear restriction estimates remains the same in the presence of exactly one zero principal curvature, while the expected range for linear restriction estimates gets smaller in the presence of (at least) one zero principal curvature. More precisely, the bilinear version of the restriction estimate $R^*_{\text{Cone}^{n-1}}(\infty \mapsto p)$ is known to hold for $p > 2(n+2)/n$ (cf. Exercise 3.32). On the other hand, $2(n+2)/n \leq 2(n-1)/(n-2)$ precisely when $n = 3, 4$.

A key difference between the cone and the paraboloid is the way in which they are rescaled. After a rotation, the cone $\text{Cone}^{n-1}$ can be identified with a subset of

$$\text{Cone}^{n-1}_* = \left\{ (\xi_1, \ldots, \xi_n) : \xi_n = \frac{\xi_1^2 + \cdots + \xi_{n-2}^2}{2\xi_{n-1}} \right\}.$$

Each cross section $\xi_{n-1} = C$ is an elliptic paraboloid that can be rescaled using parabolic rescaling. More precisely, the affine functions ($\xi_0 \in \mathbb{R}^{n-2}$, $\delta > 0$)

$$\mathbb{R}^{n-2} \times \mathbb{R} \times \mathbb{R} \ni (\xi, \xi_{n-1}, \xi_n) \mapsto \left( \frac{\xi - \xi_{n-1}\xi_0}{\delta}, \xi_{n-1}, \frac{\xi_n - \xi_0 \cdot \xi + \frac{\xi_{n-1}}{2}|\xi_0|^2}{\delta^2} \right)$$

map $\mathbb{C}\mathrm{one}_*^{n-1}$ to itself. Note that there is no stretching in the $\xi_{n-1}$ direction corresponding to the zero principal curvature. These maps are called *Lorentz transformations*.

**Exercise 4.8** (The Restriction Conjecture for $\mathbb{C}\mathrm{one}^2$ and $\mathbb{C}\mathrm{one}^3$) Use the result in Exercise 3.32 and Lorentz transformations to prove that $R^*_{\mathbb{C}\mathrm{one}^2}(\infty \mapsto p)$ holds for $p > 4$ and $R^*_{\mathbb{C}\mathrm{one}^3}(\infty \mapsto p)$ holds for $p > 3$.

Hint: Consider the subset of $\mathbb{C}\mathrm{one}_*^{n-1}$ corresponding to $-1 \leq \xi_1, \ldots, \xi_{n-2} \leq 1$ and $1 \leq \xi_{n-1} \leq 2$. Do a Whitney decomposition in the variables $\xi_1, \ldots, \xi_{n-2}$.

**Exercise 4.9** This exercise proves a weaker version of Theorem 4.6 without using a Whitney decomposition.

Let $n \geq 3$. Assume that

$$\sup_{0 < D \leq \frac{1}{2}} BR^*_{\mathbb{P}^{n-1}}(\infty \times \infty \mapsto p, D) < \infty$$

for some $2n/(n - 1) < p \leq 4$. Then the linear restriction estimate $R^*_{\mathbb{P}^{n-1}}(\infty \mapsto p)$ holds.

Hint: For each $R \gg 1$, let $C_R$ be the smallest constant such that

$$\|Ef\|_{L^p(B_R)} \leq C_R$$

holds for all balls $B_R$ and all $f: [-1, 1]^{n-1} \to \mathbb{C}$ with $\|f\|_\infty = 1$. Prove that $C_R \leq C + 1/2C_R$, for some $C$ independent of $R$. Cover the diagonal $\{(\xi, \xi): \xi \in [-1, 1]^{n-1}\}$ with $\sim D^{1-n}$ cubes of side length $D$. Use earlier arguments to estimate the contribution from these cubes by $C_p D^{\epsilon_p} C_R$, and the hypothesis to estimate the off-diagonal contribution by some $C_D$. Choose $D$ small enough.

**Exercise 4.10** Fix two disjoint cubes $\Omega_1, \Omega_2$ in $[-1, 1]^{n-1}$. Assume that for each $f$ supported on $\Omega_1 \cup \Omega_2$ we have

$$\||E_{\Omega_1} f E_{\Omega_2} f|^{\frac{1}{2}}\|_{L^p(\mathbb{R}^n)} \lesssim (\|f\|_{L^q(\Omega_1)} \|f\|_{L^q(\Omega_2)})^{1/2}.$$

Prove that this inequality continues to hold (with a different implicit constant) if the pair $(\Omega_1, \Omega_2)$ is replaced with any other pair of disjoint cubes in $\mathbb{R}^{n-1}$.

# 5

# Kakeya and Square Function Estimates

In the previous chapters, we have seen that the extension operator can be decomposed using wave packets that are spatially concentrated inside tubes. We have also demonstrated how controlling bilinear tube interactions has led to the proof of sharp bilinear restriction estimates. In this chapter, we will search for similar connections in the linear setting.

In the first section, we briefly introduce a few equivalent quantitative estimates from the hierarchy of Kakeya conjectures. Our presentation will be neither in-depth nor comprehensive. A more detailed treatment of many of these topics can be found in chapter 22 of [**83**].

In the last two sections, we discuss the relevance of square functions and Kakeya-type estimates for the pursuit of the Restriction Conjecture.

## 5.1 A Few Kakeya-Type Conjectures

A *Kakeya* or *Besicovitch* set in $\mathbb{R}^n$ is a set containing a unit line segment in every direction. A trivial example is any unit ball. It has been known for about a century that there are Kakeya sets of zero Lebesgue measure. While invisible to the naked eye, these sets cannot be too small, and one may wonder what is the ideal microscope that detects them.

The following stands as one of the most fascinating conjectures in Geometric Measure Theory.

**Conjecture 5.1** (Kakeya set conjecture) *Each Kakeya set in $\mathbb{R}^n$ has Hausdorff dimension $n$.*

The conjecture has been verified by Davies in [**42**] when $n = 2$, but is open when $n \geq 3$. A slightly weaker version of the conjecture asserts that Kakeya sets should have a full Minkowski dimension. This amounts to

$$\lim_{\delta \to 0} \frac{\log |K_\delta|}{\log \delta} = 0,$$

where $K_\delta$ is the $\delta$-neighborhood of the Kakeya set $K$.

We will continue to use the notation $\lesssim$ to hide powers of logarithms of various scales (e.g., $\delta, R$).

**Definition 5.2** An $(N_1, N_2)$-tube is a cylinder with radius $N_1$ and length $N_2 \gg N_1$. Its eccentricity is $N_2 N_1^{-1}$ and its direction (orientation) is given by the direction of the central axis. Two $(N_1, N_2)$-tubes of arbitrary orientations will be called congruent.

The Kakeya set conjecture is a fairly easy consequence of the following quantitative conjecture. We will check this fact shortly for the Minkowski dimension.

**Conjecture 5.3** (Kakeya maximal operator conjecture) *Let $\mathcal{T}$ be a collection of congruent tubes in $\mathbb{R}^n$ with eccentricity $\delta^{-1}$ and $\delta$-separated directions (in particular, there is at most one tube in each of roughly $\delta^{1-n}$ directions). Then for $n/(n-1) \le r \le \infty$,*

$$\left\| \sum_{T \in \mathcal{T}} 1_T \right\|_r \lesssim \left( \sum_{T \in \mathcal{T}} |T| \right)^{\frac{1}{r}} \delta^{\frac{n}{r} - (n-1)}. \tag{5.1}$$

Considering the family consisting of one tube centered at the origin for each direction shows that logarithmic losses are necessary at the endpoint $r = n/(n-1)$. Exercise 5.9 verifies (5.1) for this particular configuration of tubes.

Inequality (5.1) is scale invariant, the implicit constant encoded by $\lesssim$ only depends on the eccentricity $\delta^{-1}$, not on the parameters $N_1, N_2$ of the tubes. The requirement that the directions of the tubes are $\delta$-separated is important for the validity of (5.1). The more subtle inequalities for families of tubes with direction multiplicity are called *X-ray estimates*. See [78] and [120] as well as Proposition 5.6.

Note that (5.1) is trivial when $r = \infty$. Thus, using Hölder's inequality, the full range in (5.1) follows if the endpoint case $r = n/(n-1)$ is proved

$$\left\| \sum_{T \in \mathcal{T}} 1_T \right\|_r \lesssim \left( \sum_{T \in \mathcal{T}} |T| \right)^{\frac{1}{r}}. \tag{5.2}$$

Since (5.2) is trivially true when $r = 1$, we can equivalently phrase Conjecture 5.3 as saying that (5.2) holds for all $1 \le r \le n/(n-1)$.

We will make two observations, to get a better understanding of what kind of information (5.2) encodes. Let us first assume that $r$ is an integer and that (5.2) holds for *some* family $\mathcal{T}$. An equivalent way to write this is

$$\sum_{T_1 \in \mathcal{T}} \cdots \sum_{T_r \in \mathcal{T}} |T_1 \cap \cdots \cap T_r| \lesssim \sum_{T \in \mathcal{T}} |T|.$$

This says that, ignoring small (logarithmic) losses, the diagonal contribution from $T_1 = \cdots = T_r$ dominates the $r$ fold. There are only a small number of instances where such estimates with $r \in \mathbb{N}$ are relevant (e.g., Proposition 5.6 and Exercise 5.12), and simple counting arguments can be used to address them. The arguments get significantly more complicated when one deals with noninteger values of $r$.

Let us also observe that a simple application of Hölder's inequality shows that the validity of (5.2) for some particular $r \in (1, n/(n-1)]$ would imply the fact that the tubes in $\mathcal{T}$ are essentially pairwise disjoint, in the sense that

$$\sum_{T \in \mathcal{T}} |T| \lesssim |\cup_{T \in \mathcal{T}} T|. \tag{5.3}$$

This in turn immediately implies that all Kakeya sets $K$ in $\mathbb{R}^n$ have Minkowski dimension $n$. Indeed, $K_\delta$ contains a set $\cup_{T \in \mathcal{T}} T$, with $\mathcal{T}$ consisting of $\sim \delta^{1-n}$ $(\delta, 1)$-tubes T, and (5.3) shows that

$$|\log |K_\delta|| \lesssim \log \log \left( \frac{1}{\delta} \right).$$

Conjecture 5.3 was proved by Córdoba when $n = 2$; as shown in Proposition 5.6, but is open in higher dimensions. When $n \geq 3$, it has been verified by Wolff [121] for $r \geq (n+2)/n$, and improvements in high dimensions have been obtained by Katz and Tao in [74]. In the recent work [126], building on [63], Zahl improved Wolff's result when $n = 4$.

Part of the progress on Conjecture 5.3 has been achieved by analyzing two equivalent formulations that we describe next. The systematic study of these maximal functions was initiated by Bourgain in his very influential paper [22]. Among other things, he proved that Kakeya sets in $\mathbb{R}^3$ have the Hausdorff dimension at least $7/3$.

For $f : \mathbb{R}^n \to \mathbb{R}$ and $0 < \delta < 1$, we define the *Kakeya maximal function* $f_\delta^* : \mathbb{S}^{n-1} \to [0, \infty)$

$$f_\delta^*(\omega) = \sup_{T \| \omega} \frac{1}{|T|} \int_T |f|,$$

where the supremum is taken over all $(\delta, 1)$-tubes in $\mathbb{R}^n$ whose central axis is parallel to $\omega$.

We also define the *Nikodym maximal function* $f_\delta^{**} : \mathbb{R}^n \to [0, \infty)$

$$f_\delta^{**}(x) = \sup_{x \in T} \frac{1}{|T|} \int_T |f|,$$

where the supremum is taken over all $(\delta, 1)$-tubes in $\mathbb{R}^n$ containing $x$.

**Conjecture 5.4** (Kakeya maximal function conjecture) *We have for each* $p \geq n$

$$\|f_\delta^*\|_{L^p(\mathbb{S}^{n-1})} \lesssim \|f\|_{L^p(\mathbb{R}^n)}.$$

**Conjecture 5.5** (Nikodym maximal function conjecture) *We have for each* $p \geq n$

$$\|f_\delta^{**}\|_{L^p(\mathbb{R}^n)} \lesssim \|f\|_{L^p(\mathbb{R}^n)}.$$

Interpolation with $p = \infty$ shows that the validity of both conjectures would follow if they were proved for $p = n$.

The equivalence of Conjectures 5.4 and 5.5 was proved in [**110**]. The equivalence between Conjectures 5.3 and 5.4, as well as the connection between partial progress on them, is discussed in [**83**].

Let us now briefly discuss the progress on Conjecture 5.1. It has been recently proved in [**73**] that Kakeya sets in $\mathbb{R}^3$ must have Hausdorff dimension at least $5/2 + \epsilon_0$, for some very small $\epsilon_0$. The best result in $\mathbb{R}^4$ is due to Zahl [**126**], and to Katz and Tao [**74**] in dimensions five and higher.

We point out that an inequality of the form

$$\left\| \sum_{T \in \mathcal{T}} 1_T \right\|_r \lesssim \left( \sum_{T \in \mathcal{T}} |T| \right)^{\frac{1}{r}} \delta^{-s}$$

implies that the Hausdorff dimension of Kakeya sets is at least $n - sr'$; see [**107**].

The following result illustrates a simple counting argument.

**Proposition 5.6** ([**41**]) *Consider a collection $V$ of $\delta$-separated directions in the plane. Let $\mathcal{T}$ be a collection of congruent tubes in $\mathbb{R}^2$ with eccentricity $\delta^{-1}$ and with at most $m$ tubes in each direction $\mathbf{v} \in V$. Then*

$$\left\| \sum_{T \in \mathcal{T}} 1_T \right\|_2 \lesssim \log(\delta^{-1})^{\frac{1}{2}} m^{\frac{1}{2}} \left( \sum_{T \in \mathcal{T}} |T| \right)^{\frac{1}{2}}.$$

*In particular, Conjecture 5.3 holds true in $\mathbb{R}^2$.*

*Proof* The angle between any two rectangles $T_1, T_2 \in \mathcal{T}$ is $\sim j\delta$, for some $0 \leq j \lesssim \delta^{-1}$. It is easy to see that in this case we have the estimate

$$|T_1 \cap T_2| \lesssim \frac{|T_1|}{j+1}.$$

We may thus write

$$\sum_{T_1 \in \mathcal{T}} \sum_{T_2 \in \mathcal{T}} |T_1 \cap T_2| \lesssim m \sum_{T_1 \in \mathcal{T}} \sum_{j=1}^{O(\delta^{-1})} j^{-1} |T_1| \lesssim m \log(\delta^{-1}) \sum_{T_1 \in \mathcal{T}} |T_1|.$$

$\square$

This argument can be applied to show that

$$\left\| \sum_{T \in \mathcal{T}} 1_T \right\|_2 \lesssim \left( \sum_{T \in \mathcal{T}} |T| \right)^{\frac{1}{2}}$$

holds for each family of congruent tubes in $\mathbb{R}^n$, with $\delta$-separated directions restricted to a "reasonable" curve on $\mathbb{S}^{n-1}$. One such example arises with the tubes normal to $\mathbb{C}\mathrm{one}^2$. In this case, the curve is a circle on $\mathbb{S}^2$. Exercise 5.12 shows that a slightly stronger inequality holds if tubes are replaced with planks.

We next show how restriction estimates imply inequalities with Kakeya-type flavor.

**Proposition 5.7** *Assume that $R^*_{\mathbb{S}^{n-1}}(p \mapsto p)$ holds for some $p > 2n/(n-1)$. Then for each collection $\mathcal{T}$ of congruent tubes in $\mathbb{R}^n$ with eccentricity $\delta^{-1}$ and $\delta$-separated directions, we have*

$$\left\| \sum_{T \in \mathcal{T}} 1_T \right\|_r \lesssim \left( \sum_{T \in \mathcal{T}} |T| \right)^{\frac{1}{r}} \delta^{\frac{2n}{r} - 2(n-1)}, \qquad (5.4)$$

*with $r = p/2$.*

*Proof* Call $E$ the extension operator for $\mathbb{S}^{n-1}_+$ and let $\psi(\xi) = \sqrt{1 - |\xi|^2}$. Let $R_0 = \delta^{-2}$. The scale invariance of (5.4) allows us to assume that all tubes in $\mathcal{T}$ are $(R_0^{1/2}, R_0)$-tubes. We can certainly also assume locality, that is, all tubes are inside a horizontal strip of height $2R_0$.

Let us recall the computation from Example 1.8. Fix $R$ a bit larger than $R_0$, $R = CR_0$, with $C = O(1)$. Let $\Upsilon \not\equiv 0$ be a nonnegative Schwartz function supported on $[-1, 1]^{n-1}$. Consider a finitely overlapping cover $\mathcal{Q}$ of $B_{n-1}(0, 1/100)$ with cubes $\omega$ with side length $R^{-1/2}$. For each $\omega \in \mathcal{Q}$ and each cube $q \subset \mathbb{R}^{n-1}$ with side length $R^{1/2}$, we let

$$\Upsilon_{q,\omega}(\xi) := R^{\frac{n-1}{2}} \Upsilon(R^{1/2}(\xi - c_\omega)) e(-c_q \cdot (c_\omega - \xi)).$$

Then (1.7) shows that

$$|E\Upsilon_{q,\omega}(x)| \sim 1$$

whenever $x$ is in the tube

$$T_{q,\omega} := \{(\bar{x}, x_n) : |(\bar{x} - c_q) + \nabla\psi(c_\omega)x_n| \ll R^{1/2}, \ |x_n| \ll R\}.$$

Due to locality, by translating, rotating, and splitting the tubes into $O(1)$ many subcollections, and by choosing $R$ large enough, we may also assume that each $T \in \mathcal{T}$ is inside $T_{q,\omega}$, for some $\omega = \omega_T \in \mathcal{Q}$ and some $q = q_T$, with the function $T \mapsto \omega_T$ being injective. Note that if $x \in T$

$$|E\Upsilon_{q_T,\omega_T}(x)| \sim 1. \tag{5.5}$$

Let now $(r_T)_{T \in \mathcal{T}}$ be any subset of the Rademacher sequence on $[0, 1)$. For each $t$, consider the function

$$f_t(\xi) = \sum_{T \in \mathcal{T}} r_T(t)\Upsilon_{q_T,\omega_T}(\xi).$$

Using our hypothesis $R^*_{\mathbb{S}^{n-1}}(p \mapsto p)$ for each $f_t$ implies that

$$\int_{\mathbb{R}^n} \left| \sum_{T \in \mathcal{T}} r_T(t) E\Upsilon_{q_T,\omega_T}(x) \right|^p dx \lesssim \int_{[-1,1]^{n-1}} \left| \sum_{T \in \mathcal{T}} r_T(t)\Upsilon_{q_T,\omega_T}(\xi) \right|^p d\xi,$$

uniformly over $t$. Integrating with respect to $t$, using Fubini and Theorem 1.35 leads to

$$\int_{\mathbb{R}^n} \left( \sum_{T \in \mathcal{T}} |E\Upsilon_{q_T,\omega_T}(x)|^2 \right)^{p/2} dx \lesssim \int_{[-1,1]^{n-1}} \left( \sum_{T \in \mathcal{T}} |\Upsilon_{q_T,\omega_T}(\xi)|^2 \right)^{p/2} d\xi.$$

Due to (5.5), we have

$$\left\| \sum_{T \in \mathcal{T}} 1_T \right\|_{\frac{p}{2}}^{\frac{p}{2}} \lesssim \int_{\mathbb{R}^n} \left( \sum_{T \in \mathcal{T}} |E\Upsilon_{q_T,\omega_T}(x)|^2 \right)^{p/2} dx.$$

Also, since the supports of functions $\Upsilon_{q_T,\omega_T}$ have bounded overlap, we may write

$$\int_{[-1,1]^{n-1}} \left( \sum_{T \in \mathcal{T}} |\Upsilon_{q_T,\omega_T}(\xi)|^2 \right)^{p/2} d\xi \lesssim |\mathcal{T}| R^{\frac{n-1}{2}(p-1)}.$$

Combining the last two inequalities proves (5.4). $\qquad\square$

This argument shows that we may replace the assumption $R^*_{\mathbb{S}^{n-1}}(p \mapsto p)$ with $R^*_{\mathcal{M}}(p \mapsto p)$, for any $\mathcal{M}$ whose unit normal vectors cover a subset of $\mathbb{S}^{n-1}$ with nonempty interior (e.g., if $\mathcal{M}$ has nonzero Gaussian curvature).

As observed earlier, (5.4) implies that the Hausdorff dimension of Kakeya sets in $\mathbb{R}^n$ is at least

$$d_{p,n} := \frac{2p - n(p-2)}{p-2}. \tag{5.6}$$

In particular, since $d_{2n/(n-1),n} = n$, the Restriction Conjecture 1.10 is stronger than the Kakeya set conjecture.

Note also that the strongest restriction estimate we proved so far (Corollary 4.7) only produces the lower bound $d_{2(n+2)/n,n} = 2$ on the Hausdorff dimension of Kakeya sets in $\mathbb{R}^n$. This lower bound also follows trivially from Proposition 5.6, since the Hausdorff dimension of Kakeya sets is a nondecreasing function of $n$. This suggests that, in spite of the rather delicate arguments with tubes that we used in order to prove $R^*_{\mathbb{P}^{n-1}}(\infty \mapsto 2(n+2)/n; \epsilon)$, this restriction estimate does not encode genuinely three-dimensional Kakeya-type information.

We call a *hairbrush* any collection $\mathcal{H}$ of $(\delta, 1)$-tubes in $\mathbb{R}^n$ with $\delta$-separated directions intersecting a fixed $(\delta, 1)$-tube $T_0$ (see Figure 5.1). The next exercise shows that the tubes in a hairbrush are essentially pairwise disjoint. In particular, every Kakeya set whose unit line segments in all directions intersect

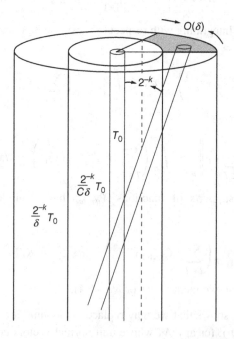

Figure 5.1 Hairbrush: one tube in some family $\mathcal{H}_{k,l}$.

a fixed line, has full Minkowski dimension. The stronger fact that (5.2) holds for $\mathcal{T} = \mathcal{H}$ at the endpoint $r = n/(n-1)$ is proved in lemma 23.4 in [**83**].

**Exercise 5.8** (a) Let $\mathcal{T}$ be a collection of $(\delta, 1)$-tubes in $\mathbb{R}^n$ with $\delta$-separated directions, which lie inside the $O(\delta)$-neighborhood of a two-dimensional plane. Prove that

$$\sum_{T \in \mathcal{T}} |T| \lesssim \left( \log \frac{1}{\delta} \right) | \cup_{T \in \mathcal{T}} T|.$$

(b) Prove that for each hairbrush $\mathcal{H}$

$$\sum_{T \in \mathcal{H}} |T| \lessapprox | \cup_{T \in \mathcal{H}} T|.$$

Hint for (b): For each $\delta \lesssim 2^{-k} \lesssim 1$, write

$$\mathcal{H} = \cup_k \mathcal{H}_k$$

where $\mathcal{H}_k$ consists of the tubes in $\mathcal{H}$ making an angle $\theta \in [2^{-k}, 2^{-k+1})$ with $T_0$. Split $\mathcal{H}_k = \cup_l \mathcal{H}_{k,l}$, with each $\mathcal{H}_{k,l}$ an essentially planar family. Let $C = O(1)$ be sufficiently large. For each $k$ and $l$, apply (a) to the segments of the tubes in $\mathcal{H}_{k,l}$ lying at distance $\geq C^{-1} 2^{-k}$ from $T_0$. Also, argue that these segments are essentially pairwise disjoint as $l$ varies.

**Exercise 5.9** Let $\mathcal{T}$ be a collection of $(\delta, 1)$-tubes in $\mathbb{R}^n$ containing the origin and with $\delta$-separated directions. Prove that for each $\delta \leq |x| \leq 1$, we have

$$\sum_{T \in \mathcal{T}} 1_T(x) \lesssim |x|^{1-n}.$$

Conclude that (5.1) holds if $\mathcal{T}$ has cardinality $\sim \delta^{1-n}$.

**Exercise 5.10** (Bilinear-to-linear reduction for Kakeya estimates) Let $p \geq n/(n-1)$ and $A \leq 0$. Assume that the inequality

$$\left\| \left( \sum_{T_1 \in \mathcal{T}_1} 1_{T_1} \sum_{T_2 \in \mathcal{T}_2} 1_{T_2} \right)^{\frac{1}{2}} \right\|_p \lesssim \delta^A$$

holds for each family $\mathcal{T}_1, \mathcal{T}_2$ of $(\delta, 1)$-tubes in $\mathbb{R}^n$ with $\delta$-separated directions within each family, and separation $\sim 1$ between the directions of each $T_1 \in \mathcal{T}_1$ and $T_2 \in \mathcal{T}_2$. Prove that

$$\left\| \sum_{T \in \mathcal{T}} 1_T \right\|_p \lessapprox \delta^A$$

for each family $\mathcal{T}$ of $(\delta, 1)$-tubes with $\delta$-separated directions.

Hint: Use a quasitriangle inequality in $L^{p/2}$ to control $\|\sum_{T\in\mathcal{T}} 1_T\|_p$ by bilinear expressions involving pairs of families of tubes with directions separated by some $\delta \lesssim 2^{-k} \lesssim 1$. Use rescaling $(\bar{x}, x_n) \mapsto (2^k \bar{x}, x_n)$ to reduce to the case when the separation is $\sim 1$.

**Exercise 5.11** Prove (5.1) for $r = 2$ and $n \geq 2$.

**Exercise 5.12** (Kakeya-type inequality for planks) Cover the $1/R$-neighborhood of the cone $\mathbb{Cone}^2$ with roughly $R^{1/2}$ rectangular boxes $\tau$ of dimensions $1, R^{-1/2}, R^{-1}$. For each such $\tau$, a dual rectangular box $P$ (we will refer to it as "plank") has the same axes as $\tau$ and dimensions $1, R^{1/2}, R$.

(a) Prove that for each two distinct planks $P_1, P_2$

$$|P_1 \cap P_2| \lesssim R^{\frac{1}{2}}\theta^{-2},$$

where $\theta$ is the angle between the axes corresponding to the long sides of $P_1, P_2$.

(b) Consider a family $\mathcal{P}$ that has at most $m$ planks corresponding to each $\tau$. Prove that

$$\left\|\sum_{P\in\mathcal{P}} 1_P\right\|_3 \lesssim m^{\frac{2}{3}}\left(\sum_{P\in\mathcal{P}} |P|\right)^{\frac{1}{3}}.$$

(c) Adapt the argument from Proposition 5.7 to show that the estimate $R^*_{\mathbb{Cone}^2}(p \mapsto p)$ for $p > 4$ implies the weaker $L^2$ estimate

$$\left\|\sum_{P\in\mathcal{P}} 1_P\right\|_2 \lesssim_\epsilon R^\epsilon\left(\sum_{P\in\mathcal{P}} |P|\right)^{\frac{1}{2}},$$

for each family $\mathcal{P}$ that has at most one plank corresponding to each $\tau$.

**Exercise 5.13** Prove that the Kakeya set conjecture in $\mathbb{R}^{n+1}$ implies the Kakeya set conjecture in $\mathbb{R}^n$.

**Exercise 5.14** Consider a collection $V$ of $\delta$-separated directions in the plane, with essentially maximal cardinality $\sim \delta^{-1}$. Let $\mathcal{T}$ be a collection of $(\delta, 1)$-tubes in $\mathbb{R}^2$, pointing in directions from $V$. Assume that there are $\sim m$ tubes in each direction $\mathbf{v} \in V$, so in particular $|\mathcal{T}| \sim m\delta^{-1}$. Let

$$A_r = \left\{x \in \mathbb{R}^2 : \sum_{T\in\mathcal{T}} 1_T(x) \geq r\right\}.$$

Prove that for each $r \geq 1$

$$|A_r| \lesssim \delta^2 \left(\frac{|\mathcal{T}|}{r}\right)^2.$$

## 5.2 Square Function Estimates

We will start the analysis in this section by mentioning two fundamental classical results. For $F \colon \mathbb{R}^n \to \mathbb{C}$ and $U \subset \mathbb{R}^n$, recall that $\mathcal{P}_U F$ is the associated Fourier projection. The following theorem is an example of square root cancellation, closely related to Theorem 1.36.

**Theorem 5.15** (Littlewood–Paley Theorem) *Let $\mathcal{R}_{LP}$ be the collection of all rectangular boxes $R = I_1 \times \cdots \times I_n$ in $\mathbb{R}^n$, with each $I_i$ of the form $[2^{k_i}, 2^{k_i+1}]$ or $[-2^{k_i+1}, -2^{k_i}]$ for some $k_i \in \mathbb{Z}$. Then for each $1 < p < \infty$ and each $F \in L^p(\mathbb{R}^n)$, we have the equivalence of norms*

$$
C_1 \left\| \left( \sum_{R \in \mathcal{R}_{LP}} |\mathcal{P}_R F|^2 \right)^{\frac{1}{2}} \right\|_{L^p(\mathbb{R}^n)} \le \|F\|_{L^p(\mathbb{R}^n)} \le C_2 \left\| \left( \sum_{R \in \mathcal{R}_{LP}} |\mathcal{P}_R F|^2 \right)^{\frac{1}{2}} \right\|_{L^p(\mathbb{R}^n)},
$$

*with $C_1, C_2$ depending only on $p, n$.*

In this section, we will search for a "curved" analog of the Littlewood–Paley equivalence, where dyadic boxes are replaced with boxlike neighborhoods of manifolds. While we choose to present the results for $\mathbb{P}^{n-1}$, they have natural analogs for $\mathbb{S}^{n-1}$, too.

It turns out that in the range $2 \le p < \infty$, one half of the equivalence has very little to do with curvature, and is in fact a consequence of the following classical theorem.

**Theorem 5.16** *Let $\mathcal{Q}$ be a partition of $\mathbb{R}^n$ into congruent cubes. Then for each $2 \le p < \infty$, we have*

$$
\left\| \left( \sum_{Q \in \mathcal{Q}} |\mathcal{P}_Q F|^2 \right)^{\frac{1}{2}} \right\|_{L^p(\mathbb{R}^n)} \lesssim \|F\|_{L^p(\mathbb{R}^n)}.
$$

*The implicit constant is independent of $\mathcal{Q}$ and $F$. The range of $p$ is sharp.*

When $n = 1$ this is due to Carleson, [36]. The higher-dimensional analog is implicit in [39]. The extension to arbitrary families of pairwise disjoint intervals is due to Rubio de Francia [96]. The generalization to higher dimensions – with intervals replaced with rectangular boxes with arbitrary side lengths but fixed orientation – appeared in [80].

**Definition 5.17** A set $\theta \subset \mathbb{R}^n$ is called an almost rectangular box if there is a genuine rectangular box $R_\theta$ such that $C^{-1} R_\theta \subset \theta \subset R_\theta$ for some $C = O(1)$.

If $R_\theta$ is a cube, $\theta$ will be called an almost cube.

For each $R \geq 1$, let us partition $[-1, 1]^{n-1}$ into almost cubes $\omega$ with diameter $\sim R^{-1/2}$. This generates a partition $\Theta(R^{-1})$ of the $1/R$-neighborhood of $\mathbb{P}^{n-1}$

$$\mathcal{N}(R^{-1}) = \{(\xi, |\xi|^2 + t) \colon |tR| < 1\}$$

into almost rectangular boxes $\theta$ of the form

$$\mathcal{N}_\omega(R^{-1}) = (\omega \times \mathbb{R}) \cap \mathcal{N}(R^{-1}).$$

**Corollary 5.18** *Assume that $\widehat{F}$ is supported on $\mathcal{N}(R^{-1})$. Then for $2 \leq p < \infty$, we have*

$$\left\| \left( \sum_{\theta \in \Theta(R^{-1})} |\mathcal{P}_\theta F|^2 \right)^{\frac{1}{2}} \right\|_{L^p(\mathbb{R}^n)} \lesssim \|F\|_{L^p(\mathbb{R}^n)}.$$

*Proof* Note that

$$\mathcal{P}_\theta F = \mathcal{P}_{\omega \times \mathbb{R}} F$$

if $\theta = \mathcal{N}_\omega(R^{-1})$. The corollary now follows from Theorem 5.16 and Fubini's theorem. See also Exercise 5.25. $\qquad\qquad\square$

Let us now investigate the other half of the "curved" Littlewood–Paley equivalence. As we shall soon see, proving the following conjecture is expected to be very difficult.

**Conjecture 5.19** (Reverse square function conjecture) *Assume that $\widehat{F}$ is supported on $\mathcal{N}(R^{-1})$. Then for $p = 2n/(n-1)$ we have*

$$\|F\|_{L^p(\mathbb{R}^n)} \gtrsim \left\| \left( \sum_{\theta \in \Theta(R^{-1})} |\mathcal{P}_\theta F|^2 \right)^{\frac{1}{2}} \right\|_{L^p(\mathbb{R}^n)}. \tag{5.7}$$

It seems plausible that (5.7) would hold even with an implicit constant independent of $R$. When $n = 2$, this was indeed proved to be the case in Proposition 3.3. The conjecture is wide open for $n \geq 3$. It turns out that if true, this conjecture would imply other similar estimates using an interesting type of interpolation that we describe in Proposition 5.20. See also Exercise 5.24 for some limitations of the method.

**Proposition 5.20** (Interpolation) *Assume that the inequality (5.7) holds true for some $p_1, p_2$ with $2 \leq p_1 < p_2 \leq 2n/(n-1)$. Assume also that the Nikodym maximal function estimate*

$$\|f_\delta^{**}\|_{L^q(\mathbb{R}^n)} \lesssim \|f\|_{L^q(\mathbb{R}^n)} \tag{5.8}$$

*holds for $q$ equal to the conjugate exponent $(p_i/2)'$, for both $i = 1$ and $i = 2$. Then (5.7) holds true for all $p_1 < p < p_2$.*

*Proof* Each $\theta \in \Theta(R^{-1})$ lies inside a rectangular box $R_\theta$ with dimensions comparable to $R^{-1/2}, \ldots, R^{-1/2}, R^{-1}$. Let $\varphi$ be a smooth function satisfying

$$1_{[-1,1]^n} \leq \varphi \leq 1_{[-2,2]^n},$$

$$|\widehat{\varphi}| \lesssim \chi.$$

Let $\varphi_\theta$ be its affinely rescaled version satisfying

$$1_{R_\theta} \leq \varphi_\theta \leq 1_{2R_\theta}$$

and

$$|\widehat{\varphi_\theta}| \lesssim \frac{1}{|T_\theta|} \chi_{T_\theta}, \tag{5.9}$$

where $T_\theta$ is a box that is dual to $R_\theta$ and contains the origin.

Define the smooth projection operator

$$\tilde{\mathcal{P}}_\theta(F)(x) = \int_{\mathbb{R}^n} \widehat{F}(\xi) \varphi_\theta(\xi) e(x \cdot \xi) \, d\xi. \tag{5.10}$$

The main step is to prove that inequality (5.7) for any given $2 \leq p \leq 2n/(n-1)$ together with (5.8) for $q = (p/2)'$ imply that the estimate

$$\left\| \sum_{\theta \in \Theta(R^{-1})} \tilde{\mathcal{P}}_\theta F_\theta \right\|_{L^p(\mathbb{R}^n)} \lessapprox \left\| \left( \sum_{\theta \in \Theta(R^{-1})} |F_\theta|^2 \right)^{\frac{1}{2}} \right\|_{L^p(\mathbb{R}^n)} \tag{5.11}$$

holds true for arbitrary functions $F_\theta$. Once we manage to prove this, we can then invoke vector-valued interpolation for the linear operator

$$\{F_\theta\}_{\theta \in \Theta(R^{-1})} \mapsto \sum_{\theta \in \Theta(R^{-1})} \tilde{\mathcal{P}}_\theta F_\theta$$

to conclude that (5.11) holds for each $p_1 < p < p_2$. But this in turn implies that (5.7) holds in the same range, simply by testing (5.11) with $F_\theta = \mathcal{P}_\theta F$ and using that $\tilde{\mathcal{P}}_\theta \mathcal{P}_\theta = \mathcal{P}_\theta$. It is worth noting that (5.11) is false even for one single $\theta$, if the smooth projection $\tilde{\mathcal{P}}_\theta$ is replaced with $\mathcal{P}_\theta$. This is due to the fact that the boundary of $\theta$ is curved. See [**48**].

Let us now assume that (5.7) holds for some $p \in [2, 2n/(n-1)]$, and let us write $q = (p/2)'$. The proof of (5.11) will rely on four observations. First, note that

$$\left\| \sum_{\theta \in \Theta(R^{-1})} \tilde{\mathcal{P}}_\theta F_\theta \right\|_{L^p(\mathbb{R}^n)} \lessapprox \left\| \left( \sum_{\theta \in \Theta(R^{-1})} |\tilde{\mathcal{P}}_\theta F_\theta|^2 \right)^{\frac{1}{2}} \right\|_{L^p(\mathbb{R}^n)}. \tag{5.12}$$

This can be seen by splitting $\Theta(R^{-1})$ into $O(1)$ subcollections $\Theta'$, and applying (5.7) to each $\Theta'$, with a slightly smaller value of $R$ and with $F = \sum_{\theta \in \Theta'} \tilde{\mathcal{P}}_\theta F_\theta$.

Second, let

$$M_{R,k} f(x) = \sup_{x \in T} \frac{1}{|T|} \int_T |f|,$$

where the supremum is taken over all $(2^k R^{1/2}, 2^k R)$-tubes $T$ in $\mathbb{R}^n$ containing $x$. Rescaling (5.8) shows that

$$\|M_{R,k} f\|_q \le C_R \|f\|_q$$

with $C_R \lesssim 1$, uniformly over $k$. Note also that due to the decay of $\chi_T$, we may write for each tube $T$

$$\frac{1}{|T|} \chi_T(x) \lesssim \sum_{k \ge 0} 2^{-10kn} \left( \frac{1}{|2^k T|} 1_{2^k T}(x) \right). \tag{5.13}$$

Using this and recalling (5.9), we get

$$|f| * |\widehat{\varphi_\theta}|(x) \lesssim \sum_{k \ge 0} 2^{-10kn} M_{R,k} f(x),$$

uniformly over $x, \theta$. We conclude that

$$\| \max_\theta |f| * |\widehat{\varphi_\theta}| \|_q \lesssim \|f\|_q. \tag{5.14}$$

Third, for arbitrary functions $G \colon \mathbb{R}^n \to \mathbb{C}$ and $H \colon \mathbb{R}^n \to [0, \infty)$, we have the following weighted $L^2$ inequality

$$\int |\tilde{\mathcal{P}}_\theta G(x)|^2 H(x)\, dx \le \int \left( \int |G(x - y)| |\widehat{\varphi_\theta}(y)|\, dy \right)^2 H(x)\, dx$$

$$\lesssim \int |G|^2 * |\widehat{\varphi_\theta}|(x) H(x)\, dx \quad \text{(Hölder's inequality)}$$

$$= \int |G^2(x)| H * |\widehat{\varphi_\theta}|(x)\, dx.$$

Fourth, combining this with (5.14) allows us to write

$$\left\| \left( \sum_{\theta \in \Theta(R^{-1})} |\tilde{\mathcal{P}}_\theta F_\theta|^2 \right)^{\frac{1}{2}} \right\|_{L^p(\mathbb{R}^n)}^2 = \sup_{\substack{H \ge 0 \\ \|H\|_q = 1}} \int \sum_{\theta \in \Theta(R^{-1})} |\tilde{\mathcal{P}}_\theta F_\theta|^2 H$$

$$\lesssim \sup_{\substack{H \ge 0 \\ \|H\|_q = 1}} \int \left( \sum_{\theta \in \Theta(R^{-1})} |F_\theta|^2 \right) \max_\theta H * |\widehat{\varphi_\theta}|$$

$$\lesssim \left\| \left( \sum_{\theta \in \Theta(R^{-1})} |F_\theta|^2 \right)^{\frac{1}{2}} \right\|_{L^p(\mathbb{R}^n)}^2.$$

Finally, combining this with (5.12) finishes the proof of (5.11).          □

We can now prove the following analog of the Littlewood–Paley theorem in the curved setting, conditional to the validity of Conjecture 5.19. It is worth mentioning that it is expected that $\lesssim\atop\approx$ can be replaced with the scale-independent symbol $\lesssim$ in the double inequality (5.15).

**Corollary 5.21** *Assume that Conjecture 5.19 holds and that $\widehat{F}$ is supported on $\mathcal{N}(R^{-1})$. Then for $2 \le p \le 2n/(n-1)$ we have*

$$\left\|\left(\sum_{\theta\in\Theta(R^{-1})} |\mathcal{P}_\theta F|^2\right)^{\frac{1}{2}}\right\|_{L^p(\mathbb{R}^n)} \underset{\approx}{\lesssim} \|F\|_{L^p(\mathbb{R}^n)} \underset{\approx}{\lesssim} \left\|\left(\sum_{\theta\in\Theta(R^{-1})} |\mathcal{P}_\theta F|^2\right)^{\frac{1}{2}}\right\|_{L^p(\mathbb{R}^n)}. \tag{5.15}$$

*Let $\tilde{\mathcal{P}}_\theta$ be as in (5.10). Then for $2n/(n+1) \le p \le 2$, we have*

$$\left\|\left(\sum_{\theta\in\Theta(R^{-1})} |\tilde{\mathcal{P}}_\theta F|^2\right)^{\frac{1}{2}}\right\|_{L^p(\mathbb{R}^n)} \underset{\approx}{\lesssim} \|F\|_{L^p(\mathbb{R}^n)} \underset{\approx}{\lesssim} \left\|\left(\sum_{\theta\in\Theta(R^{-1})} |\mathcal{P}_\theta F|^2\right)^{\frac{1}{2}}\right\|_{L^p(\mathbb{R}^n)}. \tag{5.16}$$

*Proof* Note that the case $p = 2$ in (5.15) is immediate due to orthogonality. It was proved in [**35**] that Conjecture 5.19 implies the Nikodym maximal function Conjecture 5.5. Combining these with Proposition 5.20 and Corollary 5.18 proves (5.15).

Let us now verify (5.16) for $2n/(n+1) \le p \le 2$. First, we invoke a duality argument to write

$$\left\|\left(\sum_{\theta\in\Theta(R^{-1})} |\tilde{\mathcal{P}}_\theta F|^2\right)^{\frac{1}{2}}\right\|_p$$

$$= \sup_{\|g_\theta(x)\|_{l^2(\theta)}\equiv 1} \sup_{\|H\|_{p'}=1} \int \sum_{\theta\in\Theta(R^{-1})} \tilde{\mathcal{P}}_\theta F(x)\overline{g_\theta(x)H(x)}\,dx$$

$$= \sup_{\|g_\theta(x)\|_{l^2(\theta)}\equiv 1} \sup_{\|H\|_{p'}=1} \int F(x)\overline{\sum_{\theta\in\Theta(R^{-1})} \tilde{\mathcal{P}}_\theta(g_\theta H)(x)}\,dx$$

$$\underset{\approx}{\lesssim} \|F\|_p \sup_{\|g_\theta(x)\|_{l^2(\theta)}\equiv 1} \sup_{\|H\|_{p'}=1} \left\|\left(\sum_{\theta\in\Theta(R^{-1})} |g_\theta H|^2\right)^{\frac{1}{2}}\right\|_{p'} \quad \text{(by (5.11))}$$

$$= \|F\|_p.$$

Note that this argument does not require any restriction on the support of $\widehat{F}$.

Also, if $\widehat{F}$ is supported on $\mathcal{N}(R^{-1})$ then $F = \sum_{\theta \in \Theta(R^{-1})} \mathcal{P}_\theta F$ and thus

$$\|F\|_p = \sup_{\|G\|_{p'}=1} \left| \sum_{\theta \in \Theta(R^{-1})} \int \mathcal{P}_\theta(F)\overline{G} \right|$$

$$= \sup_{\|G\|_{p'}=1} \left| \sum_{\theta \in \Theta(R^{-1})} \int \mathcal{P}_\theta(F)\overline{\mathcal{P}_\theta(G)} \right|$$

$$\leq \sup_{\|G\|_{p'}=1} \left\| \left( \sum_{\theta \in \Theta(R^{-1})} |\mathcal{P}_\theta F|^2 \right)^{\frac{1}{2}} \right\|_p \left\| \left( \sum_{\theta \in \Theta(R^{-1})} |\mathcal{P}_\theta G|^2 \right)^{\frac{1}{2}} \right\|_{p'}$$

$$\lesssim \left\| \left( \sum_{\theta \in \Theta(R^{-1})} |\mathcal{P}_\theta F|^2 \right)^{\frac{1}{2}} \right\|_p .$$

In the last inequality, we used (5.15) with $p$ and $F$ replaced with $p'$ and $G$.   $\square$

**Exercise 5.22** Show that (5.7) is false for $p > 2n/(n-1)$.

Hint: Test (5.7) with some $F$ such that $\widehat{F}$ is a smooth approximation of $1_{\mathcal{N}(R^{-1})}$. Arrange that $|\mathcal{P}_\theta F| \approx 1_{T_\theta}/|T_\theta|$ for some tube $T_\theta$ containing the origin. Then use Exercise 5.9.

The $L^4$ biorthogonality argument from Section 3.2 has a natural extension to other $L^p$ spaces, with $p$ even integer. We refer to this as *multiorthogonality*.

**Exercise 5.23** Use multiorthogonality to prove that

$$\|F\|_{L^p(\mathbb{R}^n)} \lesssim \left\| \left( \sum_{R \in \mathcal{R}_{LP}} |\mathcal{P}_R F|^2 \right)^{\frac{1}{2}} \right\|_{L^p(\mathbb{R}^n)}$$

holds for all even numbers $p$. Use the interpolation technique from the proof of Proposition 5.20 to extend this inequality to all $p \geq 2$.

The next exercise shows that the interpolation technique from the proof of Proposition 5.20 is not applicable in the range $p > 2n/(n-1)$. We illustrate this with $p = \infty$.

**Exercise 5.24** A simple application of the Cauchy–Schwarz inequality shows that for each $F: \mathbb{R}^n \to \mathbb{C}$ with the Fourier transform supported inside $\mathcal{N}(R^{-1})$, we have

$$\|F\|_{L^\infty(\mathbb{R}^n)} \lesssim R^{\frac{n-1}{4}} \left\| \left( \sum_{\theta \in \Theta(R^{-1})} |\mathcal{P}_\theta F|^2 \right)^{\frac{1}{2}} \right\|_{L^\infty(\mathbb{R}^n)}.$$

Construct functions $F_\theta$ with

$$\left\| \left( \sum_{\theta \in \Theta(R^{-1})} |F_\theta|^2 \right)^{\frac{1}{2}} \right\|_{L^\infty(\mathbb{R}^n)} = 1$$

and a smooth $\varphi$ supported on $[-2, 2]^n$, so that the smooth projections $\tilde{\mathcal{P}}_\theta$ satisfy

$$\left\| \sum_{\theta \in \Theta(R^{-1})} \tilde{\mathcal{P}}_\theta F_\theta \right\|_{L^\infty(\mathbb{R}^n)} \sim R^{\frac{n-1}{2}}.$$

Hint: Choose $\varphi$ so that $\widehat{\varphi} \geq 1$ on $[-100, 100]^n$. Let $T_\theta$ be a rectangular box dual to $R_\theta$, with dimensions $\sim R^{1/2}, \ldots, R^{1/2}, R$, such that $0 \in 10T_\theta$ and all $T_\theta$ are pairwise disjoint. Choose $F_\theta$ such that $|F_\theta| = 1_{T_\theta}$.

**Exercise 5.25** (Cylindrical reverse square function estimate) Let $\mathcal{S}$ be a collection of pairwise disjoint open sets $S$ in $\mathbb{R}^{n-1}$. Let $C_{\mathcal{S}, p}$ be the smallest constant such that

$$\|F\|_{L^p(\mathbb{R}^{n-1})} \leq C_{\mathcal{S}, p} \left\| \left( \sum_{S \in \mathcal{S}} |\mathcal{P}_S F|^2 \right)^{\frac{1}{2}} \right\|_{L^p(\mathbb{R}^{n-1})}$$

holds for all $F$ with the Fourier transform supported on $\cup_{S \in \mathcal{S}} S$. Let $\tilde{S} = S \times \mathbb{R}$. Prove that

$$\|F\|_{L^p(\mathbb{R}^n)} \leq C_{\mathcal{S}, p} \left\| \left( \sum_{S \in \mathcal{S}} |\mathcal{P}_{\tilde{S}} F|^2 \right)^{\frac{1}{2}} \right\|_{L^p(\mathbb{R}^n)}$$

for all $F$ with Fourier transform supported on $\cup_{S \in \mathcal{S}} \tilde{S}$.

## 5.3 Square Functions and the Restriction Conjecture

In this section, we explore the connection between Conjecture 1.10, Conjecture 5.3, and Conjecture 5.19. We start by proving a weighted Kakeya-type estimate for families containing multiple parallel tubes.

**Lemma 5.26** *For $p \geq 1$ and $\delta \leq 1$ we let*

$$C_{\delta, p} = \sup_{\mathcal{T}} \left\| \sum_{T \in \mathcal{T}} 1_T \right\|_p,$$

*where the supremum is over all collections $\mathcal{T}$ of $(\delta^{-1}, \delta^{-2})$-tubes with $\delta$-separated directions.*

*For each $\mathbf{v}$ in a set $V$ of $\delta$-separated directions, consider a family $\mathcal{T}_{\mathbf{v}}$ consisting of $(\delta^{-1}, \delta^{-2})$-tubes in the direction $\mathbf{v}$. For each $T \in \cup_{\mathbf{v} \in V} \mathcal{T}_{\mathbf{v}}$, let $w_T \in [0, \infty)$ satisfy for each $\mathbf{v}$*

$$\sum_{T \in \mathcal{T}_{\mathbf{v}}} w_T \leq M.$$

*Then*

$$\left\| \sum_{T \in \cup_{\mathbf{v} \in V} \mathcal{T}_{\mathbf{v}}} w_T 1_T \right\|_p \leq M C_{\delta, p}.$$

*Proof* We may assume that $M = 1$. For each $\mathbf{v}$, let $\{S_T : T \in \mathcal{T}_{\mathbf{v}}\}$ be a collection of pairwise disjoint subsets of $[0, 1]$ with $|S_T| = w_T$. Using Hölder's inequality and our hypothesis gives the desired estimate

$$\left\| \sum_{T \in \cup_{\mathbf{v} \in V} \mathcal{T}_{\mathbf{v}}} w_T 1_T \right\|_p^p = \int_{\mathbb{R}^n} \left( \int_0^1 \sum_{T \in \cup_{\mathbf{v} \in V} \mathcal{T}_{\mathbf{v}}} 1_T(x) 1_{S_T}(t) \, dt \right)^p dx$$

$$\leq \int_0^1 \left[ \int_{\mathbb{R}^n} \left( \sum_{T \in \cup_{\mathbf{v} \in V} \mathcal{T}_{\mathbf{v}}} 1_T(x) 1_{S_T}(t) \right)^p dx \right] dt$$

$$\leq (C_{\delta, p})^p. \qquad \square$$

As we have seen many times before, $Ef$ is the sum of spatially localized wave packets with distinct oscillatory phases. The role of any reverse square function estimate is to sharply quantify all the different phase cancellations, and thus to reduce the analysis of $E$ to a completely nonoscillatory problem. The details are presented in the following theorem.

**Theorem 5.27** *Assume that the Kakeya maximal operator conjecture 5.3 and the reverse square function conjecture 5.19 hold true. Then the Restriction Conjecture 1.10 holds for $\mathbb{P}^{n-1}$.*

*Proof* Using $\epsilon$ removal arguments (see, e.g., [110]), it will suffice to prove that $R^*_{\mathbb{P}^{n-1}}(\infty \mapsto 2n/(n-1); \epsilon)$ holds for each $\epsilon$. Using Proposition 1.27, this will follow if we prove that

$$\|\widehat{F}\|_{L^{\frac{2n}{n-1}}(\mathbb{R}^n)} \lesssim_\epsilon R^{-1+\epsilon} \|F\|_{L^\infty(\mathcal{N}(R^{-1}))}$$

holds for each $F$ supported on $\mathcal{N}(R^{-1})$. Fix such an $F$ and assume $\|F\|_{L^\infty(\mathcal{N}(R^{-1}))} = 1$.

Recall that each $\theta \in \Theta(R^{-1})$ is contained inside a rectangular box $T_\theta$ with dimensions comparable to $R^{-1/2}, \ldots, R^{-1/2}, R^{-1}$. Let us consider a wave packet decomposition on $T_\theta$ (Exercise 2.7)

$$\widehat{F1_\theta} = \sum_{T \in \mathcal{T}_\theta} a_T W_T.$$

Here $T$ ranges over a tiling $\mathcal{T}_\theta$ of $\mathbb{R}^n$ with rectangular boxes dual to $T_\theta$. Also,

$$|W_T(x)| \lesssim \frac{1}{|T|^{1/2}} \chi_T(x)$$

and

$$\sum_{T \in \mathcal{T}_\theta} |a_T|^2 \sim \|F1_\theta\|_2^2 \lesssim R^{-\frac{n+1}{2}}. \tag{5.17}$$

Combining this with Conjecture 5.19 leads to the initial estimate

$$\|\widehat{F}\|_{L^{\frac{2n}{n-1}}(\mathbb{R}^n)} \lessapprox \left\| \left( \sum_{\theta \in \Theta(R^{-1})} |\widehat{F1_\theta}|^2 \right)^{\frac{1}{2}} \right\|_{L^{\frac{2n}{n-1}}(\mathbb{R}^n)}$$

$$\lesssim R^{-\frac{n+1}{4}} \left\| \sum_{\theta \in \Theta(R^{-1})} \left( \sum_{T \in \mathcal{T}_\theta} |a_T| \chi_T \right)^2 \right\|_{L^{\frac{n}{n-1}}(\mathbb{R}^n)}^{\frac{1}{2}}$$

$$\lesssim R^{-\frac{n+1}{4}} \left\| \sum_{T \in \cup_\theta \mathcal{T}_\theta} |a_T|^2 \chi_T \right\|_{L^{\frac{n}{n-1}}(\mathbb{R}^n)}^{\frac{1}{2}} \quad \text{(by Cauchy–Schwarz)}.$$

Recall (5.13)

$$\chi_T(x) \lesssim \sum_{k \geq 0} 2^{-10kn} 1_{2^k T}(x).$$

On the other hand, Lemma 5.26, Conjecture 5.3, and (5.17) show that

$$\left\| \sum_{T \in \cup_\theta \mathcal{T}_\theta} |a_T|^2 1_{2^k T} \right\|_{L^{\frac{n}{n-1}}(\mathbb{R}^n)} \lessapprox 2^{k(n-1)} R^{n-1} R^{-\frac{n+1}{2}}.$$

The combination of the last two inequalities and the triangle inequality leads to

$$\left\| \sum_{T \in \cup_\theta \mathcal{T}_\theta} |a_T|^2 \chi_T \right\|_{L^{\frac{n}{n-1}}(\mathbb{R}^n)} \lessapprox R^{n-1} R^{-\frac{n+1}{2}}.$$

We conclude that

$$\|\widehat{F}\|_{L^{\frac{2n}{n-1}}(\mathbb{R}^n)} \lesssim R^{-1}. \qquad \square$$

**Remark 5.28** Combining Theorem 5.27 with Proposition 3.3 and Proposition 5.6 leads to yet another proof of the Restriction Conjecture for the parabola.

**Remark 5.29** As mentioned earlier, it was proved in [**35**] that the reverse square function conjecture 5.19 implies the Kakeya maximal operator Conjecture 5.3. We can thus conclude that Conjecture 5.19 is stronger than the Restriction Conjecture 1.10.

It seems natural to conjecture that if $\widehat{F}$ is supported on $\mathcal{N}(R^{-1})$ and $p \geq 2n/(n-1)$, we have

$$\|F\|_{L^p(\mathbb{R}^n)} \lesssim_\epsilon R^{\frac{n-1}{4}-\frac{n}{2p}+\epsilon} \left\| \left( \sum_{\theta \in \Theta(R^{-1})} |\mathcal{P}_\theta F|^2 \right)^{\frac{1}{2}} \right\|_{L^p(\mathbb{R}^n)}. \qquad (5.18)$$

The exponent of $R$ is suggested by informally interpolating between the conjectured result at the endpoint $p = 2n/(n-1)$ and the trivial result at $p = \infty$. The next exercise verifies (5.18) in the range $p \geq 2(n+1)/(n-1)$. See also Exercise 7.25, which proves (5.18) in the whole range when $n = 2$.

**Exercise 5.30**   (i) Prove that if $p \geq 2(n+1)/(n-1)$ and $F$ is as before, then for each ball $B_{R^{1/2}}$ with radius $R^{1/2}$, we have

$$\|F\|_{L^p(B_{R^{1/2}})} \lesssim_\epsilon R^{\frac{n-1}{4}-\frac{n}{2p}+\epsilon} \left\| \left( \sum_{\theta \in \Theta(R^{-1})} |\mathcal{P}_\theta F|^2 \right)^{\frac{1}{2}} \right\|_{L^p(\chi_{B_{R^{1/2}}})}.$$

(ii) Conclude that (5.18) holds if $p \geq 2(n+1)/(n-1)$.
Hint for (i): Use a local version of the Stein–Tomas theorem.

**Exercise 5.31** (Bochner–Riesz conjecture) Let $2n/(n+1) \leq p \leq 2n/(n-1)$ and let $F \in L^p(\mathbb{R}^n)$, with no restriction on the Fourier transform. Prove that Conjecture 5.19 implies

$$\left\| \sum_{\theta \in \Theta(R^{-1})} \tilde{\mathcal{P}}_\theta F \right\|_{L^p(\mathbb{R}^n)} \lesssim \|F\|_{L^p(\mathbb{R}^n)}.$$

Conclude that the case $n = 2$ of this conjecture is in fact a theorem.

# 6

# Multilinear Kakeya and Restriction Inequalities

So far in this book we have seen linear and bilinear restriction estimates. It is natural to expand our investigation to higher orders of multilinearity. It turns out that the main question can be formulated rather elegantly as follows.

**Conjecture 6.1** (*k-linear restriction for hypersurfaces*) *Let* $1 \leq k \leq n$. *For each* $1 \leq j \leq k$, *let* $\mathcal{M}_j = \{(\xi, \psi_j(\xi)) : \xi \in U_j\}$ *be a smooth hypersurface in* $\mathbb{R}^n$ *with extension operator* $E^{\mathcal{M}_j}$. *Assume that the* $\mathcal{M}_j$ *have everywhere positive principal curvatures and are* $v$-*transverse, in the sense that for each unit normals* $\mathbf{n}_j$ *to* $\mathcal{M}_j$, *the volume of the parallelepiped they span satisfies*

$$|\mathbf{n}_1 \wedge \cdots \wedge \mathbf{n}_k| > v. \tag{6.1}$$

*If* $1/q < (n-1)/2n$, $1/q \leq (n+k-2)/[(n+k)p']$ *and* $1/q \leq (n-k)/[(n+k)p'] + (k-1)/(n+k)$ *then for all* $f_j \in L^p(U_j)$

$$\left\| \left( \prod_{j=1}^{k} E^{\mathcal{M}_j} f_j \right)^{\frac{1}{k}} \right\|_{L^q(\mathbb{R}^n)} \lesssim_v \left( \prod_{j=1}^{k} \|f_j\|_{L^p(U_j)} \right)^{\frac{1}{k}}.$$

*Moreover, when* $k = n$, *the restriction* $1/q < (n-1)/2n$ *can be relaxed to* $1/q \leq (n-1)/2n$.

The case $k = 1$ was stated before as Conjecture 1.7 for $\mathbb{S}^{n-1}$ and $\mathbb{P}^{n-1}$. We have made significant progress on the case $k = 2$ in Theorem 3.1. None of these cases is completely settled in dimensions higher than 2. However, in this chapter we will see that we can give an almost complete answer when $k = n$, in the sense that we will prove local versions in the full expected range. In fact, the heart of the matter is to prove the expected estimate for the endpoint $(q, p) = (2n/(n-1), 2)$. We will achieve this in Theorem 6.24. The local estimates in the full range are then obtained by combining this with Hölder's

105

inequality and interpolation; see Exercise 6.31. It is only in this $n$-linear setting that the endpoint result corresponding to $p = 2$ solves the full problem.

We will refer to the $n$-linear inequalities as *multilinear* estimates. They have a few remarkable features. To describe the first one, note that there are two types of requirements in Conjecture 6.1, transversality and curvature. The first one cannot be relaxed, but the second one becomes less and less stringent when $k$ gets bigger. Indeed, we have already seen that when $k = 1$, the range of restriction estimates is very sensitive to whether or not the manifold has zero Gaussian curvature. On the other hand, we have also seen that this sensitivity is seriously weakened when $k = 2$. For example, the cone and paraboloid of the same dimension are expected to satisfy identical bilinear restriction estimates. However, we have proved in Section 3.5 that there is a key difference between the cylinder and the cone, in spite of them being locally isometric. Suffice to say that describing a general enough curvature-type condition that guarantees bilinear estimates in the range prescribed by Conjecture 6.1 turns out to be a rather delicate matter. See [4].

There are similar expectations that the $k$-linear theory is insensitive to the presence of $k - 1$ zero principal curvatures, though the exact description of the ideal curvature-type condition is elusive when $2 \leq k < n$. Quite remarkably however, when $k = n$, the curvature assumption in Conjecture 6.1 can be completely removed.

As in the bilinear case, working with $L^2$ in the multilinear setting will prove to be a huge advantage throughout the argument, because of the availability of orthogonality methods in this context. A second noteworthy feature of the $L^2$-based multilinear estimates ($p = 2$) is that the corresponding range for $q$ transcends the bilinear barrier $2(n + 2)/n$ (cf. Theorem 3.1); in fact, it goes all the way up to $2n/(n - 1)$, which is the endpoint for the linear restriction. When combined with a suitable multilinear-to-linear argument, this will lead to further progress on the linear restriction problem. We will explore this in the next chapter, as well as in Section 13.9.

Conjectures similar to Conjecture 6.1 may in principle be formulated for manifolds with larger codimension than one, though it is in general difficult to determine the correct substitute for principal curvatures (see, however, Exercise 6.34). We will confine ourselves to studying the analogous multilinear case, where we can completely dispense with any curvature assumption. We caution that this does not necessarily mean $k = n$; each codimension has a smallest value of $k$ at which the multilinear phenomenon is registered. To emphasize this aspect, we will replace $k$ with $m$ in the remainder of this chapter. In this generality, the trademark feature of multilinearity is the fact that it guarantees the largest possible range of restriction estimates; cf. Exercise 6.31. One of the reasons we chose to present the multilinear theory for arbitrary codimensions is its relevance to the decoupling inequalities that will be explored later in the book.

In Section 3.2 we have already seen a proof of the multilinear restriction inequality in two dimensions, for separated subsets of the parabola. The argument there made special use of biorthogonality, an efficient tool for proving inequalities of a bilinear nature. There does not seem to be an analogous multiorthogonality-type approach to restriction estimates with higher orders of multilinearity. This observation will motivate the new approach presented in this chapter, one that will in particular yield an alternative to the Córdoba–Fefferman geometric argument for $\mathbb{P}^1$.

There are three important steps in the proof of the endpoint multilinear restriction estimate. In Section 6.1, we prove this estimate in the particular case when the manifolds are affine spaces. We observe that in this flat case, there is an equivalent formulation that quantifies the transverse intersections of rectangular boxes with fixed orientations. Then, in Section 6.3 we extend this inequality to the case when the orientations are allowed to be slightly perturbed. Finally, in Section 6.4 we combine this multilinear Kakeya-type inequality with a certain bootstrapping argument to solve the multilinear case of Conjecture 6.1 for manifolds of arbitrary codimension, up to $\epsilon$-losses.

## 6.1 The Brascamp–Lieb Inequality

For $1 \leq j \leq m$, let $V_j$ be $d$-dimensional linear subspaces of $\mathbb{R}^n$ and let $l_j : \mathbb{R}^n \to V_j$ be surjective linear transformations. We impose no restriction on $m$; in particular, we allow it to be larger than $n$. Let us define the multilinear functional

$$\Lambda(f_1, \ldots, f_m) = \int_{\mathbb{R}^n} \prod_{j=1}^{m} f_j(l_j(x)) \, dx$$

for $f_j : V_j \to \mathbb{C}$. Each $V_j$ will be equipped with the $d$-dimensional Lebesgue measure and with the subspace metric inherited from $\mathbb{R}^n$.

The cornerstone of multilinear restriction theory is the Brascamp–Lieb inequality. We recall the following formulation from [11]. The reader is referred to [12] for a history of this inequality and its variants.

**Theorem 6.2** *Given a vector* $\mathbf{p} = (p_1, \ldots, p_m)$ *with all entries* $p_j \geq 1$, *we have that*

$$\sup_{\substack{f_j \in L^{p_j}(V_j) \\ 1 \leq j \leq m}} \frac{|\Lambda(f_1, \ldots, f_m)|}{\prod_{j=1}^{m} \|f_j\|_{L^{p_j}(V_j)}} < \infty \tag{6.2}$$

*if and only if*

$$\frac{n}{d} = \sum_{j=1}^{m} \frac{1}{p_j} \tag{6.3}$$

*and the following transversality condition is satisfied*

$$\dim(V) \leq \sum_{j=1}^{m} \frac{\dim(l_j(V))}{p_j}, \text{ for every linear subspace } V \subset \mathbb{R}^n. \tag{6.4}$$

Let us consider a few examples.

**Example 6.3** When $V_j = \mathbb{R}^n$ and $l_j$ is the identity, (6.2) is a reformulation of Hölder's inequality.

**Example 6.4** When $n = 2d$, $m = 3$, $V_1 = V_2 = V_3 = \mathbb{R}^d$, $l_1(x, y) = x$, $l_2(x, y) = y$, $l_3(x, y) = x - y$, (6.2) is Young's inequality for convolutions.

When all $p_j$ are equal to some $p$, the transversality condition (6.4) becomes

$$\frac{d \dim(V)}{n} \leq \frac{1}{m} \sum_{j=1}^{m} \dim(l_j(V)). \tag{6.5}$$

Testing this with one-dimensional spaces $V$ forces the restriction

$$n \leq md. \tag{6.6}$$

For any given $d < n$, there is a largest value of $m$, call it $m_{d,n}$, such that (6.5) fails to hold (for some $V$) for each choice of at most $m_{d,n} - 1$ pairs $(V_j, l_j)$, but is on the other hand satisfied by some choice of $m_{d,n}$ such pairs. See Exercise 6.11. From (6.6), we know that $m_{d,n} \geq n/d$. This lower bound is sharp when $d = 1$; see Exercise 6.10. More generally, it is sharp when $n$ is divisible by $d$. Indeed, if we let $m = n/d$ and $V_1, \ldots, V_m$ are $d$-dimensional linear spaces such that $\mathbb{R}^n$ coincides with equal to the orthogonal sum $V_1 \oplus \cdots \oplus V_m$, then it is immediate that (6.5) is satisfied with $l_j$ the orthogonal projection onto $V_j$.

However, the lower bound (6.6) is not always sharp. For example, when $d = n - 1$ we must have $m_{n-1,n} \geq n$, since each one-dimensional space $V$ with $l_j(V) = \{0\}$ for some $j$ will violate (6.5) if $m < n$. In fact, the next example shows that $m_{n-1,n} = n$.

**Example 6.5** (The Loomis–Whitney inequality) Let us consider the case when $m = n \geq 2$, $d = n - 1$, $V_j$ are hyperplanes with normal vectors $\mathbf{n}_j$ and $l_j$ are the orthogonal projections $\pi_j$ onto $V_j$. Let us check that the transversality condition (6.5) is equivalent with the linear independence of the $\mathbf{n}_j$:

$$|\mathbf{n}_1 \wedge \cdots \wedge \mathbf{n}_n| > 0.$$

If $\mathbf{n}_j$ are linearly dependent, then $V = \mathrm{span}(\mathbf{n}_1, \ldots, \mathbf{n}_n)$ has dimension $\leq n-1$ and satisfies $\dim(\pi_j(V)) < \dim(V)$ for each $j$. It is immediate that (6.5) fails for this $V$. On the other hand, for each $V$ we have $\dim(\pi_j(V)) = \dim(V) - 1$ if $\mathbf{n}_j \in V$ and $\dim(\pi_j(V)) = \dim(V)$ otherwise. If $\mathbf{n}_j$ are linearly independent, the first scenario can hold for at most $\dim(V)$ values of $j$. With these observations in mind, the verification of (6.5) is immediate.

This shows that the use of the term "transversality" in (6.5) is consistent with its earlier use in Conjecture 6.1. Under this requirement, we will next show how to reduce the Brascamp–Lieb inequality

$$\int_{\mathbb{R}^n} \prod_{j=1}^n f_j(\pi_j(x))\, dx \lesssim \prod_{j=1}^n \|f_j\|_{L^{n-1}(V_j)} \qquad (6.7)$$

to a simpler inequality that we can easily check.

Let $\mathbf{e}_j$ be the $j$th standard unit vector in $\mathbb{R}^n$. Let us denote by

$$\pi_j^*(x_1, \ldots, x_n) = (x_1, \ldots, x_{j-1}, 0, x_{j+1}, \ldots, x_n)$$

the orthogonal projection onto the canonical hyperplane $\mathbf{e}_j^\perp$. Let $L$ be the invertible linear transformation on $\mathbb{R}^n$ such that $L(\mathbf{e}_j) = \mathbf{n}_j$ for each $1 \leq j \leq n$. Note that for each $x \in \mathbb{R}^n$ and $j$, we have $\pi_j(Lx) = \pi_j(L(\pi_j^*(x)))$. Given a function $f_j$ on $V_j$, define $g_j$ on $\mathbf{e}_j^\perp$ as $g_j = f_j(\pi_j \circ L)$. The restriction of the linear transformations $\pi_j \circ L : \mathbf{e}_j^\perp \to V_j$ is nonsingular, since its kernel is trivial. It thus follows that $\|f_j\|_{L^{n-1}(V_j)} \sim_L \|g_j\|_{L^{n-1}(\mathbf{e}_j^\perp)}$. Moreover, a linear change of variables shows that

$$\int_{\mathbb{R}^n} \prod_{j=1}^n f_j(\pi_j(x))\, dx = \det[L] \int_{\mathbb{R}^n} \prod_{j=1}^n f_j(\pi_j \circ L(x))\, dx$$

$$= \det[L] \int_{\mathbb{R}^n} \prod_{j=1}^n g_j(\pi_j^*(x))\, dx.$$

These observations show that (6.7) will follow if we prove the following Loomis–Whitney inequality:

$$\int_{\mathbb{R}^n} \prod_{j=1}^n g_j(\pi_j^*(x))\, dx \leq \prod_{j=1}^n \|g_j\|_{L^{n-1}(\mathbb{R}^{n-1})}. \qquad (6.8)$$

We identify each $\mathbf{e}_j^\perp$ with $\mathbb{R}^{n-1}$ and drop the zero entry in the definition of $\pi_j^*$.

Let us see why (6.8) holds, using induction on $n$. We may assume that all $g_j$ are positive. The case $n = 2$ is immediate. Assume that the Loomis–Whitney inequality holds for $n-1$ functions. Then two applications of Hölder's inequality lead to

$$\int_{\mathbb{R}^n} \prod_{j=1}^n g_j(\pi_j^*(x))\,dx$$

$$\leq \int_{\mathbb{R}^{n-1}} g_1(x_2,\ldots,x_n) \prod_{j=2}^n \left( \int_{\mathbb{R}} g_j^{n-1}(\pi_j^*(x_1,\ldots,x_n))\,dx_1 \right)^{\frac{1}{n-1}} dx_2\ldots dx_n$$

$$\leq \|g_1\|_{L^{n-1}(\mathbb{R}^{n-1})} \left[ \int_{\mathbb{R}^{n-1}} h_2(x_3,\ldots,x_n)\ldots h_n(x_2,\ldots,x_{n-1})\,dx_2\ldots dx_n \right]^{\frac{n-2}{n-1}},$$

where

$$h_j(x_2,\ldots,x_{j-1},x_{j+1},\ldots,x_n)$$

$$= \left( \int_{\mathbb{R}} g_j^{n-1}(x_1,\ldots,x_{j-1},x_{j+1},\ldots,x_n)\,dx_1 \right)^{\frac{1}{n-2}}.$$

Using the induction hypothesis, we may dominate the last term by

$$\|g_1\|_{L^{n-1}(\mathbb{R}^{n-1})} \left( \prod_{j=2}^n \|h_j\|_{L^{n-2}(\mathbb{R}^{n-2})} \right)^{\frac{n-2}{n-1}} = \prod_{j=1}^n \|g_j\|_{L^{n-1}(\mathbb{R}^{n-1})}.$$

Throughout the remainder of this chapter, we will restrict attention to the special case when $l_j = \pi_j$ are the orthogonal projections onto $V_j$. For future use, we reformulate the Brascamp–Lieb inequality using the $L^2$ norm.

**Theorem 6.6** *Consider an $m$-tuple $\mathbb{V} = (V_1,\ldots,V_m)$ of $d$-dimensional linear spaces in $\mathbb{R}^n$ and let $q \leq 2m$. Let $\mathrm{BL}(\mathbb{V},q)$ be the smallest constant such that*

$$\left\| \left( \prod_{j=1}^m (g_j \circ \pi_j) \right)^{\frac{1}{m}} \right\|_{L^q(\mathbb{R}^n)} \leq \mathrm{BL}(\mathbb{V},q) \left( \prod_{j=1}^m \|g_j\|_{L^2(V_j)} \right)^{\frac{1}{m}} \qquad (6.9)$$

*holds for all $g_j \in L^2(V_j)$.*

*Then $\mathrm{BL}(\mathbb{V},q) < \infty$ if and only if*

$$q = \frac{2n}{d}$$

*and*

$$\dim(V) \leq \frac{n}{md} \sum_{j=1}^m \dim(\pi_j(V)), \text{ for every linear subspace } V \subset \mathbb{R}^n. \qquad (6.10)$$

*Proof* Apply Theorem 6.2 with all $p_j$ equal to $2m/q$ and $f_j = g_j^{q/m}$. $\qquad\square$

From now on, we will simplify notation by writing

$$\mathrm{BL}(\mathbb{V}) := \mathrm{BL}\left(\mathbb{V}, \frac{2n}{d}\right).$$

Let us now write (6.9) in a more familiar fashion, as a multilinear restriction/extension inequality. To simplify the computations, we restrict attention to $d = n - 1$. Let $V$ be a hyperplane containing the origin, with orthogonal projection $\pi$. We will assume that $\mathbf{e}_n \notin V$, so that $V$ is the graph of $\xi \mapsto \xi \cdot \mathbf{v}$ for some $\mathbf{v} \in \mathbb{R}^{n-1}$. Recall the extension operator associated with $V$

$$E^V h(\bar{x}, x_n) = \int h(\xi) e(\xi \cdot \bar{x} + x_n \xi \cdot \mathbf{v}) \, d\xi \, .$$

Given $f$ on $V$, define $g$ on $\mathbb{R}^{n-1}$ via the relation $g(\bar{x}) = f(\pi(\bar{x}, 0))$. Let us check that for each $x \in \mathbb{R}^n$

$$f(\pi(x)) = E^V(\widehat{g})(x).$$

Write using the Fourier inversion formula

$$E^V(\widehat{g})(\bar{x}, x_n) = \int_{\mathbb{R}^{n-1}} \int_{\mathbb{R}^{n-1}} f(\pi(\bar{y}, 0)) e(\xi \cdot (\bar{x} - \bar{y} + x_n \mathbf{v})) \, d\xi \, d\bar{y}$$

$$= f(\pi(\bar{x} + x_n \mathbf{v}, 0)).$$

It remains to notice that $\pi(\bar{x} + x_n \mathbf{v}, 0) = \pi(\bar{x}, x_n)$, since the vector $(-\mathbf{v}, 1)$ is normal to $V$.

Now, given $n$ hyperplanes $V_j$ with linearly independent normal vectors $\mathbf{n}_j$, we can assume due to the rotation invariant nature of (6.9) that they are all graphs of affine functions. Then, using Plancherel's identity, (6.9) with $q = 2n/(n-1)$ can be written as a multilinear extension inequality

$$\left\| \left( \prod_{j=1}^n E^{V_j} g_j \right)^{\frac{1}{n}} \right\|_{L^{\frac{2n}{n-1}}(\mathbb{R}^n)} \lesssim \left( \prod_{j=1}^n \|g_j\|_{L^2(\mathbb{R}^{n-1})} \right)^{\frac{1}{n}}. \tag{6.11}$$

The striking feature of this inequality is that it holds in the complete absence of curvature. We will soon see that its curved analog also holds, and that the proof in this generality will significantly dwell on the flat case discussed previously.

**Remark 6.7** (Compact vs. noncompact support) Let $V_1, V_2$ be distinct hyperplanes through the origin in $\mathbb{R}^n$. Exercise 3.13 shows that

$$\left\| \left( \prod_{j=1}^2 E^{V_j} g_j \right)^{\frac{1}{2}} \right\|_{L^4(\mathbb{R}^n)} \lesssim \left( \prod_{j=1}^2 \|g_j\|_{L^2(Q_j)} \right)^{\frac{1}{2}}$$

for each $g_1, g_2$ supported on a cubes $Q_j \subset \mathbb{R}^{n-1}$. However, Theorem 6.6 shows that the global inequality (without the compact support assumption)

$$\left\| \left( \prod_{j=1}^{2} E^{V_j} g_j \right)^{\frac{1}{2}} \right\|_{L^4(\mathbb{R}^n)} \lesssim \left( \prod_{j=1}^{2} \|g_j\|_{L^2(\mathbb{R}^{n-1})} \right)^{\frac{1}{2}}$$

is false if $n \geq 3$, since $4 \neq 2n/(n-1)$.

**Exercise 6.8** Give an alternative proof of (6.8) for $n = 3$, along the following lines. Assume first that all $g_j$ are supported on $[0, 1]^2$ and replace them with their Fourier series. Use rescaling to reduce the case of arbitrarily large compact support to this case.

**Exercise 6.9** Prove that if $\mathbb{V} = (V_1, \ldots, V_n)$ consists of hyperplanes $V_j$ with linearly independent unit normal vectors $\mathbf{n}_j$, then

$$\mathrm{BL}(\mathbb{V}) \lesssim |\mathbf{n}_1 \wedge \cdots \wedge \mathbf{n}_n|^{-\frac{1}{2n}}.$$

**Exercise 6.10** Let $\mathbb{V} = (V_1, \ldots, V_n)$ consist of one-dimensional vector spaces in $\mathbb{R}^n$ spanned by linearly independent unit vectors $\mathbf{n}_1, \ldots, \mathbf{n}_n$. Prove that

$$\mathrm{BL}(\mathbb{V}) \lesssim |\mathbf{n}_1 \wedge \cdots \wedge \mathbf{n}_n|^{-\frac{1}{2n}}.$$

**Exercise 6.11** Prove that $m_{d,n} < \infty$ for each $d < n$.
Hint: Consider all $\binom{n}{d}$ $d$-dimensional linear spaces spanned by the basis elements $\mathbf{e}_i$, equipped with the orthogonal projections $l_j = \pi_j$.

**Exercise 6.12** Let $O$ be a bounded open set in $\mathbb{R}^n$ with volume $v$. Let $v_j$ be the $n-1$ dimensional area of the projection of $O$ on the coordinate hyperplane $\mathbf{e}_j^\perp$. Use the Loomis–Whitney inequality to prove that

$$v \leq \left( \prod_{j=1}^{n} v_j \right)^{\frac{1}{n-1}}.$$

## 6.2 Plates and Joints

Let us fix $n, d$, and an $m$-tuple $\mathbb{V} = (V_1, \ldots, V_m)$ of $d$-dimensional linear spaces in $\mathbb{R}^n$ satisfying the transversality condition (6.10).

Let $\mathcal{P}_j$ be the collection of all $n - d$ dimensional affine spaces in $\mathbb{R}^n$ that are parallel to $V_j^\perp$. We will denote by $P_{j,a}$, $a \in \mathbb{N}$, a typical element of $\mathcal{P}_j$. Also, we allow multiplicity within each family, so it may happen that $P_{j,a} = P_{j,a'}$ for some $a \neq a'$.

For $W \geq 1$, we will call the $W$-neighborhood $P_{j,a,W}$ of $P_{j,a}$ a *plate*. The fact that we allow these plates to be infinitely long in $n - d$ orthogonal directions will lead to more elegant arguments, and will produce superficially stronger results.

We begin with the following equivalent formulation of Theorem 6.6.

**Proposition 6.13** *Let* $\mathrm{BL}^*(\mathbb{V})$ *be the smallest constant for which the inequality*

$$\int_{\mathbb{R}^n} \prod_{j=1}^{m} \left( \sum_{a=1}^{A_j} 1_{P_{j,a,W}} \right)^{\frac{n}{dm}} \leq \mathrm{BL}^*(\mathbb{V})^{\frac{2n}{d}} W^n \prod_{j=1}^{m} A_j^{\frac{n}{dm}} \qquad (6.12)$$

*holds true for each family* $\mathcal{P}'_j := \{ P_{j,a} \colon 1 \leq a \leq A_j \} \subset \mathcal{P}_j$ *with possible multiplicity and each* $W > 0$. *Then* $\mathrm{BL}^*(\mathbb{V}) \sim \mathrm{BL}(\mathbb{V})$, *and the similarity constants are independent of* $\mathbb{V}$.

*Proof* We first show that $\mathrm{BL}^*(\mathbb{V}) \lesssim \mathrm{BL}(\mathbb{V})$. Each $P_{j,a}$ coincides with $\pi_j^{-1}(v_{j,a})$, for some $v_{j,a} \in V_j$. Apply Theorem 6.6 to $V_j$, using

$$g_j = \left( \sum_{a=1}^{A_j} 1_{B_{V_j}(v_{j,a},W)} \right)^{1/2}. \qquad (6.13)$$

It suffices to note that

$$g_j \circ \pi_j = \left( \sum_{a=1}^{A_j} 1_{P_{j,a,W}} \right)^{1/2}, \qquad (6.14)$$

and that

$$\| g_j \|_{L^2(V_j)} \sim A_j^{\frac{1}{2}} W^{\frac{d}{2}}. \qquad (6.15)$$

It is worth realizing that the proof of this estimate is not sensitive to whether or not a certain $P_{j,a,W}$ is repeated multiple times.

Next we prove that $\mathrm{BL}(\mathbb{V}) \lesssim \mathrm{BL}^*(\mathbb{V})$. Because of density reasons, it suffices to prove that

$$\int_{\mathbb{R}^n} \left( \prod_{j=1}^{m} g_j \circ \pi_j \right)^{\frac{2n}{md}} \lesssim \mathrm{BL}^*(\mathbb{V})^{\frac{2n}{d}} \left( \prod_{j=1}^{m} \| g_j \|_{L^2(V_j)} \right)^{\frac{2n}{md}}$$

holds for all $g_j$ as in (6.13), for arbitrary $W$. Again, we allow multiplicity within each family. But this inequality follows as in the preceding, using (6.14) and (6.15). $\qquad \square$

Let us explore a discrete consequence. Note that the transversality condition (6.10) forces the intersection of the spaces $V_j^{\perp}$ to consist only of the origin. Consequently, the intersection of any $m$ spaces $P_{j,a_j} \in \mathcal{P}_j$ with $1 \le j \le m$ is either empty or it consists of just one point. The value of the limit

$$\lim_{W \to 0} \frac{|\bigcap_{j=1}^m P_{j,a_j}, W|}{W^n}$$

is equal to 0 in the first case and to some $c_V > 0$ in the second case.

Inequality (6.12) is scale invariant; if it holds for some $W$, a rescaling argument shows that it holds for arbitrary $W$. Letting $W \to 0$ in (6.12) leads to the discrete inequality

$$\int_{\mathbb{R}^n} \prod_{j=1}^m \left( \sum_{a=1}^{A_j} 1_{P_{j,a}}(x) \right)^{\frac{n}{dm}} d\mu(x) \lesssim_V \prod_{j=1}^m A_j^{\frac{n}{dm}},$$

where $d\mu$ is the counting measure on $\mathbb{R}^n$.

In particular, when $P_{j,a}$ are distinct for each $j$, the quantity on the left counts the number of $m$-tuples of plates $(P_1, \ldots, P_m) \in \mathcal{P}_1' \times \cdots \times \mathcal{P}_m'$ with nonempty intersection. A more direct proof of this inequality in the case $d = n - 1$ is explored in the next exercise. We will later find an essentially sharp upper bound on the number of joints associated with arbitrary families of lines. This is a beautiful application of the polynomial method, explored in Exercise 8.11.

**Exercise 6.14** (Joints) For $1 \le j \le n$, let $\mathcal{L}_j$ be a finite collection of lines in $\mathbb{R}^n$ with fixed direction $\mathbf{v}_j$. Assume that $\mathbf{v}_j$ are linearly independent. A joint is an $n$-tuple $(L_1, \ldots, L_n) \in \mathcal{L}_1 \times \cdots \times \mathcal{L}_n$ of intersecting lines.

Prove that the number of joints is at most $(\prod_{j=1}^n |\mathcal{L}_j|)^{\frac{1}{n-1}}$. Show that this upper bound can be sharp.

Hint: Reduce to the case when $\mathbf{v}_j = \mathbf{e}_j$ and then apply the Cauchy–Schwarz inequality.

## 6.3 The Multilinear Kakeya Inequality

The Grassmannian $\mathcal{G}(l, \mathbb{R}^n)$ is the collection of all $l$-dimensional linear spaces in $\mathbb{R}^n$. It becomes a compact metric space when equipped with the distance

$$\rho_{\mathcal{G}(l, \mathbb{R}^n)}(V_1, V_2) = \|\pi_1 - \pi_2\|,$$

where $\pi_1, \pi_2$ are the orthogonal projections associated with $V_1, V_2$, and their difference is measured in the operator norm. The distance between two translates of $V_1, V_2$ will be understood to be the same as the one between $V_1$ and $V_2$. We will sometimes refer to $\rho_{\mathcal{G}(l, \mathbb{R}^n)}(V_1, V_2)$ as the *angle* between $V_1$ and $V_2$.

Throughout this section, we fix a closed (hence compact) product set $V = V_1 \times \cdots \times V_m \subset \mathcal{G}(d, \mathbb{R}^n)^m$ such that

$$\mathrm{BL}(V) := \sup_{V' \in V} \mathrm{BL}(V') < \infty.$$

We will now denote by $\mathcal{P}_j$ the collection of all translates of $V^\perp$ with $V \in V_j$.

We now present a very important generalization of Proposition 6.13, which was proved in [13]. The case $d = n - 1$ of this theorem was a ground-breaking result first proved by Bennett et al. in [13], which has ignited essentially all developments presented in this chapter. A simpler argument was introduced later by Guth in [59], and our presentation here follows his approach.

**Theorem 6.15** (Multilinear Kakeya, first version) *For each $\epsilon > 0$ there is $C_{\epsilon,V} < \infty$ such that the inequality*

$$\int_{Q_{SW}} \prod_{j=1}^{m} \left( \sum_{a=1}^{A_j} 1_{P_{j,a,W}} \right)^{\frac{n}{dm}} \leq C_{\epsilon,V} S^\epsilon W^n \prod_{j=1}^{m} A_j^{\frac{n}{dm}} \qquad (6.16)$$

*holds true for each family $\mathcal{P}'_j := \{P_{j,a} : 1 \leq a \leq A_j\} \subset \mathcal{P}_j$ with possible multiplicity, each $W > 0$, each $S \geq 1$, and each cube $Q_{SW}$ in $\mathbb{R}^n$ with side length $SW$.*

Inequality (6.16) is scale invariant; it may be reduced to the case $W = 1$. Note its local nature and the $S^\epsilon$ term. This loss was removed by Guth [60] in the case $d = n - 1$ to produce a global version, and by Zhang [127] in the case of arbitrary $d$. Their methods are very different from the ones presented in this section and involve elements of algebraic topology.

It is common to reformulate (6.16) using plates with finite volume. For $S \geq 1$, let $\mathcal{T}_{j,S,W}$ be the collection of all rectangular boxes (plates) $T$ with $n - d$ sides parallel to some $P \in \mathcal{P}_j$ and each having length $SW$, and with the remaining $d$ sides of length $W$. Theorem 6.15 is easily seen to be equivalent to the following inequality.

**Theorem 6.16** (Multilinear Kakeya, second version) *For all finite collections $\mathcal{T}_j \subset \mathcal{T}_{j,S,W}$ with possible multiplicity and for each $c_T \geq 0$, we have*

$$\left\| \prod_{j=1}^{m} \left( \sum_{T \in \mathcal{T}_j} c_T 1_T \right)^{\frac{1}{m}} \right\|_{L^{\frac{n}{d}}(\mathbb{R}^n)} \lesssim_{\epsilon,V} S^\epsilon W^d \prod_{j=1}^{m} \left( \sum_{T \in \mathcal{T}_j} c_T \right)^{\frac{1}{m}}. \qquad (6.17)$$

In particular, when $d = n - 1$, combining this with Exercise 6.9 implies the following multilinear version of the endpoint linear Kakeya inequality (5.4). Quite remarkably, we do not ask for the tubes within a given family to have separated directions, but merely to have the upper bound on cardinality identical to the one enforced by the separated case.

**Corollary 6.17** *Assume that for each choice of* $(\delta, 1)$-*tubes* $T_j \in \mathcal{T}_{j, \delta^{-1}, \delta}$, $1 \leq j \leq n$, *their directions* $\mathbf{n}_j$ *satisfy*

$$|\mathbf{n}_1 \wedge \cdots \wedge \mathbf{n}_n| > \nu.$$

*For* $r = n/(n-1)$ *and* $\mathcal{T}_j \subset \mathcal{T}_{j, \delta^{-1}, \delta}$ *with possible multiplicity but cardinality* $O(\delta^{-(n-1)})$, *we have*

$$\left\| \prod_{j=1}^{n} \left( \sum_{T \in \mathcal{T}_j} 1_T \right)^{\frac{1}{n}} \right\|_{L^r(\mathbb{R}^n)} \lesssim_{\epsilon, \nu} \delta^{-\epsilon} \prod_{j=1}^{n} \left( \sum_{T \in \mathcal{T}_j} |T| \right)^{\frac{1}{rn}}.$$

**Example 6.18** Before we start the proof of Theorem 6.15, let us point out that the case $n = m = 2$, $d = 1$ is immediate. The reason this case is so simple is that any two nonparallel lines in the plane must intersect. More precisely, given two families $\mathcal{L}_1, \mathcal{L}_2$ of lines in $\mathbb{R}^2$ with $\sphericalangle(L_1, L_2) \geq \theta$ for each $L_i \in \mathcal{L}_i$, inequality (6.16) assumes the following sharp form involving their $W$-neighborhoods $L_{i, W}$:

$$\sum_{L_1 \in \mathcal{L}_1} \sum_{L_2 \in \mathcal{L}_2} |L_{1, W} \cap L_{2, W}| \leq \frac{W^2}{\sin \theta} |\mathcal{L}_1| |\mathcal{L}_2|. \tag{6.18}$$

This inequality is an immediate consequence of elementary geometry.

Exercise 6.21 extends this observation to higher dimensions.

The proof of Theorem 6.15 will rely on using Proposition 6.13 on small enough spatial balls, on which plates within a small angle from each other are essentially indistinguishable. This is quantified by the following simple geometric observation:

Given some $0 < \delta < 1$, there exists $\theta(\delta)$ so that whenever $P, P' \in \mathcal{G}(n - d, \mathbb{R}^n)$ are separated by an angle $\rho_{\mathcal{G}(n-d, \mathbb{R}^n)}(P, P') < \theta(\delta)$, we have

$$1_{P_W}(x) \leq 1_{P'_{2W}}(x)$$

for each $W > 0$ and $|x| \leq \delta^{-1} W$. See Figure 6.1.

The argument for Theorem 6.15 involves the use of many scales $W$. Define $f_{j, W} := \sum_{a=1}^{A_j} 1_{P_{j, a, W}}$.

**Lemma 6.19** *Fix* $\delta < 1$ *and* $P_j \in \mathcal{P}_j$. *Assume* $\mathcal{P}'_j = \{P_{j, a} : 1 \leq a \leq A_j\} \subset \mathcal{P}_j$ *is such that*

$$\rho_{\mathcal{G}(n-d, \mathbb{R}^n)}(P_j, P_{j, a}) \leq \theta(\delta) \tag{6.19}$$

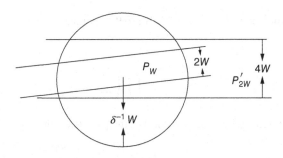

Figure 6.1 Two plates with small angle in $\mathbb{R}^2$.

*for each $P_{j,a} \in \mathcal{P}'_j$. Then for each cube $Q_S \subset \mathbb{R}^n$ with side length $S \geq \delta^{-1}W$, we have*

$$\int_{Q_S} \left( \prod_{j=1}^{m} f_{j,W} \right)^{\frac{n}{dm}} \leq C_{\mathcal{V}} \delta^n \int_{Q_S} \left( \prod_{j=1}^{m} f_{j,\delta^{-1}W} \right)^{\frac{n}{dm}},$$

*where $C_{\mathcal{V}} = C\mathrm{BL}(\mathcal{V})^{2n/d}$, for some $C$ independent of $\mathcal{V}$, $\delta$, $S$, and $W$.*

*Proof* We consider a finitely overlapping cover of $Q_S$ with balls $B$ of radius $1/100\delta^{-1}W$. For each such $B$, it suffices to prove that

$$\int_B \left( \prod_{j=1}^{m} f_{j,W} \right)^{\frac{n}{dm}} \lesssim \mathrm{BL}(\mathcal{V})^{\frac{2n}{d}} \delta^n \int_B \left( \prod_{j=1}^{m} f_{j,\delta^{-1}W} \right)^{\frac{n}{dm}}.$$

Our hypothesis (6.19) implies that for each $P_{j,a} \in \mathcal{P}'_j$, there exists a translate of $P_j$, call it $P_{j,a,B}$, so that its $2W$-neighborhood covers $P_{j,a,W}$ on $B$:

$$1_{P_{j,a,W}}(x) \leq 1_{P_{j,a,B,2W}}(x), \text{ for all } x \in B. \tag{6.20}$$

Due to (6.20), we have

$$\int_B \left( \prod_{j=1}^{m} f_{j,W} \right)^{\frac{n}{dm}} \leq \int_B \left( \prod_{j=1}^{m} \sum_{a=1}^{A_j} 1_{P_{j,a,B,2W}} \right)^{\frac{n}{dm}}.$$

The sum on the right may be restricted to those $a$ for which $P_{j,a,W}$ intersects $B$. Let $A_j(B)$ be the number of such plates. Invoking Proposition 6.13 for the families of plates $P_{j,a,B,2W}$, we get

$$\int_B \left( \prod_{j=1}^{m} \sum_{a=1}^{A_j} 1_{P_{j,a,B,2W}} \right)^{\frac{n}{dm}} \lesssim \mathrm{BL}(\mathcal{V})^{\frac{2n}{d}} W^n \prod_{j=1}^{m} A_j(B)^{\frac{n}{dm}}.$$

Since the diameter of $B$ is $1/50\delta^{-1}W$, if $P_{j,a,W}$ intersects $B$, then $1_{P_{j,a,\delta^{-1}W}}$ is identically equal to 1 on $B$. Thus we can write

$$W^n \prod_{j=1}^{m} A_j(B)^{\frac{n}{dm}} \lesssim \delta^n \int_B \prod_{j=1}^{m} \left( \sum_{a=1}^{A_j} 1_{P_{j,a,\delta^{-1}W}} \right)^{\frac{n}{dm}}.$$

When combined with the previous inequalities, this concludes the argument. $\square$

We next iterate Lemma 6.19, starting with $W = 1$. Each iteration increases the width of the plates by $\delta^{-1}$. We stop when this width is comparable to the scale of $Q_S$. At this point, the integrand is essentially constant on $Q_S$ and the integral can be related to the number of plates.

**Proposition 6.20** *Let $\delta$ and $\mathcal{P}'_j$ be as in Lemma 6.19. Then for each cube $Q_S$ with side length $S \geq 1$ in $\mathbb{R}^n$, we have*

$$\int_{Q_S} \prod_{j=1}^{m} \left( \sum_{a=1}^{A_j} 1_{P_{j,a,1}} \right)^{\frac{n}{dm}} \leq C_\mathcal{V} S^{\frac{\log C_\mathcal{V}}{\log \delta^{-1}}} \prod_{j=1}^{m} A_j^{\frac{n}{dm}}.$$

*Proof* Let $M$ be a positive integer such that $\delta^{-M+1} \leq S < \delta^{-M}$. It suffices to prove that

$$\int_{Q_{S'}} \prod_{j=1}^{m} \left( \sum_{a=1}^{A_j} 1_{P_{j,a,1}} \right)^{\frac{n}{dm}} \leq S'^{\frac{\log C_\mathcal{V}}{\log \delta^{-1}}} \prod_{j=1}^{m} A_j^{\frac{n}{dm}}$$

for $Q_{S'}$ with side length $S' = \delta^{-M}$.

Iterating Lemma 6.19 $M$ times, we get

$$\int_{Q_{S'}} \prod_{j=1}^{m} \left( \sum_{a=1}^{A_j} 1_{P_{j,a,1}} \right)^{\frac{n}{dm}} = \int_{Q_{S'}} \left( \prod_{j=1}^{m} f_{j,1} \right)^{\frac{n}{dm}} \leq C_\mathcal{V} \delta^n \int_{Q_{S'}} \left( \prod_{j=1}^{m} f_{j,\delta^{-1}} \right)^{\frac{n}{dm}}$$

$$\cdots \leq (C_\mathcal{V}\delta^n)^M \int_{Q_{S'}} \left( \prod_{j=1}^{m} f_{j,\delta^{-M}} \right)^{\frac{n}{dm}}.$$

It suffices to observe that the last quantity is dominated by

$$(C_\mathcal{V}\delta^n)^M S'^n \prod_{j=1}^{m} A_j^{\frac{n}{dm}} = S'^{\frac{\log C_\mathcal{V}}{\log \delta^{-1}}} \prod_{j=1}^{m} A_j^{\frac{n}{dm}}. \qquad \square$$

We are now ready to prove Theorem 6.15.

*Proof* (of Theorem 6.15). Rescaling, we may assume $W = 1$. Given $\epsilon > 0$, choose $\delta > 0$ small enough so that $\log C_{\mathcal{V}} / \log \delta^{-1} < \epsilon$. Using the compactness of $\mathcal{V}$, there is a number $N(\delta)$ so that we can split each family $\mathcal{P}_j$ into at most $N(\delta)$ subfamilies, each of which satisfies (6.19). Using (6.6), we find that $\int_{Q_S} \prod_{j=1}^{m} (\sum_{a=1}^{A_j} 1_{P_{j,a,1}})^{n/dm}$ is dominated by the sum of at most $N(\delta)^m$ terms of the form $\int_{Q_S} \prod_{j=1}^{m} (\sum_{a=1}^{M_j} 1_{P'_{j,a,1}})^{n/dm}$ with $M_j \leq A_j$. Moreover, each term can be bounded using Proposition 6.20 by

$$C_{\mathcal{V}} S^{\frac{\log C_{\mathcal{V}}}{\log \delta^{-1}}} \prod_{j=1}^{m} M_j^{\frac{n}{dm}} \leq C_{\mathcal{V}} S^{\epsilon} \prod_{j=1}^{m} A_j^{\frac{n}{dm}}.$$

It is now clear that (6.16) will hold with $C_{\epsilon,\mathcal{V}} = C_{\mathcal{V}} N(\delta)^m$, since $\delta$ depends only on $\epsilon$ and $\mathcal{V}$. $\square$

**Exercise 6.21** Give a simple, direct proof of Theorem 6.16 in the case $d = 1$, $m = n$, similar to the proof of (6.18), when $V_j$ are lines spanned by linearly independent vectors. This argument produces a global result.

## 6.4 The Multilinear Restriction Theorem

Let $1 \leq d \leq n - 1$ and $m \geq 1$. Given $m$ smooth functions $\psi_j \colon U_j \to \mathbb{R}^{n-d}$ with $U_j \subset \mathbb{R}^d$, consider their associated manifolds:

$$\mathcal{M}_j = \{(\xi, \psi_j(\xi)), \ \xi \in U_j\}. \tag{6.21}$$

Let $\mathbb{M} = (\mathcal{M}_1, \ldots, \mathcal{M}_m)$. It will be convenient to assume that $U_j$ are pairwise disjoint compact sets. For each $\xi \in U_j$, let $V_\xi$ be the linear space in $\mathcal{G}(d, \mathbb{R}^n)$ parallel to the tangent space to $\mathcal{M}_j$ at $(\xi, \psi_j(\xi))$. We let $\mathcal{V}_j$ be the collection of all $V_\xi$ with $\xi \in U_j$ and write $\mathcal{V} = \mathcal{V}_1 \times \cdots \times \mathcal{V}_m$. Define

$$\mathrm{BL}(\mathbb{M}) := \sup_{\mathbb{V}' \in \mathcal{V}} \mathrm{BL}(\mathbb{V}').$$

Our main assumption for $\mathbb{M}$ throughout this section will be BL-transversality.

**Definition 6.22** We will call $\mathbb{M}$ (or alternatively, the manifolds $\mathcal{M}_1, \ldots, \mathcal{M}_m$) BL-transverse if

$$\mathrm{BL}(\mathbb{M}) < \infty. \tag{6.22}$$

As observed earlier, when $\mathbb{M}$ consists of $n$ hypersurfaces in $\mathbb{R}^n$, BL-transversality coincides with transversality (as defined in (6.1)).

**Example 6.23** Let us specialize to the case when $\mathcal{M}_j$ are $n$ subsets of $\mathbb{P}^{n-1}$, which project vertically to $U_j \subset [-1, 1]^{n-1}$. A simple computation shows that (6.1) holds if and only if there is no $n - 2$ dimensional affine subspace of $\mathbb{R}^{n-1}$ intersecting each $\overline{U_j}$.

The vertical $\delta$-neighborhood on $\mathcal{M}_j$ will be denoted by

$$\mathcal{N}_{U_j}(\delta) = \{(\xi, \psi_j(\xi) + \eta) : \xi \in U_j, \ \eta \in B_{n-d}(0, \delta)\}.$$

For each $p \geq 2$ and $R \geq 1$, we let $\mathrm{MR}(p, R) = \mathrm{MR}(\mathbb{M}, p, R)$ be the smallest constant such that the inequality

$$\left\| \left( \prod_{j=1}^{m} \widehat{F_j} \right)^{\frac{1}{m}} \right\|_{L^p(\mathbb{R}^n)} \leq \mathrm{MR}(p, R) R^{-\frac{n-d}{2}} \left( \prod_{j=1}^{m} \|F_j\|_{L^2} \right)^{\frac{1}{m}} \tag{6.23}$$

holds for each $F_j : \mathcal{N}_{U_j}(R^{-1}) \to \mathbb{C}$.

The main result we will prove in this section solves Conjecture 6.1 (modulo $R^\epsilon$-losses) in the case $k = n$, and its counterpart for manifolds of arbitrary codimension. The case $d = n - 1$ has been solved in [**13**], following earlier partial results proved in [**9**] and [**14**].

**Theorem 6.24** (Multilinear restriction inequality) *We have*

$$\mathrm{MR}\left(\mathbb{M}, \frac{2n}{d}, R\right) \lesssim_{\epsilon, \mathrm{BL}(\mathbb{M})} R^\epsilon. \tag{6.24}$$

Perhaps the most striking feature of this inequality is that, unlike its linear and bilinear analogs proved earlier, its validity demands no curvature assumption on the manifolds $\mathcal{M}_j$.

It is easy to see that (6.24) implies the following inequality involving the extension operators:

$$\left\| \left( \prod_{j=1}^{m} E^{\mathcal{M}_j} f_j \right)^{\frac{1}{m}} \right\|_{L^{\frac{2n}{d}}(B_R)} \lesssim_{\epsilon, \mathrm{BL}(\mathbb{M})} R^\epsilon \left( \prod_{j=1}^{m} \|f_j\|_{L^2(U_j)} \right)^{\frac{1}{m}}. \tag{6.25}$$

To get a better understanding of this inequality, let us discuss the case of hypersurfaces, $d = n - 1$, $m = n$. It turns out that this is an endpoint result that implies the full range of $n$-linear restriction estimates in Conjecture 6.1 (modulo $R^\epsilon$-losses); see Exercise 6.31.

If $U_j$ are as in Example 6.23, we get a multilinear version of the endpoint linear restriction for the paraboloid

$$\left\| \left( \prod_{j=1}^{n} E_{U_j}^{\mathbb{P}^{n-1}} f \right)^{\frac{1}{n}} \right\|_{L^{\frac{2n}{n-1}}(B_R)} \lesssim_\epsilon R^\epsilon \left( \prod_{j=1}^{n} \|f\|_{L^2(U_j)} \right)^{\frac{1}{n}}. \tag{6.26}$$

Before we prove Theorem 6.24, we will demonstrate that there are intimate connections between the multilinear Kakeya and restriction inequalities. Following the notation from the previous section, let $\mathcal{T}_{j,R^{1/2},R^{1/2}}$ be the collection of all rectangular boxes (plates) $T$ with $d$ of the sides parallel to some $V_T \in \mathcal{V}_j$ and having length $R^{1/2}$, and the remaining $n - d$ sides of length $R$.

We will denote by $\mathrm{MK}(n/d, R) = \mathrm{MK}(\mathbb{M}, n/d, R)$ the smallest constant such that the multilinear Kakeya inequality

$$\left\| \prod_{j=1}^{m} \left( \sum_{T \in \mathcal{T}_j} c_T 1_T \right)^{\frac{1}{m}} \right\|_{L^{\frac{n}{d}}(\mathbb{R}^n)} \leq \mathrm{MK}\left(\frac{n}{d}, R\right) R^{\frac{d}{2}} \prod_{j=1}^{m} \left( \sum_{T \in \mathcal{T}_j} c_T \right)^{\frac{1}{m}}$$

holds for each set of finite collection $\mathcal{T}_j \subset \mathcal{T}_{j,R^{1/2},R^{1/2}}$ and each $c_T \geq 0$. Theorem 6.16 shows that

$$\mathrm{MK}\left(\frac{n}{d}, R\right) \lesssim_\epsilon R^\epsilon. \tag{6.27}$$

The next result is a multilinear analog of Proposition 5.7.

**Proposition 6.25** *We have*

$$\mathrm{MK}\left(\frac{n}{d}, R\right) \lesssim \mathrm{MR}\left(\frac{2n}{d}, R\right)^2, \tag{6.28}$$

*with an implicit constant independent of R. In particular, the multilinear restriction estimate (6.24) implies the corresponding multilinear Kakeya inequality (6.27).*

*Proof* Fix the collections $\mathcal{T}_j$ as just described, and let $c_T$ be positive coefficients. It is easy to see that if $\rho_{\mathcal{G}(d,\mathbb{R}^n)}(V_T, V_{T'}) \lesssim R^{-1/2}$ and $T \cap T' \neq \emptyset$, then $T \subset CT'$ with $C = O(1)$. Thus, by renormalizing the coefficients $c_T$ and changing slightly the size and orientation of the plates, we may assume that two $T, T' \in \mathcal{T}_j$ are either parallel and disjoint or else have an angle separation $\rho_{\mathcal{G}(d,\mathbb{R}^n)}(V_T, V_{T'}) \gg R^{-1/2}$.

For each $T \in \mathcal{T}_j$, let $\phi_T$ be a Schwartz function supported on $\mathcal{N}_{Q_T}(R^{-1})$, with $Q_T \subset U_j$ a cube with side length $\sim R^{-1/2}$, satisfying

$$1_T \leq |\widehat{\phi_T}|$$

and

$$\|\phi_T\|_2 \sim R^{\frac{n}{2} - \frac{d}{4}}.$$

Due to our separation assumption on the plates $T$, the functions $(\phi_T)_{T \in \mathcal{T}_j}$ can be chosen to be almost orthogonal in $L^2$. We now test (6.23) with

$$F_j(x) = F_{j,t_j}(x) = \sum_{T \in \mathcal{T}_j} (c_T)^{\frac{1}{2}} r_T(t_j) \phi_T(x)$$

and average over all $t_j$. Each $r_T$ is a distinct Rademacher function. Note that

$$\int_{[0,1]^m} \left\| \left( \prod_{j=1}^{m} \widehat{F_{j,t_j}}(x) \right)^{\frac{1}{m}} \right\|_{L^{\frac{2n}{d}}(\mathbb{R}^n)}^{\frac{2n}{d}} dt_1 \dots dt_m \sim \int_{\mathbb{R}^n} \prod_{j=1}^{m} \left( \sum_{T \in \mathcal{T}_j} c_T |\widehat{\phi_T}(x)|^2 \right)^{\frac{n}{dm}} dx$$

$$\geq \int_{\mathbb{R}^n} \prod_{j=1}^{m} \left( \sum_{T \in \mathcal{T}_j} c_T 1_T(x) \right)^{\frac{n}{dm}} dx.$$

Also , due to the almost orthogonality of the functions $\phi_T$, we have

$$\left( \prod_{j=1}^{m} \|F_{j,t_j}\|_{L^2} \right)^{\frac{2n}{md}} \sim R^{(\frac{n}{2}-\frac{d}{4})\frac{2n}{d}} \prod_{j=1}^{m} \left( \sum_{T \in \mathcal{T}_j} c_T \right)^{\frac{n}{md}}$$

for each $t_j \in [0,1]$. Integrating this over $t_1, \dots, t_m \in [0,1]$ and combining it with the last inequality and with the definition (6.23) of $\mathrm{MR}(2n/d, R)$ finishes the proof of (6.28). □

Let $\mathrm{ME}(p, R) = \mathrm{ME}(\mathbb{M}, p, R)$ be the smallest constant such that the following inequality holds for all collections $\Lambda_j \subset \mathcal{M}_j$ consisting of $1/R$-separated points $\lambda \in \mathcal{M}_j$, each $a_\lambda \in \mathbb{C}$, and each ball $B_R$ in $\mathbb{R}^n$ with radius $R$:

$$\left\| \left( \prod_{j=1}^{m} \sum_{\lambda \in \Lambda_j} a_\lambda e(\lambda \cdot x) \right)^{\frac{1}{m}} \right\|_{L^p(B_R)} \leq \mathrm{ME}(p, R) R^{\frac{d}{2}} \left( \prod_{j=1}^{m} \|a_\lambda\|_{l^2(\Lambda_j)} \right)^{\frac{1}{m}}.$$

Reasoning as in Proposition 1.29 leads to the following useful observation. See also Exercise 3.34.

**Proposition 6.26** *For each $p \geq 1$, we have*

$$\mathrm{ME}(p, R) \lesssim \mathrm{MR}(p, R).$$

At the heart of the proof of the multilinear inequality (6.24) lies the following two-scale inequality.

**Proposition 6.27** *We have*

$$\mathrm{MR}\left( \frac{2n}{d}, R \right) \lesssim \mathrm{ME}\left( \frac{2n}{d}, R^{\frac{1}{2}} \right) \mathrm{MK}\left( \frac{n}{d}, R \right)^{\frac{1}{2}}.$$

*Proof* Assume that each $F_j$ is supported on $\mathcal{N}_{U_j}(R^{-1})$. Cover $\mathcal{N}_{U_j}(R^{-1})$ with pairwise disjoint almost rectangular boxes $\mathcal{N}_Q(R^{-1})$, where each $Q$ is a cube with side length $R^{-1/2}$.

Combining wave packet decompositions (Exercise 2.7) for each $\widehat{F_j 1_Q}$ leads to the representation

$$\widehat{F_j} = \sum_{T \in \mathcal{T}_j} c_T W_T.$$

Here $\mathcal{T}_j \subset \mathcal{T}_{j, R^{1/2}, R^{1/2}}$ and

$$\sum_{T \in \mathcal{T}_j} |c_T|^2 \sim \|F_j\|_2^2. \tag{6.29}$$

Also, $W_T$ is $L^2$ normalized, has (inverse) Fourier transform supported inside the slight enlargement $C\mathcal{N}_{Q_T}(R^{-1})$, and decays rapidly outside $T$:

$$W_T(x) \lesssim \frac{1}{|T|^{1/2}} \chi_T(x).$$

We will make a few harmless approximations in order to simplify the technicalities and to isolate the main ideas of the proof. The interested reader may consult [**8**] for a fully rigorous version of this argument.

The first approximation consists of replacing $W_T(x)$ with $1/|T|^{1/2} 1_T(x) e(-\lambda_T \cdot x)$, where $\lambda_T = (\xi_T, \psi_j(\xi_T))$ and $\xi_T$ is the center of $Q_T$.

Let $\mathcal{B}$ be a finitely overlapping cover of $\mathbb{R}^n$ with balls $B$ with radius $R^{1/2}$. Denote by $\mathcal{T}_j(B)$ the plates in $\mathcal{T}_j$ that intersect $B$. As a second approximation, we will replace $W_T(x)1_B(x)$ with $|T|^{-1/2}1_B(x)e(-\lambda_T \cdot x)$ if $T \in \mathcal{T}_j(B)$ and with $0$ otherwise. This will allow us to use an exponential sum estimate on each $B$, as follows:

$$\left\| \left( \prod_{j=1}^m \widehat{F_j} \right)^{\frac{1}{m}} \right\|_{L^{\frac{2n}{d}}(\mathbb{R}^n)}^{\frac{2n}{d}} \leq \sum_{B \in \mathcal{B}} \left\| \left( \prod_{j=1}^m \widehat{F_j} \right)^{\frac{1}{m}} \right\|_{L^{\frac{2n}{d}}(B)}^{\frac{2n}{d}}$$

$$\approx \sum_{B \in \mathcal{B}} \left\| \prod_{j=1}^m \left( \sum_{T \in \mathcal{T}_j(B)} \frac{c_T}{|T|^{1/2}} e(-\lambda_T \cdot x) \right)^{\frac{1}{m}} \right\|_{L^{\frac{2n}{d}}(B)}^{\frac{2n}{d}}$$

$$\leq \mathrm{ME}\left( \frac{2n}{d}, R^{\frac{1}{2}} \right)^{\frac{2n}{d}} R^{\frac{n}{2}} R^{\frac{n}{2} - \frac{n^2}{d}} \sum_{B \in \mathcal{B}} \left( \prod_{j=1}^m \|c_T\|_{l^2(\mathcal{T}_j(B))} \right)^{\frac{2n}{md}}.$$

Note that for each $T \in \mathcal{T}_j(B)$ we have $1_B \leq 1_{10T}$, so the last expression can be dominated by

$$\text{ME}\left(\frac{2n}{d}, R^{\frac{1}{2}}\right)^{\frac{2n}{d}} R^{n-\frac{n^2}{d}} \sum_{B\in\mathcal{B}} \frac{1}{|B|} \int_B \prod_{j=1}^{m} \left(\sum_{T\in\mathcal{T}_j} |c_T|^2 1_{10T}\right)^{\frac{n}{md}}$$

$$\lesssim \text{ME}\left(\frac{2n}{d}, R^{\frac{1}{2}}\right)^{\frac{2n}{d}} R^{\frac{n}{2}-\frac{n^2}{d}} \left\|\left(\prod_{j=1}^{m}\sum_{T\in\mathcal{T}_j} |c_T|^2 1_{10T}\right)^{\frac{1}{m}}\right\|_{L^{\frac{n}{d}}(\mathbb{R}^n)}^{\frac{n}{d}}.$$

Invoking multilinear Kakeya and then (6.29), this is further dominated by

$$\lesssim \text{MK}\left(\frac{n}{d}, R\right)^{\frac{n}{d}} \text{ME}\left(\frac{2n}{d}, R^{\frac{1}{2}}\right)^{\frac{2n}{d}} R^{\frac{n}{2}-\frac{n^2}{d}} R^{\frac{n}{2}} \prod_{j=1}^{m}\left(\sum_{T\in\mathcal{T}_j}|c_T|^2\right)^{\frac{n}{md}}$$

$$\sim \text{MK}\left(\frac{n}{d}, R\right)^{\frac{n}{d}} \text{ME}\left(\frac{2n}{d}, R^{\frac{1}{2}}\right)^{\frac{2n}{d}} R^{n-\frac{n^2}{d}} \prod_{j=1}^{m}\|F_j\|_2^{\frac{2n}{md}}.$$

It suffices now to combine all these inequalities and raise to the power $d/2n$. $\qquad\square$

We can now prove Theorem 6.24.

*Proof* The argument will feature a very simple and elegant induction on scales. Combining Propositions 6.26 and 6.27, we get an inequality involving the scales $R$ and $R^{1/2}$:

$$\text{MR}\left(\frac{2n}{d}, R\right) \leq C\,\text{MR}\left(\frac{2n}{d}, R^{\frac{1}{2}}\right) \text{MK}\left(\frac{n}{d}, R\right)^{\frac{1}{2}}, \qquad (6.30)$$

with $C$ independent of $R$.

Let $l \in \mathbb{N}$ be the smallest integer such that $R^{1/2^l} \leq 2$. Note that $l \sim \log\log R$. Iterating (6.30) $l$ times, we find the following multiscale inequality:

$$\text{MR}\left(\frac{2n}{d}, R\right) \leq C^l\,\text{MR}\left(\frac{2n}{d}, R^{\frac{1}{2^l}}\right) \prod_{s=0}^{l-1} \text{MK}\left(\frac{n}{d}, R^{\frac{1}{2^s}}\right)^{\frac{1}{2}}.$$

At this point, we rely on the multilinear Kakeya inequality (6.27) to write for each $\epsilon > 0$:

$$\text{MR}\left(\frac{2n}{d}, R\right) \lesssim C^l \prod_{s=0}^{l-1} (C_\epsilon R^{\frac{\epsilon}{2^s}})^{\frac{1}{2}} \lesssim (CC_\epsilon^{\frac{1}{2}})^{O(\log\log R)} R^\epsilon.$$

This quantity is easily seen to be $O(R^{2\epsilon})$, and this closes the argument. $\qquad\square$

Combining (6.24) and Proposition 6.26 with $p = 2n/d$ leads to the following exponential sum estimate.

**Corollary 6.28** (Multilinear square root cancellation) *For each $1 \leq p \leq 2n/d$ and each collection $\Lambda_j$ of $1/R$-separated points on $\mathcal{M}_j$,*

$$\left\| \left( \prod_{j=1}^{m} \sum_{\lambda \in \Lambda_j} a_\lambda e(\lambda \cdot x) \right)^{\frac{1}{m}} \right\|_{L_\sharp^p(B_R)} \lesssim_\epsilon R^\epsilon \left( \prod_{j=1}^{m} \|a_\lambda\|_{l^2(\Lambda_j)} \right)^{\frac{1}{m}}. \tag{6.31}$$

**Remark 6.29** Apart from the $R^\epsilon$ term, this inequality is sharp in an average sense, that is, if we average the left-hand side over random choices of signs for $a_\lambda$. This should be contrasted with Exercise 6.32.

Let us now restrict attention to hypersurfaces, $d = n - 1$. The reader may wonder whether an argument like the one described in this section can be used to give a different proof of the bilinear estimate in Theorem 3.1, or more generally, of some other $L^2$-based $k$-linear restriction estimate with $k < n$ (i.e., $p = 2$ and $q \geq 2(n + k)/(n + k - 2)$ in Conjecture 6.1). The short answer is "no," and here is a brief justification. The argument we have presented here relies on applying the following inequality with $q = 2n/(n - 1)$ (and $m = n$)

$$\left\| \left( \prod_{j=1}^{m} \sum_{\lambda \in \Lambda_j} a_\lambda e(\lambda \cdot x) \right)^{\frac{1}{m}} \right\|_{L_\sharp^q(B)} \lesssim \mathrm{MR}(q, R^{1/2}) \left( \prod_{j=1}^{m} \|a_\lambda\|_{l^2(\Lambda_j)} \right)^{\frac{1}{m}}, \tag{6.32}$$

on each ball $B$ of radius $R^{1/2}$. Since a posteriori we find that

$$\mathrm{MR}\left( \frac{2n}{n - 1}, R^{1/2} \right) \lesssim R^\epsilon,$$

(6.32) is essentially sharp on each $B$, as observed in Remark 6.29. Thus, the argument has no losses in this case.

Assume now that $\mathbb{M} = (\mathcal{M}_1, \ldots, \mathcal{M}_k)$ consists of hypersurfaces as in Conjecture 6.1, with $k < n$. One may repeat the argument and apply the same inequality (6.32) on balls $B$ of radius $R^{1/2}$, this time with $m = k$ and $q \geq 2(n + k)/(n + k - 2)$. The problem is that $R^{n-1/4}$ is much bigger than $|B|^{1/q}$ when $q > 2n/(n - 1)$ and thus (6.32) is worse than a square root cancellation inequality for such $q$. In other words, it is far from being a sharp inequality (even in an average sense), and using it would lead to repeated losses throughout the argument.

**Exercise 6.30** Let $\mathbb{M} = (\mathcal{M}_1, \ldots, \mathcal{M}_m)$ with each $\mathcal{M}_j$ as in (6.21). Given $1 \leq R' \leq R$, let $\Theta_j$ be a partition of $\mathcal{N}_{U_j}(R^{-1})$ into subsets $\theta$ of the form $(q \times \mathbb{R}^{n-d}) \cap \mathcal{N}_{U_j}(R^{-1})$, with each $q$ an almost cube in $\mathbb{R}^d$ with diameter $\sim 1/R'$. Assume that $F_j$ has a Fourier transform supported on $\mathcal{N}_{U_j}(R^{-1})$.

(a) Prove that for each ball $B$ with radius $R$ we have

$$\left\| \left( \prod_{j=1}^{m} F_j \right)^{\frac{1}{m}} \right\|_{L^{\frac{2n}{d}}(B)} \lesssim \mathrm{MR}\left( \mathbb{M}, \frac{2n}{d}, R \right) \prod_{j=1}^{m} \left\| \left( \sum_{\theta \in \Theta_j} |\mathcal{P}_\theta F_j|^2 \right)^{\frac{1}{2}} \right\|_{L^{\frac{2n}{d}}(\chi_B)}^{\frac{1}{m}} .$$

(b) Conclude that if $\mathbb{M}$ is BL-transverse, we have the following global multilinear reverse square function estimate:

$$\left\| \left( \prod_{j=1}^{m} F_j \right)^{\frac{1}{m}} \right\|_{L^{\frac{2n}{d}}(\mathbb{R}^n)} \lesssim_\epsilon R^\epsilon \prod_{j=1}^{m} \left\| \left( \sum_{\theta \in \Theta_j} |\mathcal{P}_\theta F_j|^2 \right)^{\frac{1}{2}} \right\|_{L^{\frac{2n}{d}}(\mathbb{R}^n)}^{\frac{1}{m}} .$$

**Exercise 6.31** (a) Prove that if $\mathbb{M}$ satisfies (6.22) then

$$\left\| \left( \prod_{j=1}^{m} E^{\mathcal{M}_j} f_j \right)^{\frac{1}{m}} \right\|_{L^q(B_R)} \lesssim_\epsilon R^\epsilon \left( \prod_{j=1}^{m} \| f_j \|_{L^p(U_j)} \right)^{\frac{1}{m}} \tag{6.33}$$

whenever $q \geq 2n/d$ and $p' \leq d/nq$.

(b) Prove that (6.33) is false for any other choice of $1 \leq p, q \leq \infty$ and for all $\mathbb{M}$.

Hint for (a): Use (6.25), Hölder's inequality, and multilinear interpolation.

Hint for (b): Adapt the Knapp Example 1.8 to prove that $p' \leq d/nq$ is needed. To prove that $q \geq 2n/d$ is needed, test (6.33) with random functions

$$f_{j,\omega}(\xi) = \sum_{\xi_j \in \Xi_j} r_j(\omega) 1_{B_d(\xi_j, \frac{c}{R})}(\xi),$$

with $\Xi_j$ a $1/R$-separated set in $U_j$ with size $\sim R^d$, and $c > 0$ small enough.

**Exercise 6.32** For $n \geq 2$, let $\mathcal{M}_1, \ldots, \mathcal{M}_n$ be $d$-dimensional manifolds in $\mathbb{R}^n$ satisfying (6.22). Prove that (6.31) is not always sharp. More precisely, for each $1 \leq p < 2n$, prove the existence of $\alpha_p > 0$ and $\Lambda_j \subset \mathcal{M}_j$ such that

$$\left\| \left( \prod_{j=1}^{n} \sum_{\lambda \in \Lambda_j} e(\lambda \cdot x) \right)^{\frac{1}{n}} \right\|_{L^p_\sharp(B_R)} \lesssim R^{-\alpha_p} \prod_{j=1}^{n} |\Lambda_j|^{\frac{1}{2n}} .$$

Compare this with the linear case (1.19).

Hint: Choose $\Lambda_j$ in the $1/R$-neighborhood of a line and having $\sim R^{1/2}$ points. Use a linear transformation to reduce to the case when the $n$ lines are parallel to the axes. Use the fact that

$$\left\| \sum_{k=1}^{R^{\frac{1}{2}}} e(\frac{k}{R}t) \right\|_{L_{\#}^{\frac{p}{n}} ([-R,R])} \lesssim R^{\frac{1}{4}-\alpha_p}.$$

**Exercise 6.33** (Multilinear restriction for regular curves) Consider a curve in $\mathbb{R}^n$

$$\Phi(t) = (\phi_1(t), \ldots, \phi_n(t))$$

with $\phi_i \in C^1([0,1])$. Moreover, assume that

$$|\Phi'(t_1) \wedge \cdots \wedge \Phi'(t_n)| \geq \nu$$

for each $t_i$ in some fixed closed interval $I_i \subset [0,1]$.

(a) Prove that this property is satisfied for the moment curve

$$\Gamma_n = \{(t,t^2,\ldots,t^n) : t \in [0,1]\},$$

whenever $I_i$ are pairwise disjoint. Moreover, $\nu$ depends only on $n$ and on the separation between the intervals.

(b) Define the extension operator

$$E^{\Phi} f(x) = \int f(t)e(\Phi(t) \cdot x) \, dt.$$

Prove the following sharp endpoint multilinear restriction inequality

$$\left\| \left(\prod_{j=1}^{n} E^{\Phi} f_j\right)^{\frac{1}{n}} \right\|_{L^{2n}(\mathbb{R}^n)} \lesssim_{\nu} \left(\prod_{j=1}^{n} \|f_j\|_{L^2(I_j)}\right)^{\frac{1}{n}}$$

for each $f_j$ supported on $I_j$.

(c) Conclude as in Exercise 6.31 that the global estimate

$$\left\| \left(\prod_{j=1}^{n} E^{\Phi} f_j\right)^{\frac{1}{n}} \right\|_{L^q(\mathbb{R}^n)} \lesssim_{\nu} \left(\prod_{j=1}^{n} \|f_j\|_{L^p(I_j)}\right)^{\frac{1}{n}}$$

holds whenever $q \geq 2n$ and $1 - 1/p \geq n/q$.

Hint for (b): Use a change of variables to reduce matters to an $L^2$ inequality involving one function on $\mathbb{R}^n$.

The method described in the previous exercise can also be used to get optimal $L^2$-based $(n-1)$-linear restriction estimates for the moment curve.

**Exercise 6.34** (Sharp $(n-1)$-linear restriction estimates for $\Gamma_n$) Let $I_1, \ldots, I_{n-1}$ be pairwise disjoint closed subintervals of $[0,1]$.

(a) Prove that

$$\left\|\left(\prod_{j=1}^{n-1} E^{\Gamma_n} f_j\right)^{\frac{1}{n-1}}\right\|_{L^q(\mathbb{R}^n)} \lesssim \left(\prod_{j=1}^{n-1} \|f_j\|_{L^2(I_j)}\right)^{\frac{1}{n-1}}$$

holds for each $f_j$ supported on $I_j$ and each $q \geq 2(n+1)$.

(b) Prove that if $q < 2(n+1)$, the inequality is false for some functions $f_j$.

Hint for (a): Let $\Gamma_{n,I}$ be the arc on $\Gamma_n$ corresponding to $t \in I$. Prove that $\Gamma_{n,I_1} + \cdots + \Gamma_{n,I_{n-1}}$ is a hypersurface with nonvanishing Gaussian curvature. Then use Theorem 1.16.

Hint for (b): Use a Knapp-type example.

The next exercise is inspired by [6], where similar results are obtained for arbitrary orders of multilinearity, using a slightly different approach. Note that the inequality in (c) has the gain $R^{-1/4}$ over the unrestricted $L^4$ estimate with $F_1$ supported on the whole $\mathcal{N}_{U_1}(R^{-1})$. In the special cases (a) and (b), this inequality admits an easier geometric proof, similar to the proof of Proposition 3.11. The point of the alternative approach outlined in this exercise is to eliminate any use of biorthogonality. See the discussion preceding Exercise 3.45 for motivation.

**Exercise 6.35** (Improved bilinear restriction estimate in $L^4$) Fix $\nu > 0$. Let

$$\mathcal{M}_i = \{(\xi, \psi_i(\xi)), \xi \in U_i\}, \quad 1 \leq i \leq 2$$

be two smooth surfaces in $\mathbb{R}^3$. Consider a smooth curve

$$\{\Gamma(t) := (t, \gamma(t), \psi_1(t, \gamma(t))) : t \in I\}$$

on $\mathcal{M}_1$ such that we have the following transversality condition

$$|\mathbf{n}_2 \cdot \Gamma'(t)| > \nu |\Gamma'(t)| \tag{6.34}$$

for each unit normal $\mathbf{n}_2$ to $\mathcal{M}_2$ and each $t \in I$.

(a) Check that (6.34) is satisfied if $U_1$ and $U_2$ are small enough neighborhoods of $(1,0,0)$ and $(0,1,0)$, $\psi_1(\xi) = \psi_2(\xi) = |\xi|^2$, and $\gamma$ is constant.

(b) Check that (6.34) is satisfied if $\mathcal{M}_1, \mathcal{M}_2$ are angularly separated regions on the cone $\text{Cone}^2$ and $\gamma$ is linear.

(c) Adapt the argument from this section to prove that for each $F_1$ supported on the $R^{-1}$-neighborhood of $\Gamma$ and for each $F_2$ supported on $\mathcal{N}_{\mathcal{M}_2}(R^{-1})$ we have

$$\|(\hat{F}_1 \hat{F}_2)^{\frac{1}{2}}\|_{L^4(\mathbb{R}^3)} \lesssim_{\epsilon,\nu} R^{-\frac{3}{4}}(\|F_1\|_2 \|F_2\|_2)^{\frac{1}{2}}.$$

(d) Prove the necessity of (6.34) for the inequality in (c).

Hint for (c): Decompose $F_1$ using wave packets spatially concentrated inside boxes with dimensions $R \times R \times R^{1/2}$. Decompose $F_2$ into wave packets spatially concentrated in $(R^{1/2}, R)$-tubes. Use induction on the scale parameter $R$ to reduce the inequality to a simple bilinear Kakeya-type inequality involving boxes and tubes. Show that the transversality condition forces the intersection between any box and any tube arising from these decompositions to have volume $O(R^{3/2})$.

Hint for (d): Consider the case when $F_1$ is a single (boxlike) wave packet and $F_2$ is a single (tube-like) wave packet, in such a way that the tube is a subset of the box.

**Exercise 6.36** For $1 \leq k \leq n$, let $\mathcal{M}_1, \ldots, \mathcal{M}_k$ be $\nu$-transverse hypersurfaces in $\mathbb{R}^n$. Prove that

$$\left\| \left( \prod_{j=1}^{k} E^{\mathcal{M}_j} f_j \right)^{\frac{1}{k}} \right\|_{L^{\frac{2k}{k-1}}(B(0,R))} \lesssim_{\nu,\epsilon} R^{\epsilon} \left( \prod_{j=1}^{k} \|f_j\|_{L^2} \right)^{\frac{1}{k}}.$$

# 7

## The Bourgain–Guth Method

In the years immediately following the proof of the multilinear restriction theorem by Bennett, Carbery, and Tao, the experts have searched for ways to harness its power for the purpose of proving new linear restriction estimates. It became apparent that the Whitney decompositions that performed so well in translating estimates from the bilinear to the linear setup interact less efficiently with the multilinear theory.

The seminal work of Bourgain and Guth [32] came with a very satisfactory alternative solution. It introduced a new way of running the induction on scales that has since found a large number of applications. The key departure point of the new method is to replace the Whitney decomposition of the frequency domain with a uniform one. The terms in the decomposition are sorted based on their spatial contribution, as either negligible or significant. The frequency cubes that contribute significantly are seen either to be transverse or to cluster near a lower-dimensional affine subspace. The contribution in the first case is estimated using the multilinear restriction theorem, and invoking lower-dimensional theory in the second case.

The first two sections will illustrate the Bourgain–Guth method in simpler contexts, when there is no lower-dimensional contribution. In the last section, we refine the analysis to incorporate the lower-dimensional terms. We will not aim to prove new results in this chapter, but will confine ourselves to merely recovering some old ones. In Chapter 13, we will combine the Bourgain–Guth method with decoupling inequalities in order to improve the restriction estimates seen earlier in the book.

## 7.1 From Bilinear to Linear for Hypersurfaces

Our goal in this section is to introduce the uniform decomposition and to use it to reprove Theorem 4.6, which we restate as Theorem 7.1 for the reader's convenience.

**Theorem 7.1** *Let* $q > 2n/(n-1)$ *for some* $n \geq 2$. *Assume that*

$$BR^*_{\mathbb{P}^{n-1}}\left(\infty \times \infty \mapsto q, \frac{1}{2}\right) < \infty.$$

*Then the linear restriction estimate* $R^*_{\mathbb{P}^{n-1}}(\infty \mapsto q)$ *holds.*

Fix some large enough $K \in 2^{\mathbb{N}}$, to be determined later. Let $\mathcal{C}_K$ be the partition of $[-1,1]^{n-1}$ into cubes $\alpha$ with side length $2/K$. We will write $\alpha \sim \alpha'$ if $\alpha$ and $\alpha'$ are neighbors. There are at most $3^{n-1}$ neighbors for each $\alpha$.

Note that for each $f : [-1,1]^{n-1} \to \mathbb{C}$,

$$Ef(x) = \sum_{\alpha \in \mathcal{C}_K} E_\alpha f(x).$$

The following abstract lemma sets up a strategy to analyze this sum by means of a very simple dichotomy: either there are two nonadjacent cubes with significant contributions, or all significant contributors cluster near one particular cube.

**Lemma 7.2** *There are constants* $C_n$ *(independent of* $K$*) and* $C_{n,K}$ *such that given any complex numbers* $z_\alpha$ *we have*

$$\left| \sum_{\alpha \in \mathcal{C}_K} z_\alpha \right| \leq C_n \max_{\alpha \in \mathcal{C}_K} |z_\alpha| + C_{n,K} \max_{\alpha' \not\sim \alpha''} |z_{\alpha'} z_{\alpha''}|^{\frac{1}{2}}.$$

*Proof* In our applications of this lemma, it will be very important that the constant $C_n$ is independent of $K$.

The proof is very elementary. Let $\alpha^*$ be such that $|z_{\alpha^*}| = \max_\alpha |z_\alpha|$. Define

$$S_{\text{big}} = \left\{ \alpha : |z_\alpha| \geq \frac{|z_{\alpha^*}|}{K^{n-1}} \right\}.$$

There are two possibilities. Let us assume first that there is some $\alpha_0 \in S_{\text{big}}$ with $\alpha_0 \not\sim \alpha^*$. Then

$$|z_{\alpha^*} z_{\alpha_0}|^{1/2} \geq \frac{|z_{\alpha^*}|}{K^{\frac{n-1}{2}}} \geq \frac{|\sum_{\alpha \in \mathcal{C}_K} z_\alpha|}{K^{\frac{3(n-1)}{2}}}$$

and thus

$$\left| \sum_{\alpha \in \mathcal{C}_K} z_\alpha \right| \leq K^{\frac{3(n-1)}{2}} \max_{\alpha' \not\sim \alpha''} |z_{\alpha'} z_{\alpha''}|^{\frac{1}{2}}.$$

The second case is when $S_{\text{big}}$ contains only neighbors of $\alpha^*$. We then get

$$\left|\sum_{\alpha \in \mathcal{C}_K} z_\alpha\right| \leq \sum_{\alpha \sim \alpha^*} |z_\alpha| + \sum_{\alpha \not\sim \alpha^*} |z_\alpha|$$

$$\leq 3^{n-1}|z_{\alpha^*}| + K^{n-1}\frac{|z_{\alpha^*}|}{K^{n-1}}$$

$$= (3^{n-1} + 1)\max_{\alpha \in \mathcal{C}_K} |z_\alpha|. \qquad \square$$

Let us now prove Theorem 7.1. For $R \geq 1$ and $q > 2n/(n-1)$, define

$$C_{\text{lin}}(R,q) = \sup\left\{\int_{B_n(0,R)} |Ef|^q : \|f\|_{L^\infty([-1,1]^{n-1})} \leq 1\right\}.$$

We will show that

$$C_{\text{lin}}(R,q) \lesssim 1,$$

with the implicit constant independent of $R$. Fix $f$ with $\|f\|_{L^\infty([-1,1]^{n-1})} \leq 1$. We use Lemma 7.2 with $z_\alpha = E_\alpha f(x)$, replacing each maximum with a sum

$$|Ef(x)|^q \leq D_n \sum_{\alpha \in \mathcal{C}_K} |E_\alpha f(x)|^q + D_{n,K} \sum_{\alpha' \not\sim \alpha''} |E_{\alpha'} f(x) E_{\alpha''} f(x)|^{\frac{q}{2}}.$$

It is important that $D_n$ is independent of $K$. Integrate on $B_n(0,R)$

$$\int_{B_n(0,R)} |Ef|^q \leq D_n \sum_{\alpha \in \mathcal{C}_K} \int_{B_n(0,R)} |E_\alpha f|^q + D_{n,K} \sum_{\alpha' \not\sim \alpha''} \int_{B_n(0,R)} |E_{\alpha'} f E_{\alpha''} f|^{\frac{q}{2}}.$$

We will use parabolic rescaling to estimate each term in the first sum. Given $\alpha = c_\alpha + [-1/K, 1/K]^{n-1}$, changing variables leads to

$$|E_\alpha f(\bar{x}, x_n)| = K^{1-n}\left|Eg\left(\frac{\bar{x} + 2x_n c_\alpha}{K}, \frac{x_n}{K^2}\right)\right|,$$

where $g(\xi) = f(c_\alpha + \xi/K)$.

The linear transformation $(\bar{x}, x_n) \mapsto ((\bar{x} + 2x_n c_\alpha)/K, x_n/K^2)$ maps $B_n(0,R)$ to a subset of itself, if $K \gg_n 1$. Thus, after a change of variables on the spatial side we get

$$\int_{B_n(0,R)} |E_\alpha f|^q \leq K^{n+1+q(1-n)} \int_{B_n(0,R)} |Eg|^q$$

$$\leq K^{n+1+q(1-n)} C_{\text{lin}}(R,q).$$

Fix now $\alpha' \not\sim \alpha''$, so their separation is at least $2/K$. Proposition 4.2 shows that

$$\int_{\mathbb{R}^n} |E_{\alpha'} f E_{\alpha''} f|^{\frac{q}{2}} \lesssim_{K,n} 1.$$

It is important to realize that the exact dependence on $K$ of the implicit constant is irrelevant for this type of argument. We conclude that

$$\int_{B_n(0,R)} |Ef|^q \leq D_n K^{n-1} K^{n+1+q(1-n)} C_{\text{lin}}(R,q) + D'_{n,K}.$$

Since $q > 2n/(n-1)$, we can choose $K$ large enough depending on $n$ and $q$ such that

$$D_n K^{n-1} K^{n+1+q(1-n)} < \frac{1}{2}.$$

Since this inequality holds for all $|f| \leq 1$, we get the uniform inequality

$$C_{\text{lin}}(R,q) \leq 2D'_{n,K}.$$

Note that the parameter $R$ did not play a special role in the argument, other than ensuring that the constants $C_{\text{lin}}(R,q)$ are finite.

**Exercise 7.3** (a) Prove that for each $2 \leq q \leq \infty$

$$\|Ef\|_{L^q(\mathbb{R}^{n-1} \times [-1,1])} \lesssim \|f\|_{L^q([-1,1]^{n-1})}.$$

(b) Prove that for each $2 \leq q < \infty$

$$\left\| \left( \sum_{\alpha \in \mathcal{C}_{R^{1/2}}} |E_\alpha f|^q \right)^{\frac{1}{q}} \right\|_{L^q(B_n(0,R))} \lesssim R^{\frac{n}{q} - \frac{n-1}{2}} \|f\|_{L^q([-1,1]^{n-1})}.$$

**Exercise 7.4** Prove that for each $r \in \mathbb{N}$ and $q \geq 1$

$$|Ef(x)| \leq (C_n)^r \left( \sum_{\alpha \in \mathcal{C}_{K^r}} |E_\alpha f(x)|^q \right)^{\frac{1}{q}}$$

$$+ C_{n,K} \sum_{l=1}^r (C_n)^{l-1} \max_{\substack{\alpha',\alpha'' \in \mathcal{C}_{K^l} \\ \text{dist}(\alpha',\alpha'') \sim K^{-l}}} |E_{\alpha'} f(x) E_{\alpha''} f(x)|^{1/2}.$$

Use this inequality with $r$ chosen such that $K^r \sim R^{1/2}$, together with the previous exercise to reprove Theorem 7.1.

**Exercise 7.5** Let $\mathcal{D}$ be a collection of $\delta$-separated directions in the plane.

(a) If $|\mathcal{D}| \gg \left( \log (1/\delta) \right)^2$, prove that there are two subsets $\mathcal{D}_1, \mathcal{D}_2 \subset \mathcal{D}$ with the following properties:

- $|\mathcal{D}_1|, |\mathcal{D}_2| \gtrsim |\mathcal{D}|\big(\log\left(1/\delta\right)\big)^{-2}$.
- The separation $\alpha$ between $\mathcal{D}_1$ and $\mathcal{D}_2$ satisfies $\alpha \gtrsim |\mathcal{D}|\delta\big(\log\left(1/\delta\right)\big)^{-1}$.
- $\mathcal{D}_1 \cup \mathcal{D}_2$ sits inside an arc of length $\lesssim \alpha \log\left(1/\delta\right)$ on $\mathbb{S}^1_+$.

(b) Let $\mathcal{T}$ be a collection of $(\delta, 1)$-tubes pointing in directions from $\mathcal{D}$, in such a way that tubes sharing directions must necessarily be pairwise disjoint. Prove that for each $r \geq 1$

$$\left|\left\{x \in \mathbb{R}^2: \sum_{T \in \mathcal{T}} 1_T(x) > r\right\}\right| \lesssim \frac{\delta |\mathcal{T}|^2}{r^3}.$$

It is worth comparing this result with Proposition 5.6.

Hint for (a): Use a Bourgain–Guth-type argument with $K = \log\left(1/\delta\right)$.

Hint for (b): Let $A_r$ be the quantity on the left. Use (a) to argue that if $r \gg \left(\log\left(1/\delta\right)\right)^2$

$$r^2 A_r \lesssim \int \sum_{\substack{(T_1, T_2) \in \mathcal{T} \times \mathcal{T} \\ \sphericalangle(T_1, T_2) \gtrsim r\delta(\log(\frac{1}{\delta}))^{-1}}} 1_{T_1 \cap T_2}.$$

## 7.2 From Multilinear to Linear for the Moment Curve

We now extend the argument from the previous section to get the nearly optimal linear restriction estimates for the moment curve $\Gamma_n$.

**Theorem 7.6** *The inequality*

$$\|E^{\Gamma_n} f\|_{L^q(\mathbb{R}^n)} \lesssim \|f\|_{L^p([-1,1])}$$

*holds if and only if*

$$\frac{n(n+1)}{2q} + \frac{1}{p} \leq 1$$

*and*

$$q > \frac{n^2 + n + 2}{2}.$$

The necessity of the first restriction follows via a Knapp-type example; see Exercise 7.8. The necessity of the second restriction follows via a more delicate computation with $f = 1_{[-1,1]}$, appearing in [1]. See also [**34**] for a more general discussion.

We will next prove the sufficiency of the two restrictions away from the critical line. When $n = 2$, the moment curve coincides with $\mathbb{P}^1$. For $n \geq 3$, Theorem 7.6 was proved by Drury in [45], following initial progress by Prestini [92], [93] and Christ [37].

As in the previous section, we fix a large enough parameter $K \in 2^{\mathbb{N}}$ (to be chosen later) and we let $\mathcal{C}_K$ be the partition of $[-1, 1]$ into intervals $I$ of length $2/K$. We will continue to write $I \sim I'$ if $I, I'$ are neighbors.

**Lemma 7.7** *There are constants $C_n$ (independent of $K$) and $C_K$ such that given any complex numbers $z_I$, we have*

$$\left| \sum_{I \in \mathcal{C}_K} z_I \right| \leq C_n \max_{I \in \mathcal{C}_K} |z_I| + C_K \max_{I_j \not\sim I_{j'}} \left| \prod_{j=1}^{n} z_{I_j} \right|^{\frac{1}{n}},$$

*where the second maximum is taken over all pairwise nonadjacent intervals $I_1, \ldots, I_n \in \mathcal{C}_K$.*

*Proof* Define

$$S_{\text{big}} = \left\{ I \in \mathcal{C}_K : |z_I| \geq \frac{\max_{I' \in \mathcal{C}_K} |z_{I'}|}{K} \right\}.$$

There are two possibilities. Let us assume first that $S_{\text{big}}$ contains at least $3n$ intervals. Then $S_{\text{big}}$ must in fact contain $n$ intervals, no two of which are neighbors. We may write

$$\left| \prod_{j=1}^{n} z_{I_j} \right|^{\frac{1}{n}} \geq \frac{\max_{I \in \mathcal{C}_K} |z_I|}{K} \geq \frac{|\sum_{I \in \mathcal{C}_K} z_I|}{K^2}.$$

The second case is when $S_{\text{big}}$ contains fewer than $3n$ intervals. We get

$$\left| \sum_{I \in \mathcal{C}_K} z_I \right| \leq \sum_{I \in S_{\text{big}}} |z_I| + \sum_{I \notin S_{\text{big}}} |z_I|$$

$$\leq (3n - 1) \max_{I \in \mathcal{C}_K} |z_I| + \max_{I \in \mathcal{C}_K} |z_I|.$$

We may thus take $C_n = 3n$ and $C_K = K^2$. $\qquad\square$

Let us now combine this lemma with the sharp multilinear restriction estimates for curves from the previous chapter, to prove Theorem 7.6.

*Proof* We will restrict our argument to the case when $(1/p, 1/q)$ is not on the critical line, that is,

$$\frac{n(n + 1)}{2q} + \frac{1}{p} < 1.$$

and

$$q > \frac{n^2 + n + 2}{2}.$$

For $R \geq 1$ and $p, q$ as in the preceding, let us define

$$C_{\mathrm{lin}}(R, p, q) = \sup \left\{ \int_{B_n(0,R)} |E^{\Gamma_n} f|^q : \|f\|_{L^p([-1,1])} \leq 1 \right\}.$$

We will show that

$$C_{\mathrm{lin}}(R, p, q) \lesssim 1.$$

As the case $q = \infty$ is immediate, we may assume that $q < \infty$. Fix $f$ with $\|f\|_{L^p([-1,1])} \leq 1$. Write using Lemma 7.7, replacing the maximum with a sum

$$|E^{\Gamma_n} f(x)|^q \leq D_n \sum_{I \in \mathcal{C}_K} |E_I^{\Gamma_n} f(x)|^q + D_K \sum_{I_j \not\sim I_{j'}} \left| \prod_{j=1}^n E_{I_j}^{\Gamma_n} f(x) \right|^{\frac{q}{n}}.$$

As before, it is important that $D_n$ is independent of $K$. Integrate on $B_n(0, R)$

$$\int_{B_n(0,R)} |E^{\Gamma_n} f|^q \leq D_n \sum_{I \in \mathcal{C}_K} \int_{B_n(0,R)} |E_I^{\Gamma_n} f|^q + D_K \sum_{I_j \not\sim I_{j'}} \int_{\mathbb{R}^n} \left| \prod_{j=1}^n E_{I_j}^{\Gamma_n} f \right|^{\frac{q}{n}}.$$

$$(7.1)$$

We will use the following analog of parabolic rescaling for $\Gamma_n$ to estimate each term in the first sum. Given $I = c_I + [-1/K, 1/K]$, changing variables leads to

$$|E_I^{\Gamma_n} f(x_1, \ldots, x_n)| = K^{-1} |E^{\Gamma_n} g_I(L_I(x_1, \ldots, x_n))|,$$

where $g_I(\xi) = (f 1_I)(c_I + \xi/K)$ and $L_I(x_1, \ldots, x_n)$ has the $j$th entry given by $\sum_{i=j}^n x_i c_I^{i-j} / K^j$. This transformation maps $B_n(0, R)$ to a subset of itself, if $K$ is large enough. Thus, after a change of variables on the spatial side we get

$$\int_{B_n(0,R)} |E_I^{\Gamma_n} f|^q \leq K^{\frac{n(n+1)}{2} - q} \int_{B_n(0,R)} |E^{\Gamma_n} g_I|^q$$

$$\leq K^{\frac{n(n+1)}{2} - q} C_{\mathrm{lin}}(R, p, q) \|g_I\|_p^q$$

$$= K^{\frac{n(n+1)}{2} - q + \frac{q}{p}} C_{\mathrm{lin}}(R, p, q) \|f 1_I\|_p^q.$$

We can dominate the first sum in (7.1) as follows, using elementary inequalities

$$\sum_{I \in \mathcal{C}_K} \int_{B_n(0,R)} |E_I^{\Gamma_n} f|^q \leq K^{\frac{n(n+1)}{2} - q + \frac{q}{p}} C_{\text{lin}}(R, p, q) \sum_{I \in \mathcal{C}_K} \|f 1_I\|_p^q$$

$$\leq \begin{cases} K^{\frac{n^2+n+2}{2} - q} C_{\text{lin}}(R, p, q) \|f\|_{L^p([-1,1])}^q, & \text{if } p \geq q \\ K^{\frac{n^2+n}{2} - q + \frac{q}{p}} C_{\text{lin}}(R, p, q) \|f\|_{L^p([-1,1])}^q, & \text{if } p \leq q. \end{cases}$$

The exponent of $K$ is negative if $(1/p, 1/q)$ are away from the critical line, so by choosing $K$ large enough, we can make sure that

$$D_n \sum_{I \in \mathcal{C}_K} \int_{B_n(0,R)} |E_I^{\Gamma_n} f|^q < \frac{1}{2} C_{\text{lin}}(R, p, q) \|f\|_{L^p([-1,1])}^q.$$

Note also that each $p, q$ as in our theorem must necessarily satisfy the restrictions in Exercise 6.33. Applying the multilinear restriction inequality from this exercise to the moment curve shows that for each pairwise nonadjacent $I_1, \ldots, I_n \in \mathcal{C}_K$, we have

$$\int_{\mathbb{R}^n} \left| \prod_{j=1}^n E_{I_j}^{\Gamma_n} f \right|^{\frac{q}{n}} \lesssim_{K,n} \|f\|_{L^p([-1,1])}^q.$$

Putting things together, we find

$$\int_{B_n(0,R)} |E^{\Gamma_n} f|^q \leq \left( \frac{1}{2} C_{\text{lin}}(R, p, q) + D_K' \right) \|f\|_{L^p([-1,1])}^q.$$

This estimate is uniform over all choices of $f$, and we conclude that

$$C_{\text{lin}}(R, p, q) \leq 2D_K' \lesssim 1. \qquad \square$$

There are two key features that made this argument possible. One is the fact that multilinear estimates are known in the expected range of linear restriction estimates for curves. Also, using multilinear estimates proves to be enough; no lower levels of $k$-linearity are needed. This is a consequence of the fact that all $n$ pairwise disjoint arcs on $\Gamma_n$ are transverse.

Consider a curve in $\mathbb{R}^n$

$$\Phi(t) = (\phi_1(t), \ldots, \phi_n(t))$$

with $\phi_i \in C^{n+1}([-1,1])$. Moreover, assume that

$$\tau(t) = |\Phi'(t) \wedge \Phi''(t) \wedge \cdots \wedge \Phi^{(n)}(t)| \geq \nu > 0 \qquad (7.2)$$

for each $t \in [0,1]$. The quantity $\tau(t)$ is sometimes referred to as the *torsion* of the curve at $t$. In Fourier restriction, torsion plays the same role as curvature does for hypersurfaces. Curves with everywhere nonzero torsion are sometimes referred to as *regular*.

Theorem 7.6 holds true in the more general context of regular curves (see [45]), though the argument presented in this section does not easily extend to this general case because of the lack of proper rescaling.

**Exercise 7.8** Prove the necessity of the restriction $n(n+1)/2q + 1/p \le 1$ in Theorem 7.6.

Hint: Use rescaling to evaluate $\|E^{\Gamma_n} 1_{[0,\delta]}\|_{L^q(\mathbb{R}^n)}$.

**Exercise 7.9** Let $p, q$ be as in Theorem 7.6. Prove that

$$C_{\text{lin}}(R, p, q) \le D_n C_{\text{lin}}\left(\frac{nR}{K}, p, q\right) + D'_K.$$

Conclude that

$$C_{\text{lin}}(R, p, q) \lesssim_\epsilon R^\epsilon$$

if $(1/p, 1/q)$ is on the critical line.

When combined with Exercise 6.33, the next exercise shows that the multilinear restriction inequality for regular curves is similar to the one for $\Gamma_n$.

**Exercise 7.10** Consider a curve

$$\Phi(t) = (\phi_1(t), \dots, \phi_n(t))$$

with $\phi_i \in C^{n+1}([-1, 1])$.

(a) Prove that for each $t_i \in [0, 1]$ with $\Delta := \max_{1 \le i, j \le n} |t_i - t_j|$, we have

$$|\Phi'(t_1) \wedge \Phi'(t_2) \wedge \cdots \wedge \Phi'(t_n)| = \frac{\tau(t_n)}{\prod_{j=1}^{n-1} j!} \prod_{1 \le i < j \le n} |t_j - t_i| + O\left(\Delta^{\binom{n}{2}+1}\right).$$

(b) Assume that $\Phi$ satisfies (7.2). Conclude that there is some small enough $\epsilon_0$, so that whenever

$$t_1 < t_2 < \cdots < t_n$$

are such that $|t_{i+1} - t_i| \sim \epsilon$ for some $\epsilon < \epsilon_0$, we have

$$|\Phi'(t_1) \wedge \Phi'(t_2) \wedge \cdots \wedge \Phi'(t_n)| \gtrsim_\nu \epsilon^{\binom{n}{2}}.$$

Hint for (a): Write for each $i \leq n - 1$ and each $j \leq n$

$$\phi'_j(t_i) = \sum_{k=1}^{n} \frac{\phi_j^{(k)}(t_n)}{(k-1)!}(t_i - t_n)^{k-1} + O(\Delta^n).$$

## 7.3 From Multilinear to Linear for Hypersurfaces

It is now time to discuss the most subtle case of the Bourgain–Guth method, as introduced in the very influential paper [**32**]. We will seek to extend the earlier argument from the case of curves to hypersurfaces. The main new difficulty that will arise has to do with handling lower-dimensional terms. To keep the discussion as simple as possible without sacrificing the main novelty, we restrict our attention for now to the two-dimensional paraboloid. While it turns out that in this case we merely recover the known range $q > 10/3$ without improving it, extending this approach to higher dimensions as in [**32**] leads to new linear restriction estimates for the higher-dimensional paraboloid. The relation between $k$-linear and linear restriction estimates will be discussed in Chapter 13.

We denote by $E$ the extension operator for $\mathbb{P}^2$. For $q \geq 3$, let $C_{\text{lin}}(R, q)$ be the quantity

$$C_{\text{lin}}(R, q) = \sup\left\{ \int_{B(0,R)} |Ef|^q : \|f\|_{L^\infty([-1,1]^2)} \leq 1 \right\}.$$

**Theorem 7.11** *For each $q > 10/3$ we have*

$$C_{\text{lin}}(R, q) \lesssim_\epsilon R^\epsilon.$$

The necessity for the restriction $q > 10/3$ will become apparent only at the end of the argument. Before we reach that point, we will prove all our inequalities in the range $3 \leq q \leq 4$.

As in our previous applications of the Bourgain–Guth method, we will replace the Whitney decomposition with a uniform decomposition of the frequency domain. Our argument here will not make use of any bilinear restriction theory.

For $\mu \in 2^{\mathbb{N}}$, we let $\mathcal{C}_\mu$ be the partition of $[-1, 1]^2$ into squares $\alpha$ with side length $2/\mu$. In the following, $K \in 4^{\mathbb{N}}$ will be a large enough parameter whose precise value will be determined at the end of the argument. For now, we will just mention that $K$ will get larger as $q$ approaches the critical value $10/3$, but it will always be independent of $R$.

Before we present the argument in detail, let us get a bird's eye view of its main ideas. The starting point is the decomposition

$$Ef(x) = \sum_{\alpha \in \mathcal{C}_K} E_\alpha f(x), \quad x \in B(0, R).$$

Perhaps the most naive estimate available is the triangle inequality:

$$|Ef(x)| \le \|E_\alpha f(x)\|_{l^1(\mathcal{C}_K)}. \tag{7.3}$$

This of course will not lead us too far, as it destroys the fine cancellations between the terms of the sum. However, it can prove efficient when used for certain smaller contributions to the sum.

The magnitude of each term $|E_\alpha f(x)|$ can vary significantly when $x$ ranges through $B(0, R)$. We will, however, determine that it remains essentially constant when $x$ is constrained to a ball $B_K$ with much smaller radius $K$. This suggests analyzing $Ef$ separately on each ball $B_K$ from some finitely overlapping cover of $B(0, R)$. We will use a very elementary inequality (Lemma 7.13) to determine that the squares $\alpha \in S_{\mathrm{big}}(B_K)$ corresponding to the significant terms $|E_\alpha f|$ on each given $B_K$ display exactly one of the following three behaviors: they either cluster near some $\alpha^*$, or cluster inside a narrow strip, or are spread in such a way that they contain a triple determining a triangle with a "large" area.

The first scenario proves extremely favorable, and in fact allows us to strengthen (7.3) beyond even a square root cancellation, replacing the $l^1$ norm with the $l^q$ norm. In the second scenario, the significant part of $Ef$ is localized in frequency near a parabola, and the required estimate is of lower-dimensional nature. The third scenario brings to bear trilinear restriction estimates.

We initiate our detailed presentation by revisiting the critical concept of transversality.

**Definition 7.12** We will say that three subsets $S_1, S_2, S_3$ of $[-1, 1]^2$ are $\nu$-transverse if the area of the triangle determined by any $\xi_1 \in S_1, \xi_2 \in S_2, \xi_3 \in S_3$ is greater than $\nu$.

A simple computation shows that the area of the triangle determined by $\xi_1, \xi_2, \xi_3$ is comparable to $|\mathbf{n}(\xi_1) \wedge \mathbf{n}(\xi_2) \wedge \mathbf{n}(\xi_3)|$, where $\mathbf{n}(\xi)$ denotes the unit normal to $\mathbb{P}^2$ at the point $(\xi, |\xi|^2)$. Thus this definition is consistent with the earlier one (see, e.g., (6.1)). In particular, for each $q \ge 3$, (6.26) guarantees that the estimate

$$\left( \int_{B(0,R)} \left| \prod_{i=1}^{3} E_{\alpha_i} f(x) \right|^{\frac{q}{3}} dx \right)^{\frac{1}{q}} \lesssim_{\epsilon, K} R^\epsilon \|f\|_{L^\infty([-1,1]^2)} \tag{7.4}$$

holds uniformly over all choices of $f$ and of $K^{-2}$-transverse squares $\alpha_1, \alpha_2, \alpha_3 \in \mathcal{C}_K$.

Given a line $L$ in $\mathbb{R}^2$ that intersects $[-1, 1]^2$, we define the strip

$$S_L = \left\{ \xi \in \mathbb{R}^2 : \operatorname{dist}(\xi, L) \leq \frac{1000}{K} \right\}.$$

The following lemma has analogs in the previous two sections. The main twist here is its universal nature: it provides uniform estimates over all $z_\alpha$ subordinated to some fixed $Z_\alpha$. The need for such estimates is a subtle aspect of the whole theory, and it is ultimately due to the presence of lower-dimensional terms in the anatomy of $Ef$.

**Lemma 7.13** (Structure result) *Let $Z_\alpha \geq 0$ for $\alpha \in C_K$. Then at least one of the following scenarios holds.*

*(Sc1) The inequality*

$$\left| \sum_{\alpha \in C_K} z_\alpha \right| \leq 100 \max_\alpha Z_\alpha$$

*holds for each $z_\alpha \in \mathbb{C}$ satisfying $|z_\alpha| \leq Z_\alpha$.*

*(Sc2) There is a line $L$ such that*

$$\left| \sum_{\alpha \in C_K} z_\alpha \right| \leq \max_\alpha Z_\alpha + \left| \sum_{\alpha \subset S_L} z_\alpha \right|$$

*for each $z_\alpha \in \mathbb{C}$ satisfying $|z_\alpha| \leq Z_\alpha$. Moreover,*

$$\max_{\alpha \not\subset S_L} Z_\alpha \leq K^{-2} \max_\alpha Z_\alpha.$$

*(Sc3) There are three $K^{-2}$-transverse squares $\alpha_1, \alpha_2, \alpha_3$ such that*

$$\left| \sum_{\alpha \in C_K} z_\alpha \right| \leq C_K (Z_{\alpha_1} Z_{\alpha_2} Z_{\alpha_3})^{1/3}$$

*for each $z_\alpha \in \mathbb{C}$ satisfying $|z_\alpha| \leq Z_\alpha$. The constant $C_K$ only depends on $K$.*

*Proof* Define

$$S_{\text{big}} = \left\{ \alpha' \in C_K : Z_{\alpha'} \geq \frac{\max Z_\alpha}{K^2} \right\}.$$

We will analyze three distinct possibilities.

First, if $S_{\text{big}}$ contains no two squares that are separated by at least $10/K$, then $S_{\text{big}}$ certainly contains fewer than 99 squares. The inequality in (Sc1) is then easily seen to hold, invoking the triangle inequality.

We will next assume that $S_{\text{big}}$ contains two squares $\alpha_1$ and $\alpha_2$ that are separated by at least $10/K$. We choose these squares so that their distance is as large as possible. Let $L$ be the line determined by their centers.

The second possibility corresponds to the case when

$$\max_{\alpha \not\subset S_L} Z_\alpha \le K^{-2} \max_\alpha Z_\alpha.$$

The triangle inequality then shows that in this case we also have

$$\left| \sum_{\alpha \in \mathcal{C}_K} z_\alpha \right| \le \max_\alpha Z_\alpha + \left| \sum_{\alpha \subset S_L} z_\alpha \right|.$$

The last possibility is when there is $\alpha_3 \in S_{\text{big}}$ with $\alpha_3 \not\subset S_L$. A simple geometric observation using the maximality of $\text{dist}\,(\alpha_1, \alpha_2)$ shows that $\alpha_1, \alpha_2, \alpha_3$ must be $K^{-2}$-transverse. We then estimate

$$\left| \sum_{\alpha \in \mathcal{C}_K} z_\alpha \right| \le K^2 \max_\alpha Z_\alpha \le K^4 (Z_{\alpha_1} Z_{\alpha_2} Z_{\alpha_3})^{1/3}. \qquad \square$$

Given a ball $B = B(c_B, R)$ in $\mathbb{R}^3$, we introduce the weight function

$$w_B(x) = \frac{1}{\left(1 + \frac{|x - c_B|}{R}\right)^{100}}.$$

The main properties of these weights that will be used in this section are recorded in the following lemma, whose verification is left to the reader.

**Lemma 7.14** *For $r \ge 1$, let $\mathcal{B}_r$ be a finitely overlapping cover of $\mathbb{R}^3$ with balls of radius $r$. Then, uniformly over $B = B(c, R)$,*

$$\sum_{B' \in \mathcal{B}_R} w_{B'}(x) w_B(c_{B'}) \lesssim w_B(x) \lesssim \sum_{B' \in \mathcal{B}_R} 1_{B'}(x) w_B(c_{B'}). \qquad (7.5)$$

*Also*

$$\sup_{R > 1} \sup_{|x - y| \le R} \frac{w_{B(0, R)}(x)}{w_{B(0, R)}(y)} < \infty, \qquad (7.6)$$

*and for each $1 \le K \le R$*

$$\sum_{\substack{B' \in \mathcal{B}_K \\ B' \cap B(0, R) \neq \emptyset}} w_{B'}(x) \lesssim w_{B(0, R)}(x). \qquad (7.7)$$

The fact that various contributions from $Ef$ are not perfectly localized on the spatial side will only cause minor technical complications.

**Corollary 7.15** *For each $f : [-1, 1]^2 \to \mathbb{C}$, we have*

$$\int |Ef|^q w_{B(0, R)} \lesssim C_{\text{lin}}(R, q) \|f\|_\infty^q.$$

*Proof* Using (3.30), it is immediate that

$$\int_{B(c,R)} |Ef|^q \lesssim C_{\text{lin}}(R,q) \|f\|_\infty^q,$$

uniformly over $c$ and $f$. Combining this with (7.5), we write

$$\int |Ef|^q w_{B(0,R)} \lesssim \sum_{B' \in \mathcal{B}_R} w_{B(0,R)}(c_{B'}) \int_{B'} |Ef|^q$$

$$\lesssim C_{\text{lin}}(R,q) \|f\|_\infty^q \sum_{B' \in \mathcal{B}_R} w_{B(0,R)}(c_{B'})$$

$$\lesssim C_{\text{lin}}(R,q) \|f\|_\infty^q. \qquad \square$$

For the rest of the argument, we fix a finitely overlapping cover $\mathcal{B}_{K,R}$ of $B(0,R)$ with balls $B_K$ of radius $K$. Write

$$\int_{B(0,R)} |Ef|^q \le \sum_{B_K \in \mathcal{B}_{K,R}} \int_{B_K} \left| \sum_{\alpha \in \mathcal{C}_K} E_\alpha f \right|^q.$$

Let $\eta$ be a Schwartz function on $\mathbb{R}^3$ with a Fourier transform equal to 1 on $B(0,10)$ and such that $\eta(0) = 1$. Let

$$\eta_K(x) = \frac{1}{K^3} \eta\left(\frac{x}{K}\right).$$

Define also the functions

$$\theta(x) = \max_{|y| \le 1} |\eta|(y - x)$$

and

$$\theta_{B_K}(x) = \frac{1}{K^3} \theta\left(\frac{x-c}{K}\right),$$

for $B_K = B(c,K)$. Note that $1_{B_K} \le K^3 \theta_{B_K}$.

The next two propositions can be thought of as versions of the Uncertainty Principle. They quantify the fact that $|F|$ is essentially constant on balls of radius $K$, whenever $\widehat{F}$ is supported on a ball of radius $O(K^{-1})$. For example, a positive continuous function on a compact set is constant if its largest value matches its $L^1$ average, or alternatively, if the $L^1$ average matches the $L^q$ average of the function for some $q \in (1,\infty)$. In our context, it is natural to consider smooth averages.

**Proposition 7.16** *For each* $\alpha \in C_K$ *and each ball* $B_K$,

$$\sup_{x \in B_K} |E_\alpha f(x)| \le \int |E_\alpha f| \theta_{B_K}.$$

*Proof* Let $F_\alpha$ be the modulation of $E_\alpha f$, with Fourier transform supported on $B(0, 10K^{-1})$. Note that

$$F_\alpha = F_\alpha * \eta_K.$$

It follows that if $x \in B_K$,

$$|E_\alpha f(x)| = |F_\alpha(x)| \le \int |E_\alpha f(y) \eta_K(x - y)| \, dy \le \int |E_\alpha f(y)| \theta_{B_K}(y) \, dy.$$

$\square$

**Proposition 7.17** (Reverse Hölder's inequality) *For each* $q \ge 1$ *and* $\alpha \in C_K$,

$$\left( \int |E_\alpha f|^q \theta_{B_K} \right)^{1/q} \lesssim_K \int |E_\alpha f| w_{B_K}. \tag{7.8}$$

*Proof* Let $\mathcal{B}_K$ be a finitely overlapping cover of $\mathbb{R}^3$ with balls $B = B(c_B, K)$ of radius $K$. Due to the rapid decay of $\theta$, we may write

$$\theta_{B_K}(x) \le \sum_{B \in \mathcal{B}_K} 1_B(x) \sup_{y \in B} \theta_{B_K}(y) \lesssim \sum_{B \in \mathcal{B}_K} 1_B(x) [w_{B_K}(c_B)]^q. \tag{7.9}$$

We claim that it suffices to prove that

$$\left( \int_B |E_\alpha f|^q \right)^{1/q} \lesssim_K \int |E_\alpha f| w_B,$$

uniformly over $B \in \mathcal{B}_K$. Indeed, combining this with (7.5), (7.9), and the fact that the $l^1$ norm dominates the $l^q$ norm leads to the desired inequality.

Let $\psi_B$ be a Schwartz function with Fourier transform supported on $B(0, 1/K)$ and such that $1_B \le |\psi_B| \lesssim w_B$. Using Young's inequality, it follows that

$$\|E_\alpha f\|_{L^q(B)} \le \|\psi_B E_\alpha f\|_{L^q(\mathbb{R}^3)} \lesssim_K \|\psi_B E_\alpha f\|_{L^1(\mathbb{R}^3)} \lesssim \int |E_\alpha f| w_B.$$

$\square$

Tracking down constants reveals that $\lesssim_K$ may be replaced with $\lesssim K^{-3}$ in (7.8), which can thus be viewed as a reverse Hölder's inequality. However, this precise form of the constants will have no relevance for the forthcoming argument.

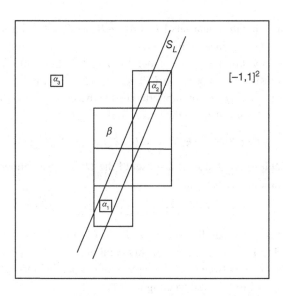

Figure 7.1 Three transverse squares $\alpha_i$ and the family $\mathcal{F}_L$.

**Remark 7.18** One can generalize the previous two propositions to show that for each function $F$ with Fourier transform supported on a rectangular box $B$, its absolute value $|F|$ is essentially constant on boxes dual to $B$. This is already apparent from the wave packet representation (2.13).

Let $\alpha \in \mathcal{C}_K$ be centered at $c_\alpha$. Our argument here only uses the fact that $|E_\alpha f|$ is essentially constant on balls of radius $K$. However, more is true. Since $\widehat{E_\alpha f}$ is supported on a box with dimensions roughly $1/K$ and $1/K^2$, $|E_\alpha f|$ is in fact essentially constant on $(K, K^2)$-tubes pointing in the direction $(-2c_\alpha, 1)$. This stronger observation was exploited in [**32**] to make further progress on the restriction problem.

For each line $L$ in $\mathbb{R}^2$ that intersects $[-1, 1]^2$, let us cover $S_L \cap [-1, 1]^2$ with a family $\mathcal{F}_L$ consisting of squares $\beta \in \mathcal{C}_{K^{1/2}}$ (see Figure 7.1).

The inequality proved in the following lemma is an example of what we will later call $l^q$ decoupling.

**Lemma 7.19** (Lower dimensional contribution) *Let $2 \le q \le 4$. For each $f$ supported on $\cup_{\beta \in \mathcal{F}_L} \beta$, we have*

$$\left( \int_{B_K} |Ef|^q \right)^{1/q} \lesssim_\epsilon K^{\frac{1}{2}(\frac{1}{2} - \frac{1}{q}) + \epsilon} \left( \sum_{\beta \in \mathcal{F}_L} \int |E_\beta f|^q w_{B_K} \right)^{1/q}.$$

*Proof* Assume that the equation of $L$ is $A\xi_1 + B\xi_2 + C = 0$, with $A^2 + B^2 = 1$. Note that $C = O(1)$, since $L$ must intersect $[-1,1]^2$. The argument follows a pattern that is now familiar to the reader. Let $\eta_{B_K}$ be a Schwartz function with the Fourier transform supported on $B(0, K^{-1})$ and satisfying $1_{B_K} \leq \eta_{B_K}$. The Fourier transform of $\eta_{B_K} Ef$ is supported inside the $O(K^{-1})$-neighborhood $\mathcal{N}$ of the $O(K^{-1/2})$-wide strip on the paraboloid

$$\{(\xi_1, \xi_2, \xi_1^2 + \xi_2^2): \text{dist}\,((\xi_1, \xi_2), L) \lesssim K^{-1/2}\}.$$

It is not difficult to see that $\mathcal{N}$ is a subset of the $O(K^{-1})$-neighborhood of the cylinder containing the parabola

$$\{(\xi_1, \xi_2, \xi_1^2 + \xi_2^2): (\xi_1, \xi_2) \in L \cap [-1,1]^2\}$$

and all lines through it parallel to $\mathbf{v} = (A, B, -2C)$. Indeed, the normal vectors $(2\xi_1, 2\xi_2, -1)$ to $\mathbb{P}^2$ at the points of this parabola are all perpendicular to $\mathbf{v}$. Moreover, the angle between $\mathbf{v}$ and the plane of the parabola is $\gtrsim 1$, uniformly over all choices of lines $L$ intersecting $[-1,1]^2$.

Combining (3.6), Corollary 5.21, and Exercise 5.25 shows that we have the following reverse square function estimate for each $2 \leq q \leq 4$:

$$\|\eta_{B_K} Ef\|_{L^q(\mathbb{R}^3)} \lesssim_\epsilon K^\epsilon \left\|\left(\sum_{\beta \in \mathcal{F}_L} |\eta_{B_K} E_\beta f|^2\right)^{\frac{1}{2}}\right\|_{L^q(\mathbb{R}^3)}.$$

Finally, it suffices to use Hölder's inequality and the fact that $(\eta_{B_K})^q \lesssim w_{B_K}$. $\qquad\square$

Note that at the heart of the previous argument was the inequality (3.6) for the parabola. This justifies referring to the contribution from $\mathcal{F}_L$ as being lower dimensional.

**Proposition 7.20** *For* $2 \leq q \leq 4$ *and* $\epsilon > 0$, *there are constants* $C_1$, $C_\epsilon$ *(independent of $K$) and $C_K$, and there is a line* $L = L(B_K)$ *for each* $B_K$ *such that*

$$\int_{B_K} |Ef|^q \leq C_1 \sum_{\alpha \in \mathcal{C}_K} \int |E_\alpha f|^q w_{B_K}$$

$$+ C_\epsilon K^{\frac{q}{4} - \frac{1}{2} + \epsilon} \sum_{\beta \in \mathcal{F}_{L(B_K)}} \int |E_\beta f|^q w_{B_K}$$

$$+ C_K \max_{\substack{\alpha_1, \alpha_2, \alpha_3 \in \mathcal{C}_K \\ K^{-2}-transverse}} \left(\prod_{i=1}^3 \int |E_{\alpha_i} f| w_{B_K}\right)^{\frac{q}{3}}.$$

*Proof* Define

$$Z_\alpha = Z_\alpha(B_K) = \left( \int |E_\alpha f|^q \theta_{B_K} \right)^{\frac{1}{q}}.$$

Proposition 7.16 shows that

$$\sup_{x \in B_K} |E_\alpha f(x)| \le Z_\alpha.$$

We will apply Lemma 7.13 with $z_\alpha = E_\alpha f(x)$, for each $x \in B_K$. If scenario (Sc1) holds, then we may write

$$\int_{B_K} |Ef|^q \le 100^q |B_K| \max_{\alpha \in \mathcal{C}_K} (Z_\alpha)^q \le C_1 \sum_{\alpha \in \mathcal{C}_K} \int |E_\alpha f|^q w_{B_K}.$$

If (Sc3) holds, then integrating the pointwise inequality and then using Proposition 7.17 leads to

$$\int_{B_K} |Ef|^q \lesssim_K \max_{\substack{\alpha_1, \alpha_2, \alpha_3 \\ K^{-2}-\text{transverse}}} \left( \prod_{i=1}^{3} \int |E_{\alpha_i} f|^q \theta_{B_K} \right)^{\frac{1}{3}}$$

$$\le C_K \max_{\substack{\alpha_1, \alpha_2, \alpha_3 \\ K^{-2}-\text{transverse}}} \left( \prod_{i=1}^{3} \int |E_{\alpha_i} f| w_{B_K} \right)^{\frac{q}{3}}.$$

It remains to analyze scenario (Sc2). Integrating again the pointwise inequality on $B_K$ leads to the initial estimate

$$\int_{B_K} |Ef|^q \lesssim \sum_{\alpha \in \mathcal{C}_K} \int |E_\alpha f|^q w_{B_K} + \int_{B_K} \left| \sum_{\alpha \subset S_L} E_\alpha f \right|^q,$$

with the implicit constant independent of $K$. Write using the triangle inequality

$$\left\| \sum_{\alpha \subset S_L} E_\alpha f \right\|_{L^q(B_K)} \le \left\| \sum_{\beta \in \mathcal{F}_L} E_\beta f \right\|_{L^q(B_K)} + \sum_{\alpha \not\subset S_L} \| E_\alpha f \|_{L^q(B_K)}.$$

Using Lemma 7.19, the first term on the right is

$$\lesssim_\epsilon K^{\frac{1}{2}(\frac{1}{2} - \frac{1}{q}) + \epsilon} \left( \sum_{\beta \in \mathcal{F}_L} \int |E_\beta f|^q w_{B_K} \right)^{\frac{1}{q}}.$$

Then

$$\sum_{\alpha \notin S_L} \|E_\alpha f\|_{L^q(B_K)} \lesssim |B_K|^{1/q} \left( \sum_{\alpha \notin S_L} Z_\alpha \right)$$

$$\leq |B_K|^{1/q} \max_{\alpha \in \mathcal{C}_K} Z_\alpha$$

$$\leq |B_K|^{1/q} \left( \sum_{\alpha \in \mathcal{C}_K} (Z_\alpha)^q \right)^{1/q}$$

$$\lesssim \left( \sum_{\alpha \in \mathcal{C}_K} \int |E_\alpha f|^q w_{B_K} \right)^{1/q}.$$

Raising to the power $q$, we conclude that subject to (Sc2), we have

$$\int_{B_K} |Ef|^q \leq C_1 \sum_{\alpha \in \mathcal{C}_K} \int |E_\alpha f|^q w_{B_K} + C_\epsilon K^{\frac{q}{4}-\frac{1}{2}+\epsilon} \sum_{\beta \in \mathcal{F}_L} \int |E_\beta f|^q w_{B_K}.$$

$\square$

We will use the following rescaling argument to estimate terms corresponding to both scale $K^{-1}$ and scale $K^{-1/2}$.

**Lemma 7.21** (Parabolic rescaling) *Let $\tau$ be a square in $[-1,1]^2$ with side length $2\delta \ll 1$. Then*

$$\int |E_\tau f|^q w_{B(0,R)} \leq \delta^{2q-4} C_{\text{lin}}(R,q) \|f\|_{L^\infty}^q.$$

*Proof* Assume that $\tau = [a_1 - \delta, a_1 + \delta] \times [a_2 - \delta, a_2 + \delta]$. We may write as before

$$|E_\tau f(x_1, x_2, x_3)| = \delta^2 |Eg(\delta(x_1 + 2a_1 x_3), \delta(x_2 + 2a_2 x_3), \delta^2 x_3)|,$$

for some $g$ with the same $L^\infty$ norm as $f$. Changing variables on the spatial side, it suffices to note that

$$w_{B(0,R)}(x_1, x_2, x_3) \leq w_{B(0,R)}(\delta(x_1 + 2a_1 x_3), \delta(x_2 + 2a_2 x_3), \delta^2 x_3)$$

if $\delta$ is small enough. Finally, use Corollary 7.15. $\square$

We next estimate the transverse contribution using the trilinear restriction estimate for the paraboloid.

**Lemma 7.22** (Transverse contribution) *For each $\epsilon > 0$, $q \geq 3$ and each $K^{-2}$-transverse $\alpha_1, \alpha_2, \alpha_3 \in \mathcal{C}_K$, we have*

$$\sum_{B_K \in \mathcal{B}_{K,R}} \left( \prod_{i=1}^3 \int |E_{\alpha_i} f| w_{B_K} \right)^{\frac{q}{3}} \lesssim_{\epsilon, K} R^\epsilon \|f\|_\infty^q.$$

*Proof* Let $w_K = w_{B(0,K)}$. Then, due to (7.6) we have

$$\int |E_{\alpha_i} f| w_{B_K} \lesssim \inf_{x \in B_K} |E_{\alpha_i} f| * w_K(x).$$

Thus, using Hölder's inequality and (3.30),

$$\left( \prod_{i=1}^{3} \int |E_{\alpha_i} f| w_{B_K} \right)^{\frac{q}{3}} \lesssim_K \int_{B_K} \prod_{i=1}^{3} (|E_{\alpha_i} f| * w_K(x))^{\frac{q}{3}} dx$$

$$\lesssim_K \int_{B_K} \prod_{i=1}^{3} \left( \int_{\mathbb{R}^3} |E_{\alpha_i} f(x + y^i)|^{\frac{q}{3}} w_K(y^i) dy^i \right) dx$$

$$= \int_{\mathbb{R}^9} \left[ \int_{B_K} \left( \prod_{i=1}^{3} |E_{\alpha_i} f_{y^i}(x)| \right)^{\frac{q}{3}} dx \right]$$

$$\times \prod_{i=1}^{3} w_K(y^i) \, dy^1 dy^2 dy^3,$$

with $f_y(\xi) = f(\xi) e(\xi \cdot \bar{y} + |\xi|^2 y_3)$ for each $y = (\bar{y}, y_3)$.

Summing over all $B_K \in \mathcal{B}_{K,R}$ and invoking (7.4) leads to the final estimate:

$$\sum_{B_K \in \mathcal{B}_{K,R}} \left( \prod_{i=1}^{3} \int |E_{\alpha_i} f| w_{B_K} \right)^{\frac{q}{3}}$$

$$\lesssim_K \int_{\mathbb{R}^9} \left[ \int_{B(0,CR)} \left| \prod_{i=1}^{3} E_{\alpha_i} f_{y^i}(x) \right|^{\frac{q}{3}} dx \right] \prod_{i=1}^{3} w_K(y^i) \, dy^1 dy^2 dy^3$$

$$\lesssim_{\epsilon,K} R^{\epsilon} \|f\|_{\infty}^{q} \int_{\mathbb{R}^9} \prod_{i=1}^{3} w_K(y^i) \, dy^1 dy^2 dy^3$$

$$\lesssim_{\epsilon,K} R^{\epsilon} \|f\|_{\infty}^{q}. \qquad \square$$

We are now in the position to prove the estimate $C_{\text{lin}}(R, q) \lesssim_{\epsilon} R^{\epsilon}$ for $q > 10/3$. We may in addition assume that $q < 4$. Let us sum the inequality in Proposition 7.20 over all balls $B_K \in \mathcal{B}_{K,R}$, using (7.7) along the way

$$\int_{B_R} |Ef|^q \leq C_2 \sum_{\alpha \in \mathcal{C}_K} \int |E_{\alpha} f|^q w_{B_R}$$

$$+ C_{2,\epsilon} K^{\frac{q}{4} - \frac{1}{2} + \epsilon} \sum_{\beta \in \mathcal{C}_{K^{\frac{1}{2}}}} \int |E_{\beta} f|^q w_{B_R}$$

$$+ C_K \sum_{\substack{\alpha_1, \alpha_2, \alpha_3 \\ K^{-2}-\text{transverse}}} \sum_{B_K \in \mathcal{B}_{K,R}} \left( \prod_{i=1}^{3} \int |E_{\alpha_i} f| w_{B_K} \right)^{\frac{q}{3}}.$$

We use Lemma 7.21 for the first two terms and Lemma 7.22 for the third one to write

$$\int_{B_R} |Ef|^q \leq C_{3,\epsilon} C_{\text{lin}}(R,q)(|\mathcal{C}_K|K^{4-2q} + |\mathcal{C}_{K^{\frac{1}{2}}}|K^{2-q}K^{\frac{q}{4}-\frac{1}{2}+\epsilon})\|f\|_\infty^q$$
$$+ C_{K,\epsilon}R^\epsilon\|f\|_\infty^q.$$

Taking the supremum over all $f \in L^\infty([-1,1]^2)$, we get

$$C_{\text{lin}}(R,q) \leq C_{3,\epsilon} C_{\text{lin}}(R,q)(K^{6-2q} + K^{\frac{5}{2}-\frac{3q}{4}+\epsilon}) + C_{K,\epsilon}R^\epsilon.$$

Note that both exponents of $K$ are negative when $q > 10/3$, if $\epsilon$ is small enough. Thus, we may choose $K$ large enough so that

$$C_{3,\epsilon}(K^{6-2q} + K^{\frac{5}{2}-\frac{3q}{4}+\epsilon}) \leq \frac{1}{2}.$$

With this choice of $K$, the main inequality becomes

$$C_{\text{lin}}(R,q) \leq \frac{1}{2}C_{\text{lin}}(R,q) + C_{K,\epsilon}R^\epsilon.$$

We conclude that $C_{\text{lin}}(R,q) \lesssim_\epsilon R^\epsilon$ for $q > 10/3$.

We remark that the only source of inefficiency in this argument is coming from the lower-dimensional contribution, which generates the term $K^{5/2-3q/4}$. This is ultimately due to the fact that the line $L(B_K)$ depends on $B_K$. In the absence of additional information about the form of this dependence, the best one can do is to use the crude inequality

$$\sum_{B_K \in \mathcal{B}_{K,R}} \sum_{\beta \in \mathcal{F}_{L(B_K)}} \int |E_\beta f|^q w_{B_K} \leq \sum_{B_K \in \mathcal{B}_{K,R}} \sum_{\beta \in \mathcal{C}_{K^{\frac{1}{2}}}} \int |E_\beta f|^q w_{B_K}.$$

**Exercise 7.23** For $\delta \ll 1$ consider three $\delta^2$-transverse squares $\alpha_1, \alpha_2, \alpha_3 \in \mathcal{C}_{\delta^{-1}}$ such that the radius of the circle determined by their centers is $\sim \delta$.

(a) Use parabolic rescaling to prove that for each $f: [-1,1]^2 \to \mathbb{C}$ and $R \gtrsim \delta^{-2}$

$$\left\|\left(\prod_{i=1}^3 E_{\alpha_i} f\right)^{\frac{1}{3}}\right\|_{L^3(B(0,R))} \lesssim_\epsilon \delta^{-\frac{1}{3}}R^\epsilon\|f\|_2. \tag{7.10}$$

(b) Combine this with the trivial estimate $R^*_{\mathbb{P}^2}(1 \mapsto \infty)$ to argue that for each $R \gtrsim 1$,

$$\left\|\left(\prod_{i=1}^3 E_{\alpha_i} f\right)^{\frac{1}{3}}\right\|_{L^q(B(0,R))} \lesssim_\epsilon \delta^{2-\frac{4}{q}}R^\epsilon\|f\|_\infty.$$

Hint for (b): Transversality is not needed in the range $R \lesssim \delta^{-2}$. The estimate

$$\|E_{\alpha_i} f\|_{L^q(B(0,R))} \lesssim \delta^{2-\frac{4}{q}} \|f\|_{\infty}$$

follows by interpolating $L^2$ and $L^{\infty}$ estimates.

Not every triple can easily be rescaled as in this exercise. The reader is invited to explore the case when $\alpha_1$ is centered at $(1, 0)$, $\alpha_2$ is centered at $(-1, 0)$ and $\alpha_3$ is at distance $\sim \delta$ from origin. A clever argument due to Ramos [95] shows that the bound $\delta^{-1/3}$ in (7.10) applies to all $\delta^2$-transverse sets, with no restriction on their sizes. This bound can be used to recover some weak linear restriction estimates, following an approach similar to the bilinear-to-linear reduction described in Chapter 4. The following exercise explores a basic version of this approach. The reader may try to further optimize the argument by exploiting orthogonality.

**Exercise 7.24** (Trilinear Whitney-type decomposition) Let

$$S = \{(\xi_1, \xi_2, \xi_3, \xi_4, \xi_5, \xi_6) \in [-1, 1]^6 : (\xi_1, \xi_2), (\xi_3, \xi_4), (\xi_5, \xi_6) \text{ are collinear}\}.$$

(a) Argue as in the proof of Corollary 4.4 that there is a collection $\mathcal{Q}$ of pairwise disjoint cubes $\alpha$ with the following properties:

(P1) $\text{dist}(\alpha, S) \sim l(\alpha)$ and

$$[-1, 1]^6 \setminus S = \bigcup_{\alpha \in \mathcal{Q}} \alpha.$$

(P2) If $\mathcal{Q}_{\delta}$ are the cubes in $\mathcal{Q}$ with side length $\delta$, then $|\mathcal{Q}_{\delta}| \lesssim \delta^{-5}$.

(P3) If $\alpha = \alpha_1 \times \alpha_2 \times \alpha_3 \in \mathcal{Q}_{\delta}$, then $\alpha_1, \alpha_2, \alpha_3$ are (at least) $\delta^2$-transverse.

(b) Use the result of Ramos to argue that for each $\alpha = \alpha_1 \times \alpha_2 \times \alpha_3 \in \mathcal{Q}_{\delta}$ and $q \geq 3$,

$$\left\| \left( \prod_{i=1}^{3} E_{\alpha_i} f \right)^{\frac{1}{3}} \right\|_{L^q(B(0,R))} \lesssim_{\epsilon} \delta^{2-\frac{4}{q}} R^{\epsilon} \|f\|_{\infty}.$$

(c) Conclude using the representation

$$(Ef)^3 = \sum_{\alpha_1 \times \alpha_2 \times \alpha_3 \in \mathcal{Q}} E_{\alpha_1} f E_{\alpha_2} f E_{\alpha_3} f$$

that the linear restriction estimate $R^*_{\mathbb{P}^2}(q \mapsto \infty; \epsilon)$ holds for each $q > 12$.

The next exercise is a follow-up to Exercises 5.30 and 6.30.

**Exercise 7.25** We use the notation from Section 5.2.

(i) Let $\Omega_1, \Omega_2$ be two pairwise disjoint cubes in $[-1,1]^{n-1}$. Prove that if $\widehat{F}$ is supported on $\mathcal{N}_{\Omega_1}(R^{-1}) \cup \mathcal{N}_{\Omega_2}(R^{-1})$ and if $F_i$ is the Fourier projection of $F$ onto $\mathcal{N}_{\Omega_i}(R^{-1})$, then for each ball $B_{R^{1/2}}$ with radius $R^{1/2}$ and each $p \geq 2(n+2)/n$ we have

$$\left\| |F_1 F_2|^{1/2} \right\|_{L^p(B_{R^{1/2}})} \lesssim_\epsilon R^{\frac{n-1}{4} - \frac{n}{2p} + \epsilon} \left\| \left( \sum_{\theta \in \Theta(R^{-1})} |\mathcal{P}_\theta F|^2 \right)^{\frac{1}{2}} \right\|_{L^p(w_{B_{R^{1/2}}})}.$$

(ii) Use the Bourgain–Guth method to conclude that if $p \geq 2(n+2)/n$ and $\widehat{F}$ is supported on $\mathcal{N}(R^{-1})$, then

$$\|F\|_{L^p(\mathbb{R}^n)} \lesssim_\epsilon R^{\frac{n-1}{4} - \frac{n}{2p} + \epsilon} \left\| \left( \sum_{\theta \in \Theta(R^{-1})} |\mathcal{P}_\theta F|^2 \right)^{\frac{1}{2}} \right\|_{L^p(\mathbb{R}^n)}.$$

**Exercise 7.26** Consider the subset $\mathbb{Cone}^{n-1}_{\mathrm{comp}}$ of the full cone

$$\mathbb{Cone}^{n-1}_* = \left\{ (\xi_1, \ldots, \xi_n) \colon \xi_n = \frac{\xi_1^2 + \cdots + \xi_{n-2}^2}{2\xi_{n-1}} \right\}$$

corresponding to $(\xi_1, \ldots, \xi_{n-2}) \in [-1,1]^{n-2}$ and $1 \leq \xi_{n-1} \leq 2$. Partition $[-1,1]^{n-2}$ into cubes $\alpha \in \mathcal{C}_{\delta^{-1}}$ with side length $\delta$, and let

$$\mathbb{Cone}^{n-1}_{\mathrm{comp}}(\alpha) = \{ (\xi, \xi_{n-1}, \xi_n) \in \mathbb{Cone}^{n-1}_{\mathrm{comp}} \colon \xi \in \alpha \}.$$

(a) Prove that for each pairwise disjoint $\alpha_1, \alpha_2, \alpha_3 \in \mathcal{C}_{\delta^{-1}}$ and each unit normal vectors $\mathbf{n}_i$ to $\mathbb{Cone}^{n-1}_{\mathrm{comp}}(\alpha_i)$, we have

$$|\mathbf{n}_1 \wedge \mathbf{n}_2 \wedge \mathbf{n}_3| > \nu_\delta,$$

for some $\nu_\delta > 0$.

(b) Use the arguments in this chapter to prove that the case $n = 5, k = 3$ of Conjecture 6.1 for $\mathbb{Cone}^4$ (see the discussion following the statement of the conjecture) implies the full range of estimates $R^*_{\mathbb{Cone}^4}(\infty \mapsto p)$, $p > 8/3$ (cf. Conjecture 1.11).

(c) Use the same kind of arguments, combined this time with the known trilinear restriction inequality for the cone $\mathbb{Cone}^2$ (Theorem 6.24), to give a different proof for the estimate $R^*_{\mathbb{Cone}^2}(\infty \mapsto p)$ in the full range $p > 4$.

# 8

## The Polynomial Method

The proof of the finite field Kakeya conjecture by Dvir [46] sparked a revolution in many areas of discrete mathematics. It introduced a robust way of counting structures (e.g., special points, lines) with the aid of polynomials. An important subsequent development came through the work [61] of Guth and Katz, which solved the Erdös distance conjecture. In that paper, the Euclidean space $\mathbb{R}^3$ is partitioned into cells determined by the zero set of a suitable polynomial. A typical cell contains a proper subset of the structures that are being counted, opening the door to an inductive approach. A critical feature of the Euclidean space exploited by Guth and Katz is its topology, which when combined with the fundamental theorem of algebra limits the number of cells that a given line can enter.

One notable feature of affine subspaces, such as points and lines, is their lack of scale. Can polynomials be used to count cubes and tubes? How about to estimate integrals involving complicated expressions? A very satisfactory answer came in the form of the resolution of the endpoint multilinear Kakeya conjecture by Guth [60].

Perhaps even more surprising is the fact that Guth also managed to tailor the polynomial method to produce significant progress on the Restriction Conjecture, a highly oscillatory problem that is not a priori formulated as a counting problem. His approach can be summarized as follows. Use a polynomial $P$ of small degree to partition the spatial ball $B(0, R)$ into cells where $|Ef|^p$ has roughly the same mass. On $B(0, R)$, $Ef$ is the sum of wave packets whose mass is concentrated inside tubes. The key is to understand how the various wave packets interact with the cells. The contribution from the tubes that lie roughly tangent to the zero set $Z(P)$ of $P$ is estimated directly, using both counting arguments for tubes and the oscillatory properties of $Ef$. The contribution from the other tubes is estimated using a well-crafted induction hypothesis, exploiting only the algebraic properties of the cells and the $L^2$ orthogonality of the wave packet decomposition.

In the first two sections, we will introduce the polynomial partitioning and explore some of its applications in the discrete setting. We will only emphasize the ones that have continuous analogs relevant to the overall scope of our investigation. Guth's book [58] offers a comprehensive treatment of the polynomial method, and a unique perspective on how it has expanded over the last 10 years to reach a hall of fame status. We will close the chapter by explaining the details of Guth's argument from [57], which uses polynomial partitioning in order to advance our understanding of the restriction problem.

# 8.1 Polynomial Partitioning

At the heart of the method we are about to describe lies the idea of simultaneously bisecting $N$ "masses" in $\mathbb{R}^n$ using a polynomial of low degree. This is possible due to a combination of topological reasons and the richness of the family of polynomials of a fixed degree. The advantage of realizing the bisection with a small degree polynomial $P$ will be exploited through the fact that any line can intersect $Z(P)$ only a small number of times.

The starting point in our presentation is the following classical result from algebraic topology.

**Theorem 8.1** (Borsuk–Ulam) *If $F: \mathbb{S}^N \to \mathbb{R}^N$ is a continuous function such that $F(\mathbf{v}) = -F(-\mathbf{v})$ for each $\mathbf{v} = (v_1, \ldots, v_{N+1}) \in \mathbb{S}^N$, then there exists $\mathbf{v} \in \mathbb{S}^N$ such that $F(\mathbf{v}) = 0$.*

This result can be used to give an elegant proof of the following "Ham Sandwich" Theorem due to Stone and Tukey [104]. We will denote by $Z(f)$ the zero set of a function $f$.

**Theorem 8.2** *Let $\mathcal{F}$ be a (real) vector space with dimension $\dim(\mathcal{F})$, consisting of real-valued continuous functions on $\mathbb{R}^n$. We denote by $f_0$ the trivial function which equals zero everywhere. For $N < \dim(\mathcal{F})$ let $\mu_1, \ldots, \mu_N$ be finite Borel measures on $\mathbb{R}^n$, such that for each $f \in \mathcal{F} \setminus \{f_0\}$ and each $j$ we have $\mu_j(Z(f)) = 0$.*

*Then there exists $f \in \mathcal{F} \setminus \{f_0\}$ such that for each $1 \leq j \leq N$ we have*

$$\mu_j(\{x \in \mathbb{R}^n : f(x) > 0\}) = \mu_j(\{x \in \mathbb{R}^n : f(x) < 0\}).$$

*Proof* Let $\mathcal{F}'$ be some $N + 1$ dimensional subspace of $\mathcal{F}$, with basis $f_1, \ldots, f_{N+1}$. We identify $\mathbb{R}^{N+1}$ with $\mathcal{F}'$ via the map

$$\mathbf{v} = (v_1, \ldots, v_{N+1}) \mapsto f^{\mathbf{v}} = \sum_{j=1}^{N+1} v_j f_j.$$

We define the function $F: \mathbb{S}^N \to \mathbb{R}^N$ with components

$$F_j(\mathbf{v}) = \mu_j(\{x \in \mathbb{R}^n : f^{\mathbf{v}}(x) > 0\}) - \mu_j(\{x \in \mathbb{R}^n : f^{\mathbf{v}}(x) < 0\}).$$

Any zero $\mathbf{v}$ of $F$ will produce the desired function $f = f^{\mathbf{v}}$. Let us verify the assumptions of the Borsuk–Ulam theorem for $F$. It is clear that $F(\mathbf{v}) = -F(-\mathbf{v})$.

Recall that the sign $\mathrm{sgn}(r)$ of a real number $r$ is the obvious element of $\{1, -1, 0\}$. To check the continuity of $F$, it suffices to prove that for each sequence $\mathbf{v}^k$ in $\mathbb{S}^N$ converging to $\mathbf{v}$ and each $j$, we have $\lim_{k \to \infty} \mu_j(A_k) = 0$, where

$$A_k = \{x \in \mathbb{R}^n : \mathrm{sgn}(f^{\mathbf{v}^k}(x)) \neq \mathrm{sgn}(f^{\mathbf{v}}(x))\}.$$

Since $f^{\mathbf{v}^k}$ converges pointwise to $f^{\mathbf{v}}$, we have

$$\cap_{K \geq 1}(\cup_{k \geq K} A_k) \subset Z(f^{\mathbf{v}}).$$

Using the continuity of $\mu_j$ and our hypothesis $\mu_j(Z(f^{\mathbf{v}})) = 0$, it follows that $\mu_j(A_k) \to 0$, as desired. $\qquad\square$

We will apply the Ham Sandwich Theorem to $\mathcal{F} = \mathrm{Poly}_D(\mathbb{R}^n)$, the collection of all polynomials on $\mathbb{R}^n$ with real coefficients and degree at most $D$. We need to make two observations. First, a simple counting argument reveals that this vector space has a fairly large dimension:

$$\dim(\mathrm{Poly}_D(\mathbb{R}^n)) = \binom{D+n}{n} \sim_n D^n. \tag{8.1}$$

Also, the zero set $Z(P)$ of any nontrivial polynomial $P$ has zero Lebesgue measure, as can be seen using a simple induction argument.

**Corollary 8.3** (Polynomial Ham Sandwich) *Let $\mu_1, \ldots, \mu_N$ be finite Borel measures on $\mathbb{R}^n$ that are absolutely continuous with respect to the Lebesgue measure. There is a nontrivial polynomial $P \in \mathrm{Poly}_D(\mathbb{R}^n)$ with $D \lesssim_n N^{1/n}$ such that for each $1 \leq j \leq N$ we have*

$$\mu_j(\{x \in \mathbb{R}^n : P(x) > 0\}) = \mu_j(\{x \in \mathbb{R}^n : P(x) < 0\}).$$

We may iterate this corollary to split a fixed mass into many equal pieces.

**Theorem 8.4** (Polynomial partitioning, continuous version, [57]) *Let* $W \in L^1(\mathbb{R}^n)$ *be a positive function. Then for each degree* $D$, *there is a nontrivial polynomial* $P \in \text{Poly}_D(\mathbb{R}^n)$ *such that* $\mathbb{R}^n \setminus Z(P)$ *is the union of* $\sim_n D^n$ *pairwise disjoint open sets* $O_i$ *(cells), such that all integrals* $\int_{O_i} W$ *are equal and, moreover, each line in* $\mathbb{R}^n$ *can intersect at most* $D + 1$ *cells* $O_i$.

*Proof* We first apply the previous corollary with $N = 1$ and $\mu_1 = W \, dx$ to find some polynomial $P_1$ of degree $D_1 = O_n(1)$ such that

$$\int_{P_1 > 0} W = \int_{P_1 < 0} W.$$

Next, we apply Corollary 8.3 with $N = 2$ and $\mu_1 = (W 1_{P_1 > 0}) \, dx$, $\mu_2 = (W 1_{P_1 < 0}) \, dx$ to get a polynomial $P_2$ of degree $D_2 = O_n(1)$ such that for each $s_1, s_2 \in \{1, -1\}$

$$\int_{\substack{\text{sgn}(P_1) = s_1 \\ \text{sgn}(P_2) = s_2}} W = 2^{-2} \int_{\mathbb{R}^n} W.$$

Repeating this $m$ times, we find a polynomial $P_m$ of degree $O_n(2^{m/n})$ such that for each $s_1, \ldots, s_m \in \{1, -1\}$

$$\int_{\substack{\text{sgn}(P_1) = s_1 \\ \vdots \\ \text{sgn}(P_m) = s_m}} W = 2^{-m} \int_{\mathbb{R}^n} W.$$

Let $P = \prod_{j=1}^{m} P_j$. Its degree is at most $C_n 2^{m/n}$. Choose $m$ such that $D/2 \le C_n 2^{m/n} \le D$. The set $\mathbb{R}^n \setminus Z(P)$ is partitioned into exactly $2^m$ open cells $O_i$ of the form

$$\{x \in \mathbb{R}^n : \text{sgn}(P_1(x)) = s_1, \ldots, \text{sgn}(P_m(x)) = s_m\}.$$

It is worth observing that each cell $O_i$ is the union of connected components of $\mathbb{R}^n \setminus Z(P)$.

Finally, let us observe that any hypothetical line $L = x_0 + \mathbb{R}v$ intersecting $D + 2$ of the cells $O_i$ would have to cross the boundary $Z(P)$ at least $D + 1$ times. We can reinterpret this as saying that the polynomial of one variable $t \mapsto P(x_0 + tv)$ has at least $D + 1$ distinct roots. Since its degree is at most $D$, this would force $P$ to be zero along the whole $L$, contradicting the fact that $L$ intersects at least one cell $O_i$. □

**Example 8.5** Let $W = 1_{[0,1]^n}$. The polynomial

$$P(x_1, \ldots, x_n) = \prod_{j=1}^{n} \prod_{m=1}^{M-1} \left( x_j - \frac{m}{M} \right)$$

has degree $D = n(M - 1)$ and $\mathbb{R}^n \setminus Z(P)$ has $M^n$ connected components $O_i$ satisfying $\int_{O_i} W = M^{-n}$.

**Definition 8.6** A point $x$ is called nonsingular for $P$ if $\nabla P(x) \neq 0$. The polynomial $P$ is called nonsingular if each point in its zero set is nonsingular for $P$.

The neighborhood of $Z(P)$ close enough to a nonsingular point is a smooth hypersurface.

**Remark 8.7** In most applications, it suffices for the integrals $\int_{O_i} W$ to be comparable, rather than equal to each other. Using a density argument, this can be achieved by using only nonsingular polynomials $P_i$ in the proof of Theorem 8.4 (cf. [57]). A product of nonsingular polynomials has the property that its nonsingular points are dense in its zero set. This additional regularity makes some of the arguments easier.

Note that the topology of $\mathbb{R}^n$ is used twice in the proof of the previous theorem. First, it is implicitly used to create the cells with equal mass, via the Borsuk–Ulam theorem. Second, it is used to argue that a line intersecting two cells must also intersect the (topological) boundary of each cell.

## 8.2 A Discrete Application of Polynomial Partitioning

Theorem 8.4 has a discrete version, proved by Katz and Guth in [61]. Its proof is very similar to its continuous version, so we omit it.

**Theorem 8.8** (Polynomial partitioning, discrete version) *Let $X$ be a finite set in $\mathbb{R}^n$. For any $D$, there is a nontrivial $P \in \text{Poly}_D(\mathbb{R}^n)$ such that $\mathbb{R}^n \setminus Z(P)$ has $\leq C_n D^n$ connected components (cells), each containing at most $C_n D^{-n}|X|$ points of $X$.*

The main difference between this result and its continuous version is that we allow some of the points in $X$ (possibly all of them) to lie inside $Z(P)$.

In preparation for the main application of polynomial partitioning in this chapter, we illustrate the applicability of the method in the discrete setting.

**Definition 8.9** (2-rich points) For a family $\mathcal{L}$ of lines in $\mathbb{R}^3$, we denote by $\mathcal{P}_2(\mathcal{L})$ the collection of 2-rich points in $\mathbb{R}^3$, those contained in at least two lines from $\mathcal{L}$.

It is easy to find families $\mathcal{L}$ for which $|\mathcal{P}_2(\mathcal{L})|$ is comparable to the trivial upper bound $|\mathcal{L}|^2$. Perhaps the easiest example is when all lines lie inside a plane. Another example consists of $|\mathcal{L}|/2$ lines of the form $z = xy_i$, $y = y_i$ ($1 \le i \le L/2$) together with $|\mathcal{L}|/2$ lines of the form $z = x_iy$, $x = x_i$ ($1 \le i \le L/2$) inside the doubly ruled hyperbolic paraboloid $z = xy$.

One of the questions in incidence geometry is to determine the maximum size of $\mathcal{P}_2(\mathcal{L})$ subject to various constraints. In both previous examples, the lines giving rise to a large collection of 2-rich points belong to the zero set of a polynomial with degree at most 2. The main result in this section illustrates the fact that if $\mathcal{L}$ is not allowed to have too many lines inside zero sets of low-degree polynomials, then $|\mathcal{P}_2(\mathcal{L})|$ is forced to get significantly smaller than $|\mathcal{L}|^2$.

**Theorem 8.10** ([57]) *For each $\epsilon > 0$, there is $D = D(\epsilon)$ such that for any collection $\mathcal{L}$ of $L$ lines in $\mathbb{R}^3$ with at most $S$ of them lying in the zero set of any nontrivial polynomial $P \in \text{Poly}_D(\mathbb{R}^3)$, we have*

$$|\mathcal{P}_2(\mathcal{L})| \le C(\epsilon, S)L^{\frac{3}{2}+\epsilon}. \tag{8.2}$$

*The constant $C(\epsilon, S)$ is independent of $L$.*

*Proof* Let $C_3$ be the constant from the statement of Theorem 8.8. For each $\epsilon > 0$, let $D = D(\epsilon)$ be large enough so that

$$4C_3(8C_3)^{\frac{3}{2}+\epsilon} \le D^{2\epsilon}.$$

Define also

$$C(\epsilon, S) = 8S^2 D(\epsilon).$$

We will prove (8.2) using induction on $L$. Note first that (8.2) is immediate if $L = 2$. Assume now that we have verified (8.2) for all families containing fewer than $L \ge 3$ lines.

Fix $\epsilon > 0$ and let $D = D(\epsilon)$ be as before. Also, fix a family $\mathcal{L}$ with $L$ lines, with at most $S$ of them lying in the zero set of any nontrivial polynomial $P \in \text{Poly}_D(\mathbb{R}^3)$.

Let $P \in \text{Poly}_D(\mathbb{R}^3)$ be a polynomial as in Theorem 8.8, relative to $X = \mathcal{P}_2(\mathcal{L})$. Call $O_i$ the corresponding cells. Note that there are two types of lines in $\mathcal{L}$. The ones contained in $Z(P)$ will be referred to as *tangent* lines. The other ones will be called *transverse* and can intersect $Z(P)$ at most $D$ times. We distinguish two scenarios.

First, let us assume that at least half of the points from $\mathcal{P}_2(\mathcal{L})$ lie in the union of the cells. Then among all cells, at least $D^3/(4C_3)$ of them contain at least $|\mathcal{P}_2(\mathcal{L})|/(4C_3D^3)$ 2-rich points each. Let us restrict attention to these cells, and call $\mathcal{L}_{cell}$ the collection of lines intersecting at least one of these cells. Since each line in $\mathcal{L}_{cell}$ is transverse, it can intersect at most $D + 1$ of these cells. Thus, a double counting argument shows that one of these cells, call it $O_{i_0}$, will be intersected by at most $8LC_3/D^2$ lines. The induction hypothesis shows that there are at most $C(\epsilon, S)\big(8LC_3/D^2\big)^{3/2+\epsilon}$ 2-rich points in $O_{i_0}$. Combining this upper bound with the earlier lower bound shows that

$$\frac{|\mathcal{P}_2(\mathcal{L})|}{4C_3D^3} \leq C(\epsilon, S)\left(\frac{8LC_3}{D^2}\right)^{\frac{3}{2}+\epsilon}.$$

Thus (8.2) will follow because of our choice of $D$.

The second scenario is when $Z(P)$ contains more than half of the 2-rich points. In this case, we estimate the cardinality of $\mathcal{P}_2(\mathcal{L})$ directly. We split the 2-rich points inside $Z(P)$ into two categories: the ones contained inside some transverse line and the ones that are only contained inside tangent lines. Since each transverse line intersects $Z(P)$ at most $D$ times, there are at most $DL$ 2-rich points of the first type. Also, it is clear that there can be at most $S^2$ 2-rich points of the second type. To conclude, there can be at most $2(DL + S^2)$ 2-rich points. Thus (8.2) is again satisfied, because of our choice of $C(\epsilon, S)$.  □

It turns out that the restriction on the maximum number of lines inside a variety can be removed if instead of counting 2-rich points we count 3-rich points that lie at the intersection of three noncoplanar (transverse) lines. Such points are particular instances of what we called *joints* in Section 6.2. Exercise 6.14 explored the sharp upper bound on the cardinality of joints under the special hypothesis that the lines have a fixed direction within each family. One of the consequences of the multilinear Kakeya inequalities from [13] is the fact that the estimate remains true if the families of lines are quantitatively transverse.

The next exercise guides the reader through a proof of the optimal bound in all dimensions, without any use of quantitative transversality. The result is due to Guth and Katz [62] in $\mathbb{R}^3$ and to Kaplan et al. [70] in higher dimensions. The very elegant and elementary argument we present here is due to Quilodrán [94].

**Exercise 8.11** (Joints, revisited) Given a family $\mathcal{L}$ of lines in $\mathbb{R}^n$, we define a joint to be a point lying at the intersection of $n$ transverse lines from $\mathcal{L}$.

(a) Prove that for any finite set $S \subset \mathbb{R}^n$ there is a nontrivial polynomial $P \in \mathrm{Poly}_D(\mathbb{R}^n)$ with degree $\leq n|S|^{1/n}$, that vanishes on $S$.
(b) Prove that if a polynomial $P$ vanishes on $n$ transverse lines that meet at $x$, then $\nabla P(x) = 0$.

(c) Prove that if $\mathcal{L}$ determines $J$ joints, then at least one of the lines in $\mathcal{L}$ contains at most $nJ^{1/n}$ of these joints.

(d) Use (c) to conclude that $\mathcal{L}$ can generate at most $(n|\mathcal{L}|)^{n/(n-1)}$ joints. The exponent $n/(n-1)$ is sharp.

Hint for (a): Prove first that there is a nontrivial polynomial in $\mathrm{Poly}_D(\mathbb{R}^n)$ that vanishes on $S$, as long as $\dim_{\mathbb{R}}(\mathrm{Poly}_D(\mathbb{R}^n)) > |S|$. Then use (8.1).

Hint for (c): Let $P$ be a nontrivial polynomial with smallest possible degree that vanishes on all the joints. Prove that if each line in $\mathcal{L}$ contains more than $nJ^{1/n}$ joints, then $P$ is forced to be trivial.

## 8.3 A New Linear Restriction Estimate

Earlier in the book, we used the bilinear restriction theorem to derive the linear restriction estimate $R^*_{\mathbb{P}^2}(\infty \mapsto p)$ for $p > 10/3$. In the remainder of this chapter, we will explain how polynomial partitioning can be used to improve this range. More precisely, we will seek to understand the following result due to Guth [57]. Let $E$ denote the extension operator for $\mathbb{P}^2$.

**Theorem 8.12** *For each $R \geq 1$, $\epsilon > 0$ and $f : [-1,1]^2 \to \mathbb{C}$, we have*

$$\int_{B(0,R)} |Ef|^{3.25} \lesssim_\epsilon R^\epsilon \|f\|_\infty^{3.25}.$$

The proof will have a subtle bilinear flavor. It will make use of the following operator

$$\mathfrak{B}_{K,A} f(x) = \min_{\tau_1,\dots,\tau_A \in \mathcal{C}_K} \max_{\substack{\tau \in \mathcal{C}_K \\ \tau \not\sim \tau_1,\dots,\tau_A}} |E_\tau f(x)|,$$

defined for all $K \in 2^{\mathbb{N}}$ and $A \geq 1$. As before, $\mathcal{C}_K$ will refer to all dyadic squares inside $[-1,1]^2$ with side length $2/K$, and the symbol $\sim$ will denote neighbors. All $\tau, \tau_i$ encountered in this chapter will be elements from $\mathcal{C}_K$. In the definition of $\mathfrak{B}_{K,A}$, we do not insist that the $\tau_i$ are distinct, as this is anyway irrelevant to value of the operator.

Note that $\mathfrak{B}_{K,A} f(x) \leq \mathfrak{B}_{K,A'} f(x)$ if $A \geq A'$, and $\mathfrak{B}_{K,A} f(x) = 0$ if $A \geq K^2$. The following two lemmas will help us compare the operator $\mathfrak{B}_{K,A}$ with more familiar quantities.

**Lemma 8.13** *For each $A \geq 1$, we have*

$$\mathfrak{B}_{K,A} f(x) \leq \max_{\tau_1 \not\sim \tau_2} |E_{\tau_1} f(x) E_{\tau_2} f(x)|^{\frac{1}{2}},$$

*with the maximum taken over all pairs of nonadjacent squares in $\mathcal{C}_K$.*

*Proof* Pick $\tau_1$ such that $|E_{\tau_1} f(x)| = \max_\tau |E_\tau f(x)|$. Pick $\tau_2 \not\sim \tau_1$ with largest possible value for $|E_{\tau_2} f(x)|$. It suffices to note that

$$\mathfrak{B}_{K,1} f(x) \leq |E_{\tau_2} f(x)| \leq |E_{\tau_1} f(x) E_{\tau_2} f(x)|^{\frac{1}{2}}. \qquad \square$$

The next result is rather immediate.

**Lemma 8.14** *If* $\tau_1, \ldots, \tau_{9A+1}$ *are distinct squares in* $\mathcal{C}_K$, *then*

$$\min_{1 \leq i \leq 9A+1} |E_{\tau_i} f(x)| \leq \mathfrak{B}_{K,A} f(x).$$

The first lemma tells us that $\mathfrak{B}_{K,A}$ can be controlled using the more familiar two-term geometric averages. This observation will prove helpful later when we will use the fact that Tao's bilinear restriction estimate is applicable to $\mathfrak{B}_{K,A}$. But what is it that makes the operator $\mathfrak{B}_{K,A} f$ more useful than $|E_{\tau_1} f E_{\tau_2} f|^{1/2}$? The answer is rather subtle and will only become apparent in the heart of the main argument. Here is a brief explanation. The function $f$ undergoes a critical decomposition $f = f_{\text{tang}} + f_{\text{trans}}$, following the application of the polynomial method. Anticipating more precise (but also slightly technical) definitions, $E f_{\text{tang}}$ and $E f_{\text{trans}}$ will correspond to the parts of $E f$ supported on tangent and transverse tubes, respectively, relative to the zero set of a certain polynomial. It turns out that one can estimate both $|E_{\tau_1} f_{\text{tang}} E_{\tau_2} f_{\text{tang}}|^{1/2}$ and $|E_{\tau_1} f_{\text{trans}} E_{\tau_2} f_{\text{trans}}|^{1/2}$, but there is no easy way to estimate the mixed term $|E_{\tau_1} f_{\text{tang}} E_{\tau_2} f_{\text{trans}}|^{1/2}$. On the other hand, the mixed-term estimate will not be needed if we work with $\mathfrak{B}_{K,A}$.

A critical observation is that the operator $\mathfrak{B}_{K,A}$ has an intrinsically (quasi-) sublinear (rather than bi-sublinear) nature. More precisely, while $\mathfrak{B}_{K,A}$ is not genuinely sublinear, once we allow the parameter $A$ to be flexible, the following related inequality will hold.

**Lemma 8.15** *For each* $A_1, A_2$ *and each* $f_1, f_2$, *we have*

$$\mathfrak{B}_{K,A_1+A_2}(f_1 + f_2)(x) \leq \mathfrak{B}_{K,A_1} f_1(x) + \mathfrak{B}_{K,A_2} f_2(x).$$

*Proof* It suffices to note that for all (not necessarily distinct) $\tau_1, \ldots, \tau_{A_1+A_2}$, we have

$$\max_{\substack{\tau \not\sim \tau_i \\ 1 \leq i \leq A_1+A_2}} |E_\tau(f_1 + f_2)(x)| \leq \max_{\substack{\tau \not\sim \tau_i \\ 1 \leq i \leq A_1}} |E_\tau f_1(x)| + \max_{\substack{\tau \not\sim \tau_i \\ A_1+1 \leq i \leq A_1+A_2}} |E_\tau f_2(x)|.$$

$$\square$$

The sole purpose of working with different values of $A$ is to guarantee this type of quasi-sublinearity. The value of $A$ will change only slightly throughout the argument, and this will not cause any harm.

We will also need the following related inequality.

**Lemma 8.16** *For each $A \geq 1$,*

$$|\mathfrak{B}_{K,A}(f)(x) - \mathfrak{B}_{K,A}(g)(x)| \leq \max_{\tau \in \mathcal{C}_K} |E_\tau f(x) - E_\tau g(x)|.$$

*Proof* Let $\tau_1, \ldots, \tau_A$ be such that

$$\mathfrak{B}_{K,A}(g)(x) = \max_{\substack{\tau \in \mathcal{C}_K \\ \tau \not\sim \tau_1, \ldots, \tau_A}} |E_\tau g(x)|.$$

Then for each $\tau' \not\sim \tau_1, \ldots, \tau_A$, we have

$$|E_{\tau'} f(x)| \leq |E_{\tau'} g(x)| + \max_{\tau \in \mathcal{C}_K} |E_\tau f(x) - E_\tau g(x)|$$

$$\leq \mathfrak{B}_{K,A}(g)(x) + \max_{\tau \in \mathcal{C}_K} |E_\tau f(x) - E_\tau g(x)|.$$

This implies the desired inequality

$$\mathfrak{B}_{K,A}(f)(x) - \mathfrak{B}_{K,A}(g)(x) \leq \max_{\tau \in \mathcal{C}_K} |E_\tau f(x) - E_\tau g(x)|.$$

$\square$

We will prove that $\mathfrak{B}_{K,A} f$ satisfies the following inequality.

**Theorem 8.17** *For each $R \geq 1$, $K \in 2^{\mathbb{N}}$, $\epsilon > 0$ and $f : [-1,1]^2 \to \mathbb{C}$, we have*

$$\int_{B(0,R)} (\mathfrak{B}_{K,A} f)^{3.25} \lesssim_{\epsilon,K} R^\epsilon \|f\|_\infty^{3.25},$$

*whenever $A \geq \epsilon^{-6}$.*

Let us now see why Theorem 8.17 implies Theorem 8.12. The proof involves a variant of the Bourgain–Guth method from the previous chapter.

**Lemma 8.18** *For each $x$, $K$, and $A$, we have*

$$|Ef(x)| \leq K^4 \mathfrak{B}_{K,A} f(x) + (9A + 1) \max_{\tau \in \mathcal{C}_K} |E_\tau f(x)|.$$

*Proof* Define

$$S_{\text{big}}(x) = \left\{ \tau \in \mathcal{C}_K : |E_\tau f(x)| \geq K^{-2} \max_{\tau' \in \mathcal{C}_K} |E_{\tau'} f(x)| \right\}.$$

We distinguish two possibilities. First, if $S_{\text{big}}(x)$ contains at least $9A + 1$ squares, then using Lemma 8.14 we may write

$$|Ef(x)| \leq K^2 \max_{\tau' \in \mathcal{C}_K} |E_{\tau'} f(x)| \leq K^4 \min_{\tau \in S_{\text{big}}(x)} |E_\tau f(x)| \leq K^4 \mathfrak{B}_{K,A} f(x).$$

On the other hand, if $S_{\text{big}}(x)$ contains at most $9A$ squares, then the triangle inequality implies that

$$|Ef(x)| \leq \sum_{\tau \in S_{\text{big}}(x)} |E_\tau f(x)| + \sum_{\tau \notin S_{\text{big}}(x)} |E_\tau f(x)|$$

$$\leq (9A + 1) \max_{\tau \in \mathcal{C}_K} |E_\tau f(x)|. \qquad \square$$

Fix now $\epsilon > 0$ and choose $\epsilon^{-6} < A \leq 2\epsilon^{-6}$. Let $K$ be large enough, to be chosen momentarily. Lemma 8.18 implies that

$$\int_{B(0,R)} |Ef|^{3.25} \leq CK^C \int_{B(0,R)} |\mathfrak{B}_{K,A} f|^{3.25} + CA^{3.25} \sum_{\tau \in \mathcal{C}_K} \int_{B(0,R)} |E_\tau f|^{3.25}.$$

Let

$$C_{\text{lin}}(R) = \sup \left\{ \int_{B(0,R)} |Ef|^{3.25} : \|f\|_{L^\infty([-1,1]^2)} \leq 1 \right\}.$$

Combining the previous inequality with Theorem 8.17 and parabolic rescaling shows that for some $C = O(1)$, we have

$$C_{\text{lin}}(R) \leq C_{\epsilon, K} R^\epsilon + CA^{3.25} K^{-\frac{1}{2}} C_{\text{lin}}(R).$$

Choose now $K$ large enough (depending only on $\epsilon$) so that $CA^{3.25} K^{-1/2} \leq 1/2$. We conclude that $C_{\text{lin}}(R) \lesssim_\epsilon R^\epsilon$, as desired.

To gauge the strength of Theorem 8.12, let us observe that it implies something nontrivial about the Kakeya set conjecture. More precisely, using (5.6), we see that Kakeya sets in $\mathbb{R}^3$ must have Hausdorff dimension at least $11/5$. We thus anticipate that the forthcoming argument will use more delicate information about tubes than was actually needed in the proof of the bilinear restriction estimate in Theorem 3.1. This information will come in the form of a sharp upper bound for the number of directions of tubes tangent to a variety.

## 8.4 The Induction Hypothesis

One of the difficulties in proving Theorem 8.17 directly is the use of $L^\infty$ norm for $f$, which is virtually insensitive to orthogonality. One of the innovations of

Guth's argument was to make an elegant reduction to $L^2$-based quantities, by proving a slightly stronger result.

We will extend our notation $C_r$ to the case when $r$ is not necessarily an integer, to denote a family of finitely overlapping squares of side length $2/r$ that cover $[-1,1]^2$.

**Theorem 8.19** *There are constants $C_0, \epsilon_0 > 0$ such that the following property $P(\epsilon)$ holds for each $K \in 2^\mathbb{N}$ and each $0 < \epsilon \le \epsilon_0$:*

$P(\epsilon)$: *For each $R \ge 1$, $A \ge A(\epsilon) := \epsilon^{-6}$ and each $f : [-1,1]^2 \to \mathbb{C}$ satisfying*

$$\max_{\omega \in C_{R^{1/2}}} \frac{1}{|\omega|} \int_\omega |f|^2 \le 1, \tag{8.3}$$

*we have*

$$\int_{B(0,R)} (\mathfrak{B}_{K,A} f)^{3.25} \le C_{\epsilon,K} R^{C_0\epsilon} \|f\|_2^{3+\epsilon},$$

*with $C_{\epsilon,K}$ independent of $R$ and $f$.*

Note that (8.3) forces $\|f\|_2 \lesssim 1$. The choice of the slightly bigger exponents $3 + \epsilon$ (as opposed to 3) and $C_0\epsilon$ (as opposed to $\epsilon$) will play a crucial role in our argument, when estimating the cellular and the transverse contributions, respectively. They will offset certain losses in the same way as in the proof of Theorem 8.10.

The threshold 3 in the exponent of $\|f\|_2$ is not definitive, but merely forced by the use of the polynomial method. This will become clear when we study the cellular contribution. However, once we choose to work with this exponent, the value 3.25 cannot be lowered any further. This limitation will only become apparent when we estimate the contribution from tangent tubes; see Exercise 8.41 and the discussion in Section 8.10.

It is rather immediate that Theorem 8.19 implies Theorem 8.17. In fact, it implies the following stronger version of it.

**Corollary 8.20** *For each $R \ge 1$, $K \in 2^\mathbb{N}$, $\epsilon > 0$ and $f : [-1,1]^2 \to \mathbb{C}$, we have*

$$\|\mathfrak{B}_{K,A} f\|_{L^{\frac{13}{4}}(B(0,R))} \lesssim_{\epsilon,K} R^\epsilon \|f\|_2^{\frac{12}{13}} \|f\|_\infty^{\frac{1}{13}},$$

*whenever $A \ge \epsilon^{-6}$.*

This inequality is false if $\mathfrak{B}_{K,A} f$ is replaced with the linear operator $Ef$, as can be easily seen using the Knapp example.

The proof in [**57**] of Theorem 8.19 was phrased as a double induction on both $R$ and $\|f\|_2$. While essentially following the same steps, we choose to recast that argument here as an induction on $\epsilon$. More precisely, Theorem 8.19 will be a consequence of the following two results. Indeed, iterating the map $\epsilon \mapsto \epsilon - \epsilon^7$ leads to arbitrarily small values of $\epsilon$.

The values of $\epsilon_0$ and $C_0$ in Theorem 8.19 will become clear at the end of the argument. Along the way, we will derive all the necessary restrictions. The first two are presented in the next lemma.

**Lemma 8.21** *Property $P(\epsilon_0)$ holds if*

$$\epsilon_0 \leq \frac{1}{4} \quad and \quad C_0 \epsilon_0 \geq \frac{3}{8}. \tag{8.4}$$

*Proof* Since $\mathfrak{B}_{K,A}f(x) \leq \max_\tau |E_\tau f(x)|$, we have $\|\mathfrak{B}_{K,A}f\|_{L^4(\mathbb{R}^3)} \lesssim_K \|f\|_2$ (Stein–Tomas theorem) and $\|\mathfrak{B}_{K,A}f\|_{L^2(B(0,R))} \lesssim_K R^{1/2}\|f\|_2$ (conservation of mass). Thus

$$\int_{B(0,R)} (\mathfrak{B}_{K,A}f)^{3.25} \lesssim_K R^{\frac{3}{8}} \|f\|_2^{3.25}.$$

The last quantity is $O\left(R^{C_0\epsilon_0}\|f\|_2^{3+\epsilon_0}\right)$, since $\|f\|_2 \lesssim 1$. $\qquad\square$

The remainder of the chapter will be devoted to the proof of the following result.

**Theorem 8.22** *If $P(\epsilon)$ holds for some $\epsilon \leq \epsilon_0$, then $P(\epsilon - \epsilon^7)$ also holds.*

Fix $\epsilon$ such that $P(\epsilon)$ holds and fix some $A \geq (\epsilon - \epsilon^7)^{-6}$. Fix also $R \geq 1$, $K \in 2^\mathbb{N}$ and some $f$ satisfying (8.3). These quantities will stay the same throughout the remainder of the chapter. To simplify notation, we will write $\mathfrak{B}_A$ for $\mathfrak{B}_{K,A}$.

We need to verify $P(\epsilon - \epsilon^7)$ for $f$, more precisely the inequality

$$\int_{B(0,R)} (\mathfrak{B}_A f)^{3.25} \leq C R^{C_0(\epsilon - \epsilon^7)} \|f\|_2^{3+\epsilon-\epsilon^7}, \tag{8.5}$$

for some $C$ independent of $f$ and $R$. It is clear that we may restrict attention to large enough values of $R$ depending (only) on $\epsilon$, as the inequality is trivially true otherwise.

Let $D = R^{\epsilon^4}$. We invoke Theorem 8.4 to find a polynomial $P$ of degree at most $D$ (in fact, a product of nonsingular polynomials, cf. Remark 8.7) such that $\mathbb{R}^3 \setminus Z(P)$ consists of $N \sim D^3$ open cells $O_i$, and so that for each $i \leq N$,

$$C_1^{-1} \int_{B(0,R)} (\mathscr{B}_A f)^{3.25} \leq D^3 \int_{O_i \cap B(0,R)} (\mathscr{B}_A f)^{3.25} \leq C_1 \int_{B(0,R)} (\mathscr{B}_A f)^{3.25}.$$

$$(8.6)$$

Here $C_1$ is a universal constant.

For the rest of the argument, we also fix the parameter $\delta = \epsilon^2$. For a scale $s$, we will denote by $\mathcal{N}_s(Z(P))$ the $s$-neighborhood of $Z(P)$. In particular, we will refer to

$$W = \mathcal{N}_{100R^{\frac{1}{2}+\delta}}(Z(P))$$

as the *wall*. Also, the interior of the cell $O_i$ is defined to be

$$O_i' = (O_i \cap B(0,R)) \setminus W.$$

We perform a wave packet decomposition of $Ef$ at scale $R$ as in Theorem 2.2

$$Ef = \sum_{T \in \mathbb{T}} Ef_T,$$

with $f_T = a_T \Upsilon_T$ and $Ef_T$ spatially concentrated near the $(R^{1/2}, R)$-tube $T$. Recall that $\Upsilon_T$ is supported on the square $\omega_T \in \Omega_R$ with side length $8R^{-1/2}$.

Moreover, let us rewrite (2.1), (2.5), and (2.6) in our context:

$$\left\| \sum_{T \in \mathcal{T}'} f_T \right\|_2^2 \lesssim R \sum_{T \in \mathcal{T}'} |a_T|^2, \quad \text{for each } \mathcal{T}' \subset \mathbb{T}, \tag{8.7}$$

$$R \sum_{\substack{T \in \mathbb{T} \\ \omega_T = \omega}} |a_T|^2 \sim \|f 1_\omega\|_2^2, \quad \text{for each } \omega \in \Omega_R, \tag{8.8}$$

$$R \sum_{T \in \mathbb{T}} |a_T|^2 \sim \|f\|_2^2. \tag{8.9}$$

Note that

$$\int_{B(0,R)} (\mathscr{B}_A f)^{3.25} \leq \sum_{i=1}^{N} \int_{O_i'} (\mathscr{B}_A f)^{3.25} + \int_{W \cap B(0,R)} (\mathscr{B}_A f)^{3.25}.$$

The sum will be referred to as *cellular contribution*, while the last integral is called the *wall contribution*.

We distinguish two cases, which will be analyzed separately in the following sections. The forthcoming argument will mimic the one behind Theorem 8.10, with lines replaced by tubes. The thrust of the polynomial partitioning is to reduce the initial restriction estimate to the case when all wave packets are spatially concentrated near the wall, a set with much smaller volume than $B(0, R)$ and with good algebraic properties.

## 8.5 Cellular Contribution

In this section we will verify (8.5) under the assumption that the cellular contribution dominates the wall contribution

$$\sum_{i=1}^{N} \int_{O_i'} (\mathcal{B}_A f)^{3.25} \geq 2NC_1^2 \int_{W \cap B(0, R)} (\mathcal{B}_A f)^{3.25}. \tag{8.10}$$

Combining this with (8.6) we find that for each $i \leq N$

$$\int_{O_i'} (\mathcal{B}_A f)^{3.25} \geq \frac{1}{2} \int_{O_i \cap B(0, R)} (\mathcal{B}_A f)^{3.25} \geq \frac{1}{2D^3 C_1} \int_{B(0, R)} (\mathcal{B}_A f)^{3.25}. \tag{8.11}$$

Recalling the definition of $R^\delta T$ from Chapter 2, we let

$$\mathbb{T}_i = \{T \in \mathbb{T} \colon R^\delta T \cap O_i' \neq \emptyset\}$$

and

$$f_i = \sum_{T \in \mathbb{T}_i} f_T.$$

The following key observation is a consequence of the fact that a line not contained in $Z(P)$ can intersect $Z(P)$ at most $D$ times. Indeed, whenever $R^\delta T$ intersects two distinct $O_i'$, its central axis intersects $Z(P)$.

**Lemma 8.23** *Each tube $T \in \mathbb{T}$ lies in at most $D + 1$ of the collections $\mathbb{T}_i$.*

We aim to use the induction hypothesis $P(\epsilon)$ for one of the functions $f_i$. In order to do so, we need to detect a particular $f_{i_0}$ with good $L^2$ properties. This will be achieved via pigeonholing, with the aid of the next two propositions.

**Proposition 8.24** *We have*

$$\sum_{i=1}^{N} \|f_i\|_2^2 \lesssim D \|f\|_2^2.$$

*Proof* The inequality follows by first using (8.7) for each $i$, then (8.9):

$$\sum_{i=1}^{N} \|f_i\|_2^2 = \sum_{i=1}^{N} \left\| \sum_{T \in \mathbb{T}_i} f_T \right\|_2^2$$

$$\lesssim R \sum_{i=1}^{N} \sum_{T \in \mathbb{T}_i} |a_T|^2$$

$$\leq (D+1) R \sum_{T \in \mathbb{T}} |a_T|^2$$

$$\lesssim D \|f\|_2^2. \qquad \square$$

**Proposition 8.25** *Given a family $\mathcal{T} \subset \mathbb{T}$, let $g = \sum_{T \in \mathcal{T}} f_T$. Then*

$$\max_{\omega \in \mathcal{C}_{R^{1/2}}} \frac{1}{|\omega|} \int_{\omega} |g|^2 \lesssim 1.$$

*Proof* It will suffice to prove that for each $\omega' \in \Omega_R$, we have

$$\left\| \sum_{T \in \mathcal{T}(\omega')} f_T \right\|_{L^2(\mathbb{R}^2)}^2 \lesssim R^{-1},$$

where $\mathcal{T}(\omega')$ are those $T \in \mathcal{T}$ with $\omega_T = \omega'$. Indeed, note that on each $\omega \in \mathcal{C}_{R^{1/2}}$, $g$ is the sum of $O(1)$ such functions, the ones corresponding to those $\omega' \in \Omega_R$ intersecting $\omega$. Combining (8.7) with (8.8), we write

$$\left\| \sum_{T \in \mathcal{T}(\omega')} f_T \right\|_{L^2(\mathbb{R}^2)}^2 \lesssim R \sum_{T \in \mathcal{T}(\omega')} |a_T|^2$$

$$\leq R \sum_{\substack{T \in \mathbb{T} \\ \omega_T = \omega'}} |a_T|^2$$

$$\lesssim \|f\|_{L^2(\omega')}^2.$$

To close the argument, we invoke (8.3). $\qquad \square$

Since $N \sim D^3$, we can pick $i_0$ satisfying

$$\|f_{i_0}\|_2 \lesssim D^{-1} \|f\|_2 \qquad (8.12)$$

and

$$\max_{\omega \in \mathcal{C}_{R^{1/2}}} \frac{1}{|\omega|} \int_{\omega} |f_{i_0}|^2 \lesssim 1. \qquad (8.13)$$

Recall that we may assume $R$ to be large enough, depending on $\epsilon$ (or in other words, on $\delta$). Thus, using (2.3), (2.6), and the Cauchy–Schwarz inequality, we get

$$\sup_{x \in O'_{i_0}} \sum_{T \in \mathbb{T} \setminus \mathbb{T}_{i_0}} |Ef_T(x)| \leq R^{-100} \|f\|_2.$$

This in turn implies that

$$\sup_{x \in O'_{i_0}} \max_{\tau \in \mathcal{C}_K} |E_\tau f(x) - E_\tau f_{i_0}(x)| \leq R^{-100} \|f\|_2.$$

Consequently, using Lemma 8.16, we deduce that

$$|\mathfrak{B}_A f(x) - \mathfrak{B}_A f_{i_0}(x)| \leq R^{-100} \|f\|_2,$$

when $x \in O'_{i_0}$. Combining this with (8.11), we conclude that

$$\int_{B(0,R)} (\mathfrak{B}_A f)^{3.25} \lesssim D^3 \int_{B(0,R)} (\mathfrak{B}_A f_{i_0})^{3.25} + \|f\|_2^{3.25}. \qquad (8.14)$$

We now invoke the hypothesis $P(\epsilon)$ for $B^{-1} f_{i_0}$ ($B$ large enough but $O(1)$), which is possible due to (8.13). Using also (8.12), we may write

$$\int_{B(0,R)} (\mathfrak{B}_A f_{i_0})^{3.25} \lesssim_{\epsilon,K} R^{C_0 \epsilon} D^{-3-\epsilon} \|f\|_2^{3+\epsilon}. \qquad (8.15)$$

We rely of course on the fact that $A \geq A(\epsilon - \epsilon^7) \geq A(\epsilon)$. At this stage, only the monotonicity of the function $\epsilon \mapsto A(\epsilon)$ is needed, but later in the argument the reason behind the specific choice of this function will be revealed.

Since $\|f\|_2 \lesssim 1$, it follows that if $\epsilon \leq 1/4$

$$\|f\|_2^{3.25} \lesssim \|f\|_2^{3+\epsilon-\epsilon^7}.$$

We may now combine (8.14) and (8.15) to state our final estimate for the cellular contribution.

**Corollary 8.26** *Assume (8.10). The estimate (8.5)*

$$\int_{B(0,R)} (\mathfrak{B}_A f)^{3.25} \lesssim_{\epsilon,K} R^{C_0(\epsilon-\epsilon^7)} \|f\|_2^{3+\epsilon-\epsilon^7}$$

*holds if $C_0 \epsilon - \epsilon^5 \leq C_0(\epsilon - \epsilon^7)$, in particular if $\epsilon \leq \epsilon_0 \leq 1/4$ and*

$$C_0 \epsilon_0^2 \leq 1. \qquad (8.16)$$

Note that the saving $\epsilon^5$ in the exponent of $R$ is due to the choice of the exponent $3 + \epsilon$ for $\|f\|_2$, which allows us to capitalize on the gain (8.12) in the $L^2$ norm of $f_{i_0}$. Note also how the gain $D^{-3}$ cancels out the loss $D^3$ from (8.11).

It is worth observing that if the cellular contribution were always dominant, then the argument in this section would suffice to prove the full Restriction Conjecture for $\mathbb{P}^2$. Indeed, there was nothing special about working with 3.25 so far, and we may have considered instead any $p > 3$. In reality, there is no reason to expect that the cellular term always dominates, and we will need to analyze the alternative scenario, when the wall contribution is dominant. The next few sections contain a careful analysis of this contribution, and will ultimately justify the need for working with the exponent $p = 3.25$.

## 8.6  Two Types of Wall Contribution

It remains to verify (8.5) under the following assumption:

$$\sum_{i=1}^{N} \int_{O_i'} (\mathfrak{B}_A f)^{3.25} \leq 2NC_1^2 \int_{W \cap B(0,R)} (\mathfrak{B}_A f)^{3.25}. \tag{8.17}$$

It is easy to see that in this case, a significant fraction of the total mass of $\mathfrak{B}_A f$ is concentrated in $W$, more precisely

$$\int_{B(0,R)} (\mathfrak{B}_A f)^{3.25} \lesssim D^3 \int_{W \cap B(0,R)} (\mathfrak{B}_A f)^{3.25}. \tag{8.18}$$

We thus need a favorable estimate for the wall contribution. We will divide this contribution into two parts.

Let us start by covering $B(0, R)$ with a family of roughly $R^{3\delta}$ finitely overlapping balls $B_j$ of radius $R^{1-\delta}$. We do this in order to exploit the induction hypothesis $P(\epsilon)$ at slightly smaller radial scales.

In the proof of Theorem 8.10, we found it useful to classify lines as being either tangent or transverse, relative to a variety. The following definition quantifies a similar dichotomy for tubes. We will denote by $C \times R^\delta T$ the dilation of $R^\delta T$ by a factor of $C$ around its center.

**Definition 8.27** (Tangent and transverse tubes) Let $j \lesssim R^{3\delta}$. We define the collection $\mathbb{T}_{j,\text{tang}}$ of tangent tubes to $Z(P)$ in $B_j$ to be those tubes $T \in \mathbb{T}$ satisfying the following two properties, for some $C_2, C_3 = O(1)$.

(Ta1) $R^\delta T \cap W \cap B_j \neq \emptyset$.

(Ta2) If $x$ is any nonsingular point of $P$ lying in $C_2 B_j \cap (C_3 \times R^\delta T)$, then the angle between the central axis $v_T$ of $T$ and the tangent plane to $Z(P)$ at $x$ is smaller than $R^{2\delta - 1/2}$.

The collection of those tubes $T \in \mathbb{T}$ with $W \cap B_j \cap R^\delta T \neq \emptyset$ that are not in $\mathbb{T}_{j,\text{tang}}$ is said to be transverse in $B_j$, and will be denoted by $\mathbb{T}_{j,\text{trans}}$.

The exact values of $C_2, C_3$ will not concern us here. The fact that $P$ is a product of nonsingular polynomials implies that the set of nonsingular points of $P$ is dense in $Z(P)$. Thus, if $C_2, C_3$ are large enough, a simple computation shows that (Ta1) forces the existence of a nonsingular point of $P$ lying in $C_2 B_j \cap (C_3 \times R^\delta T)$. Thus, each tube satisfying (Ta1) is in exactly one of the collections $\mathbb{T}_{j,\text{tang}}, \mathbb{T}_{j,\text{trans}}$.

The distinction between tangent and transverse tubes is, of course, a bit blurry. Some transverse tubes will look very close to being tangent. However, in spite of this seemingly weak definition of transversality, it will prove to be good enough to guarantee the critical property of transverse tubes, as quantified by Proposition 8.30.

A wave packet $Ef_T$ may contribute both to different cells and also to the wall, both transversely and tangentially. For example, the tube shown in Figure 8.1 lies in both $\mathbb{T}_{1,\text{tang}}$ and, $\mathbb{T}_{2,\text{trans}}$, and, moreover, it also contributes to two different cells.

A tangent tube must have a significant portion lying close to $Z(P)$. This is an elementary exercise in calculus whose proof is left to the reader.

**Proposition 8.28** *There exists a constant $C = O(1)$ such that for each tangent tube $T \in \mathbb{T}_{j,\text{tang}}$, there is a segment of $R^\delta T$ of length $\sim R^{1-\delta}$ that lies inside the slight enlargement of the wall $W$:*

$$\{x \in \mathbb{R}^3 : \text{dist}\,(x, Z(P)) \leq C R^{\frac{1}{2}+\delta}\}.$$

Define

$$f_{j,\text{tang}} = \sum_{T \in \mathbb{T}_{j,\text{tang}}} f_T,$$

$$f_{j,\text{trans}} = \sum_{T \in \mathbb{T}_{j,\text{trans}}} f_T.$$

Note that each tube $T \in \mathbb{T}$ such that $R^\delta T \cap W \cap B(0, R) \neq \emptyset$ belongs to at least one collection $\mathbb{T}_{j,\text{tang}}$ or $\mathbb{T}_{j',\text{trans}}$. The tubes for which $R^\delta T \cap W = \emptyset$ will

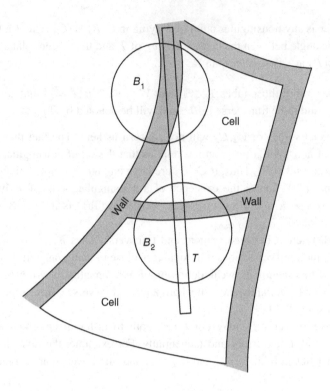

Figure 8.1 Polynomial partitioning.

contribute negligibly to $\int_W (\mathfrak{B}_A f)^{3.25}$. More precisely, assuming as before that $R$ is large enough, we have the estimate

$$\sup_{x \in B_j \cap W} \sum_{T \in \mathbb{T} \setminus (\mathbb{T}_{j,\text{tang}} \cup \mathbb{T}_{j,\text{trans}})} |Ef_T(x)| \le R^{-100} \|f\|_2.$$

Combining this with Lemma 8.16 proves that

$$\sup_{x \in B_j \cap W} |\mathfrak{B}_A f(x) - \mathfrak{B}_A (f_{j,\text{tang}} + f_{j,\text{trans}})(x)| \le R^{-100} \|f\|_2.$$

Lemma 8.15 comes in handy to separate the tangent and transverse contributions as follows:

$$\mathfrak{B}_A (f_{j,\text{tang}} + f_{j,\text{trans}})(x) \le \mathfrak{B}_1 (f_{j,\text{tang}})(x) + \mathfrak{B}_{A-1} (f_{j,\text{trans}})(x).$$

Combining these observations with (8.18), we conclude that, subject to scenario (8.17), we have the following estimate.

**Corollary 8.29**

$$\int_{B(0,R)} (\mathcal{B}_A f)^{3.25} \lesssim \|f\|_2^{3.25} + R^{3\epsilon^4} \left( \sum_{j=1}^{O(R^{3\delta})} \int_{B_j} (\mathcal{B}_1 f_{j,\text{tang}})^{3.25} \right.$$

$$\left. + \sum_{j=1}^{O(R^{3\delta})} \int_{B_j} (\mathcal{B}_{A-1} f_{j,\text{trans}})^{3.25} \right).$$

The first term is an acceptable contribution, since $\|f\|_2 \lesssim 1$. The transverse term will be estimated using the induction hypothesis. The fact that we lower the parameter $A$ to $A-1$ will prove harmless. For the tangent term, the drop from $A$ to $1$ may seem rather dramatic. However, Lemma 8.13 shows that the value $A=1$ already captures the essence of bilinearity. The tangent contribution will be evaluated directly – without using the induction hypothesis – by exploiting Tao's bilinear restriction estimate and the fact that tangent tubes cannot point in too many directions.

## 8.7 The Transverse Contribution

In this section, we prove the inequality

$$\sum_{j=1}^{O(R^{3\delta})} \int_{B_j} (\mathcal{B}_{A-1} f_{j,\text{trans}})^{3.25} \lesssim_{\epsilon, K} R^{C_0(\epsilon - \epsilon^7) - 3\epsilon^4} \|f\|_2^{3 + \epsilon - \epsilon^7}.$$

Here is the key geometric fact about the transverse tubes. The reader is referred to [57] for a proof.

**Proposition 8.30** *Each $T \in \mathbb{T}$ belongs to at most $D^{O(1)} = R^{O(\delta^2)} = R^{O(\epsilon^4)}$ families $\mathbb{T}_{j,\text{trans}}$.*

This is reminiscent of the fact that a line can intersect $Z(P)$ at most $D+1$ times. We will be content with an upper bound that is polynomial (rather than linear) in $D$.

We will use our hypothesis $P(\epsilon)$ for each $f_{j,\text{trans}}$ on the ball $B_j$. There will be a gain coming from the fact that these balls have smaller radius $R^{1-\delta}$. The fact that $B_j$ is not centered at the origin can be dealt with by invoking (3.30). We need to verify that each $f_{j,\text{trans}}$ has good $L^2$ properties.

**Proposition 8.31** *We have for some $C_{\text{trans}}$ independent of $R, f, \epsilon$*

$$\sum_j \|f_{j,\text{trans}}\|_2^2 \lesssim R^{C_{\text{trans}}\epsilon^4} \|f\|_2^2$$

*and*

$$\max_{j} \max_{\omega' \in \mathcal{C}_{\frac{1-\delta}{R^{\frac{1-\delta}{2}}}}} \frac{1}{|\omega'|} \int_{\omega'} |f_{j,\text{trans}}|^2 \lesssim 1.$$

*Proof* The proof of the first inequality is very similar to that of Proposition 8.24. We first use (8.7) for each $i$, then Proposition 8.30 followed by (8.9) for $f$

$$\sum_{j} \|f_{j,\text{trans}}\|_2^2 = \sum_{j} \left\| \sum_{T \in \mathbb{T}_{j,\text{trans}}} f_T \right\|_2^2$$

$$\lesssim R \sum_{j} \sum_{T \in \mathbb{T}_{j,\text{trans}}} |a_T|^2$$

$$\leq R^{1+O(\delta^2)} \sum_{T \in \mathbb{T}} |a_T|^2$$

$$\lesssim R^{C_{\text{trans}} \epsilon^4} \|f\|_2^2.$$

Proposition 8.25 implies that

$$\max_{\omega \in \mathcal{C}_{\frac{1}{R^{\frac{1}{2}}}}} \frac{1}{|\omega|} \int_{\omega} |f_{j,\text{trans}}|^2 \lesssim 1.$$

It now suffices to use the fact that small-scale averages dominate large-scale averages:

$$\max_{\omega' \in \mathcal{C}_{\frac{1-\delta}{R^{\frac{1-\delta}{2}}}}} \frac{1}{|\omega'|} \int_{\omega'} |f_{j,\text{trans}}|^2 \lesssim \max_{\omega \in \mathcal{C}_{\frac{1}{R^{\frac{1}{2}}}}} \frac{1}{|\omega|} \int_{\omega} |f_{j,\text{trans}}|^2. \qquad \square$$

Recall that $A \geq (\epsilon - \epsilon^7)^{-6}$. This guarantees that $A - 1 \geq \epsilon^{-6}$. Invoking $P(\epsilon)$ for $C^{-1} f_{j,\text{trans}}$ ($C = O(1)$) on each $B_j$, we get

$$\sum_{j} \int_{B_j} (\mathcal{B}_{A-1} f_{j,\text{trans}})^{3.25} \lesssim_{\epsilon, K} R^{C_0 \epsilon (1-\delta)} \sum_{j} \|f_{j,\text{trans}}\|_2^{3+\epsilon}$$

$$\lesssim_{\epsilon, K} R^{C_0(\epsilon - \epsilon^3)} \left( \sum_{j} \|f_{j,\text{trans}}\|_2^2 \right)^{\frac{3+\epsilon}{2}}$$

$$\lesssim_{\epsilon, K} R^{C_0(\epsilon - \epsilon^3)} R^{C_{\text{trans}} \frac{\epsilon^4 (3+\epsilon)}{2}} \|f\|_2^{3+\epsilon}.$$

Recalling that $\|f\|_2 \lesssim 1$, we rewrite our estimate for the transverse contribution as follows.

**Corollary 8.32** *Let $\epsilon_0$ be small enough so that*

$$C_0\epsilon_0^3 > C_0\epsilon_0^7 + \frac{C_{\text{trans}}}{2}\epsilon_0^4(3 + \epsilon_0) + 3\epsilon_0^4. \tag{8.19}$$

*Then, if $0 < \epsilon \le \epsilon_0$, we have*

$$\sum_j \int_{B_j} (\mathscr{B}_{A-1}f_{j,\text{trans}})^{3.25} \lesssim_{\epsilon,K} R^{C_0(\epsilon-\epsilon^7)-3\epsilon^4}\|f\|_2^{3+\epsilon-\epsilon^7}.$$

To recap, we have only lost the small quantity $R^{O(\epsilon^4)}$, since a given tube does not intersect the wall transversely too many times. This loss is offset by the more substantial gain $R^{C_0\epsilon^3}$, which comes from applying the induction hypothesis on balls of smaller radius $R^{1-\epsilon^2}$.

## 8.8 Tubes Tangent to a Variety

In this section, we prove that the tangent tubes $T \in \cup_j \mathbb{T}_{j,\text{tang}}$ can point in at most $R^{1/2+O(\delta)}$ distinct directions. This represents a significant gain over the full collection of tubes $\mathbb{T}$, which are allowed to point in roughly $R$ directions. In light of Proposition 8.28, it is useful to analyze tubes contained in small neighborhoods of $Z(P)$. We will lose many powers of $R^\delta$ throughout the argument in this section, but that will prove to be acceptable for later applications.

The following theorem was first proved for $n = 3$ by Guth [57], then for $n = 4$ by Zahl [126], and finally for arbitrary $n$ by Katz and Rogers [72].

**Theorem 8.33** (Few separated directions near $Z(P)$) *Let $C = O(1)$. For each polynomial $P$ of degree $D$ in $\mathbb{R}^n$ and each $L \gg 1$, the maximum number of $(1, L)$-tubes with $1/L$-separated directions lying inside $\mathcal{N}_C(Z(P)) \cap B_{CL}$ is $\lesssim_{\epsilon,D} L^{n-2+\epsilon}$, uniformly over all balls $B_{CL}$ of radius $CL$ in $\mathbb{R}^n$.*

To put this theorem into perspective, we will first show that it is a consequence of the Kakeya maximal operator conjecture 5.3. We will rely on an essentially sharp upper bound on the volumes of neighborhoods of $Z(P)$.

**Theorem 8.34** (Volumes of walls) *Let $V$ be a rectangular box in $\mathbb{R}^n$ with dimensions $R_1 \times \cdots \times R_n$ satisfying $1 \le R_1 \le R_2 \le \cdots \le R_n$. Then*

$$|\mathcal{N}_C(Z(P)) \cap V| \lesssim_C D \prod_{j=2}^{n} R_j.$$

The size of the coefficients of $P$ plays no role in this estimate; only its degree matters. The case when all $R_j$ are equal is due to Wongkew [122]. The more general version presented here was proved by Guth in [57].

**Proposition 8.35** *Assume that Conjecture 5.3 holds for some $n$. Then the maximum number of $(1, L)$-tubes in $\mathbb{R}^n$ with $1/L$-separated directions lying inside a set $X$ is $\lesssim_\epsilon L^{\epsilon-1}|X|$. In particular, Theorem 8.33 holds.*

*Proof* Recall that the validity of Conjecture 5.3 would force any family $\mathcal{T}$ of such tubes to satisfy the essential pairwise disjointness property

$$|\mathcal{T}|L^{1-\epsilon} \lesssim_\epsilon |\cup_{T \in \mathcal{T}} T|.$$

This proves the first part of the proposition. Theorem 8.33 would then follow by using Theorem 8.34 with $R_1 = \cdots = R_n = 2CL$.                    □

The proof in [72] of Theorem 8.33 for arbitrary $n$ uses deep results from algebraic and differential geometry. When $n = 3$, Guth [57] found an elementary argument that also gives an explicit upper bound that is polynomial in $D$. We reproduce his proof here.

**Theorem 8.36** *When $n = 3$, the number of directions in Theorem 8.33 is $O(D^2 L(\log L)^2)$.*

Before proving this result let us observe the following rather immediate consequence that will be used in the next section.

**Corollary 8.37** *The tubes in $T \in \cup_j \mathbb{T}_{j, \text{tang}}$ can point in at most $O(R^{1/2+O(\delta)})$ directions.*

*Proof* Let $\mathcal{T}$ be a collection of tangent tubes with distinct directions. According to Proposition 8.28, for each $T \in \mathcal{T}$, there is some $(R^{1/2+\delta}, R^{1-\delta})$-tube inside $R^\delta T$, which also lies inside $\mathcal{N}_{CR^{1/2+\delta}}(Z(P))$. Rescaling everything by $R^{1/2+\delta}$, we find a collection of smaller $(1, R^{1/2-2\delta})$-tubes lying inside $\mathcal{N}_C(Z(\tilde{P}))$, where $\tilde{P}(x) = P(R^{1/2+\delta}x)$. Note that $\tilde{P}$ and $P$ have the same degree. Moreover, the small tubes will sit inside a ball $B$ with radius $CR^{1/2-\delta}$. Cover $B$ with a family of finitely overlapping balls $B'$ having smaller radius $CR^{1/2-2\delta}$, so that each small tube lies inside some $B'$.

We split the collection of small tubes into $O(R^{7\delta})$ subcollections $\mathcal{T}'$, so that the tubes in each $\mathcal{T}'$ have directions separated by $R^{-1/2+2\delta}$, and lie inside a fixed ball $B'$. Finally, we apply Theorem 8.36 to each $\mathcal{T}'$ with $L = R^{1/2-2\delta}$, to find that there can be at most $O(D^2 R^{1/2-2\delta}(\log R)^2)$ distinct directions for each $\mathcal{T}'$.                                                       □

Let us now prove Theorem 8.36. Let $\mathcal{T}$ be a collection of tubes as in Theorem 8.33. The idea is to find a hairbrush $\mathcal{H}$ (see Exercise 5.8) containing a significant fraction of the tubes from $\mathcal{T}$. Since these tubes are on the one hand (essentially) pairwise disjoint and on the other hand concentrated inside a wall with small volume, we will end up getting a good estimate on the cardinality of $\mathcal{H}$, and consequently also on the cardinality of $\mathcal{T}$.

Letting $|\mathcal{T}| = \beta L$, we need to prove that $\beta \lesssim D^2 (\log L)^2$. This is immediate if $\beta \lesssim D$, so we will assume that $\beta \gg D$.

Define the auxiliary set

$$S = \{(Q, T_1, T_2) \colon Q = \mathbf{n} + [0,1]^3, \mathbf{n} \in \mathbb{Z}^3, T_1, T_2 \in \mathcal{T} \text{ and } T_1 \cap T_2 \cap Q \neq \emptyset\}.$$

We will start by proving a lower bound for $|S|$. Using Hölder's inequality and Theorem 8.34 with $R_1 = R_2 = R_3 = 2CL$, we get

$$\beta L^2 \sim \left\| \sum_{T \in \mathcal{T}} 1_T \right\|_1$$

$$\leq \left\| \sum_{T \in \mathcal{T}} 1_T \right\|_2 |\mathcal{N}_C(Z(P)) \cap B_{CL}|^{1/2}$$

$$\lesssim D^{\frac{1}{2}} L \left\| \sum_{T \in \mathcal{T}} 1_T \right\|_2.$$

Combining this with the obvious inequality

$$\left\| \sum_{T \in \mathcal{T}} 1_T \right\|_2^2 \leq |S|$$

shows that $|S| \gtrsim D^{-1} \beta^2 L^2$.

For $1/L \lesssim 2^{-k} \lesssim 1$, we let

$$S_k = \{(Q, T_1, T_2) \in S \colon \sphericalangle(T_1, T_2) \in [2^{-k}, 2^{-k+1})\}.$$

Let also

$$S_* = \{(Q, T, T) \in S\},$$

and note that $|S_*| \lesssim \beta L^2$. Since $\beta \gg D$, it follows that

$$\left| \bigcup_k S_k \right| = |S \setminus S_*| \gtrsim D^{-1} \beta^2 L^2.$$

Pigeonholing, we find some $k$ such that

$$|S_k| \gtrsim \frac{\beta^2 L^2}{D \log L}.$$

It further follows that there is $T_1 \in \mathcal{T}$ such that

$$|\{(Q,T_2): (Q,T_1,T_2) \in \mathcal{S}_k\}| \gtrsim \frac{\beta L}{D \log L}.$$

Denote by $\mathcal{H}$ the hairbrush consisting of those $T_2$ that contribute to the left-hand side. Since for each $T_2 \in \mathcal{H}$ we have $|T_1 \cap T_2| \lesssim 2^k$, it follows that

$$|\mathcal{H}| \gtrsim \frac{\beta L}{2^k D \log L}.$$

To summarize, we have identified a hairbrush containing a significant fraction of $\mathcal{T}$.

Now, reasoning as in Exercise 5.8, we get an explicit bound

$$|\cup_{T_2 \in \mathcal{H}} T_2| \gtrsim L(\log L)^{-1}|\mathcal{H}|.$$

Also, since the tubes in $\mathcal{H}$ lie inside a cylinder with radius $O(2^{-k}L)$ around $T_1$, using again Theorem 8.34, this time with $R_1 = R_2 = O(2^{-k}L)$ and $R_3 = CL$, we get

$$|\cup_{T_2 \in \mathcal{H}} T_2| \lesssim D2^{-k}L^2.$$

Combining the last three inequalities leads to the desired conclusion $\beta \lesssim D^2(\log L)^2$.

**Exercise 8.38** Use Theorem 8.34 to prove that for each $\delta \leq 1$ and each ball $B_1$ of radius 1 in $\mathbb{R}^n$, we have

$$|\mathcal{N}_\delta(Z(P)) \cap B_1| \lesssim_D \delta.$$

Conclude that the Hausdorff dimension of $Z(P)$ is at most $n - 1$.

## 8.9 Controlling the Tangent Contribution

We now combine the geometric information obtained in the previous section with Tao's bilinear restriction inequality, to provide the following estimate for the tangent term

$$\sum_{j=1}^{O(R^{3\delta})} \int_{B_j} (\mathcal{B}_1 f_{j,\text{tang}})^{3.25} \lesssim_{\epsilon,K} R^{C_0(\epsilon-\epsilon^7)-3\epsilon^4} \|f\|_2^{3+\epsilon-\epsilon^7}.$$

In fact, we will see that each $B_j$ may be replaced with $B(0,CR)$.

**Theorem 8.39** *Assume* $g: [-1, 1]^2 \to \mathbb{C}$ *is supported on the union of* $M$ *squares* $\omega \in \mathcal{C}_{R^{1/2}}$ *and that it satisfies (8.3), that is,*

$$\max_{\omega \in \mathcal{C}_{R^{1/2}}} \frac{1}{|\omega|} \int_\omega |g|^2 \leq 1.$$

*Assume that* $\tau_1, \tau_2 \in \mathcal{C}_K$ *are not neighbors. Then for each* $\eta > 0$, $3 < p \leq 10/3$ *and* $0 \leq \Delta < p - 3$, *we have*

$$\|(E_{\tau_1} g E_{\tau_2} g)^{\frac{1}{2}}\|^p_{L^p(B(0,R))} \lesssim_{\eta, K} R^{\eta + 4 + \frac{\Delta}{2} - \frac{5p}{4}} M^{\frac{p-3}{2}} \|g\|_2^{3+\Delta}.$$

*Proof* The bilinear restriction Theorem 3.1 shows that for each $\eta > 0$,

$$\|(E_{\tau_1} g E_{\tau_2} g)^{\frac{1}{2}}\|_{L^{\frac{10}{3}}(B(0,R))} \lesssim_{\eta, K} R^\eta \|g\|_2.$$

Interpolating this with the trivial inequality

$$\|(E_{\tau_1} g E_{\tau_2} g)^{\frac{1}{2}}\|_{L^2(B(0,R))} \lesssim R^{\frac{1}{2}} \|g\|_2$$

leads to the bound

$$\|(E_{\tau_1} g E_{\tau_2} g)^{\frac{1}{2}}\|^p_{L^p(B(0,R))} \lesssim_{\eta, K} R^{\eta + \frac{5}{2} - \frac{3p}{4}} \|g\|_2^p. \tag{8.20}$$

Our hypothesis implies that $\|g\|_2 \lesssim (MR^{-1})^{1/2}$. This allows us to write

$$\|g\|_2^p \lesssim \|g\|_2^{3+\Delta} (MR^{-1})^{\frac{p-3-\Delta}{2}} \leq \|g\|_2^{3+\Delta} (MR^{-1})^{\frac{p-3}{2}} R^{\frac{\Delta}{2}}. \tag{8.21}$$

The desired conclusion follows by combining (8.20) and (8.21). □

We will apply this theorem with $p = 3.25$, $M = O(R^{1/2+O(\delta)})$, $\eta = \delta$, $\Delta = \epsilon - \epsilon^7$ and $g = f_{j,\text{tang}}$. It is here that the necessity of working with $p = 3.25$ is finally revealed, as the exponent

$$4 - \frac{5p}{4} + \frac{p-3}{4}$$

of $R$ is zero precisely for this value of $p$.

The hypotheses of the theorem are easily seen to hold for these choices of $p, M, \eta, \Delta$, and $g$ due to Proposition 8.25 and Corollary 8.37. Combining this theorem with Lemma 8.13, we get the following satisfactory estimate for the tangent contribution.

**Corollary 8.40** *There is a constant* $C_{\text{tang}}$ *such that*

$$\sum_{j=1}^{O(R^{3\delta})} \int_{B_j} (\mathfrak{B}_1 f_{j,\text{tang}})^{3.25} \lesssim_{\delta, K} R^{C_{\text{tang}}\delta + \frac{\epsilon - \epsilon^7}{2}} \|f\|_2^{3+\epsilon-\epsilon^7}. \tag{8.22}$$

*In particular, if $\epsilon \le \epsilon_0 \le 1/4$ and, in addition, $\epsilon_0$ is small enough so that*

$$C_{\text{tang}}\epsilon_0^2 + 3\epsilon_0^4 \le \left(C_0 - \frac{1}{2}\right)(\epsilon_0 - \epsilon_0^7), \tag{8.23}$$

*then*

$$\sum_{j=1}^{O(R^{3\delta})} \int_{B_j} (\mathfrak{B}_1 f_{j,\text{tang}})^{3.25} \lesssim_{\epsilon,K} R^{C_0(\epsilon-\epsilon^7)-3\epsilon^4} \|f\|_2^{3+\epsilon-\epsilon^7}.$$

**Exercise 8.41** (a) Given any $M \lesssim R^{1/2}$ and $p \ge 3$, construct a function $g$ as in the Theorem 8.39 such that

$$\|(E_{\tau_1}g E_{\tau_2}g)^{\frac{1}{2}}\|_{L^p(B(0,R))}^p \gtrsim_K R^{4-\frac{5p}{4}} M^{\frac{p-3}{2}} \|g\|_2^3$$

and, moreover, all tubes in the wave packet decomposition at scale $R$ of $g$ are tangent to a plane.

In particular, when $M \sim R^{1/2}$ and $p < 3.25$, we have

$$\|(E_{\tau_1}g E_{\tau_2}g)^{\frac{1}{2}}\|_{L^p(B(0,R))}^p \gtrsim_K R^{\eta_p} \|g\|_2^3,$$

for some $\eta_p > 0$.

(b) Prove that for each $p \ge 3$,

$$\sum_j \int_{B_j} (\mathfrak{B}_1 f_{j,\text{tang}})^p \lesssim_{\delta,K} R^{C_{\text{tang}}\delta} \|f\|_2^2.$$

Hint for (a): Use Rademacher functions.

## 8.10 Putting Things Together

Let $\epsilon_0, C_0$ satisfy (8.4), (8.16), (8.19), and (8.23). It is easy to see that these restrictions can be simultaneously fulfilled. The estimate (8.5) follows by combining Corollaries 8.26, 8.29, 8.32, and 8.40. This closes the proof of Theorem 8.22, and thus also of Theorem 8.19.

Let us conclude with a few remarks. The two exponents appearing in Theorem 8.19 are $p = 3.25$ and the weight $w = 3 + \epsilon$. The fact that the weight is greater than 3 allowed us to close the induction $P(\epsilon) \implies P(\epsilon - \epsilon^7)$ in the case when the cellular contribution dominates. Part (a) of Exercise 8.41 shows that we only have a favorable estimate for the tangent term when $p \ge 3.25$.

In order to be able lower the value of 3.25, the weight $w$ needs to be chosen smaller than 3. Part (b) of the same exercise shows that if 2 is used in place of 3, then the tangent contribution remains favorable even at the critical exponent $p = 3$. This observation has led to a further improvement in the restriction estimate for the paraboloid. In [116], Wang uses a more careful analysis of the cellular–transverse–tangent trichotomy that relies on a multistep iteration and on additional geometric information about tubes. Her approach proves the restriction estimate $R^*_{\mathbb{P}^2}(\infty \mapsto p)$ for $p > 3 + 3/13$.

The following conjecture seems plausible, and if proved, it would completely solve the Restriction Conjecture for $\mathbb{P}^2$.

**Conjecture 8.42** *For each $R \geq 1$, $K \in 2^{\mathbb{N}}$, $\epsilon > 0$ and $f : [-1,1]^2 \to \mathbb{C}$, we have*

$$\|\mathfrak{B}_{K,A}f\|_{L^3(B(0,R))} \lesssim_{\epsilon, K} R^\epsilon \|f\|_2^{\frac{2}{3}} \|f\|_\infty^{\frac{1}{3}},$$

*whenever $A \gtrsim_\epsilon 1$.*

It is worth comparing Guth's argument with the Bourgain–Guth approach to the Restriction Conjecture described in Section 7.3. The tangent contribution in the former argument is somewhat analogous to the lower-dimensional contribution from the latter argument. However, the spatial localization is better in the first case. Indeed, the tangent tubes live inside a narrow region of $\mathbb{R}^3$. This restriction forces (via Corollary 8.37) the frequencies associated with the tangent tubes to be essentially localized inside some $R^{-1/2}$-wide (possibly curved) strip in $[-1,1]^2$. On the other hand, in the case of the Bourgain–Guth lower-dimensional contribution, this strip is always the neighborhood of a line segment, and the spatial tubes are scattered over potentially large regions in $\mathbb{R}^3$. Another notable feature of Guth's argument is that it avoids the use of the Bennett–Carbery–Tao trilinear restriction estimate.

Exercise 8.43 proposes an alternative proof to (8.22), without using the bilinear restriction Theorem 3.1. It also explores a proof of a slightly weaker version of Theorem 3.1 that essentially only relies on the polynomial partitioning and on two-dimensional estimates. This type of argument is part of a far-reaching scheme. For each level of $k$-linearity, there is a corresponding operator $\mathfrak{B}$ that is similar in spirit with but behaves a bit better than geometric averages of $k$-transverse terms. In [56], Guth proved that Conjecture 6.1 holds with $p = 2$ for these better behaved operators. This in turn can be used to make progress on the (linear) Restriction Conjecture in higher dimensions.

**Exercise 8.43** (a) Prove that for each $j$ and each cube $q$ in $\mathbb{R}^3$ with side length $R^{1/2}$ and such that $q \cap B_j \cap W \neq \emptyset$, the tubes in $\mathbb{T}_{j,\text{tang}}$ that intersect $q$ are nearly coplanar. More precisely, their directions are within $R^{-1/2+O(\delta)}$ from a fixed plane.

(b) Use (a) to prove (8.22) without any appeal to Theorem 3.1.

(c) Use (a) to prove that for each $j$ and each $f$,

$$\int_{B_j \cap W} (\mathfrak{B}_1 f_{j,\text{tang}})^{\frac{10}{3}} \lesssim_{\delta, K} R^{O(\delta)} \|f\|_2^{\frac{10}{3}}.$$

(d) Adapt the argument from this chapter to prove the inequality

$$\|\mathfrak{B}_{K,A} f\|_{L^{\frac{10}{3}}(B(0,R))} \lesssim_{\epsilon, K} R^{\epsilon} \|f\|_2,$$

for $A \gtrsim_\epsilon 1$.

Hint for (b), (c): Use the bilinear restriction estimate for the parabola to prove an $L^4$ estimate on each $q$, then sum over all $q$ from a finitely overlapping cover of $B_j \cap W$.

# 9

## An Introduction to Decoupling

The revolutionary paper of Wolff [**119**] was a majestic swansong. In it he introduced a powerful Fourier analytic tool, which we now call *decoupling*, and demonstrated its extraordinary potential for applications.

The subsequent work of Łaba–Wolff [**79**], Łaba–Pramanik [**76**], Pramanik–Seeger [**91**], and Garrigós–Seeger [**51**], [**52**] has expanded the applicability of the new method and the range of techniques involved in proving decouplings. Another important milestone was the paper of Bourgain [**20**], which brought to bear multilinear restriction theory and introduced a stunning new round of applications to exponential sums. All these results have paved the way for a more recent flurry of essentially sharp results, leading to an unanticipated breadth of applications.

We will start by introducing and motivating decoupling and will then describe the general toolbox that will be used throughout the remaining chapters.

### 9.1 The General Framework

Let $(x_j)_{j=1}^N$ be $N$ elements of a normed space $(X, \| \cdot \|_X)$. In this generality, the triangle inequality

$$\left\| \sum_{j=1}^N x_j \right\|_X \leq \sum_{j=1}^N \|x_j\|_X$$

is the best estimate available for the norm of the sum of $x_j$. When combined with the Cauchy–Schwarz inequality, it leads to

$$\left\| \sum_{j=1}^N x_j \right\|_X \leq N^{\frac{1}{2}} \left( \sum_{j=1}^N \|x_j\|_X^2 \right)^{1/2} .$$

Choosing $X = L^1$ and disjointly supported functions with equal $L^1$ norms shows that this inequality can be sharp.

However, if $X$ is a Hilbert space and if $x_j$ are pairwise orthogonal, then we have the stronger inequality

$$\left\| \sum_{j=1}^{N} x_j \right\|_X \leq \left( \sum_{j=1}^{N} \|x_j\|_X^2 \right)^{1/2}. \tag{9.1}$$

The fact that this is actually an equality will be less important to us. An example that is ubiquitous in Fourier analysis is when $X = L^2(\mathbb{R}^n)$ and $x_j$ are functions with disjointly supported Fourier transforms. It is natural to ask whether there is an analogous phenomenon in $L^p(\mathbb{R}^n)$ when $p \neq 2$, in the absence of Hilbert space orthogonality. We will see that the answer is "yes" in some rather general framework, when $p \geq 2$. The reader is invited to check Exercise 9.8 for the necessity of the restriction $p \geq 2$.

We will continue to denote by $\mathcal{P}_U$ the Fourier projection operator associated with the set $U \subset \mathbb{R}^n$.

**Definition 9.1** Let $p \geq 2$. For a family $\mathcal{S}$ consisting of (possibly infinitely many) pairwise disjoint sets $S_i \subset \mathbb{R}^n$, let $\mathrm{Dec}(\mathcal{S}, p)$ be the smallest constant for which the inequality

$$\|F\|_{L^p(\mathbb{R}^n)} \leq \mathrm{Dec}(\mathcal{S}, p) \left( \sum_i \|\mathcal{P}_{S_i} F\|_{L^p(\mathbb{R}^n)}^2 \right)^{1/2}$$

holds for each $F$ with Fourier transform supported on $\cup S_i$.

We will refer to this inequality as $l^2(L^p)$ decoupling, or simply $l^2$ decoupling, when the value of $p$ is clear from the context. The number $\mathrm{Dec}(\mathcal{S}, p)$ will be called the $l^2(L^p)$ decoupling constant of $\mathcal{S}$.

It is immediate that $\mathrm{Dec}(\mathcal{S}, 2) = 1$. Let us also take note of the trivial upper bound, when $\mathcal{S}$ is finite

$$\mathrm{Dec}(\mathcal{S}, p) \leq |\mathcal{S}|^{\frac{1}{2}}.$$

Typically, we will consider infinitely many such families $\mathcal{S}$ and will be interested in proving better upper bounds for $\mathrm{Dec}(\mathcal{S}, p)$ in terms of the cardinality $|\mathcal{S}|$. The ideal scenario – we will sometimes call it *genuine* $l^2$ decoupling – is when

$$\mathrm{Dec}(\mathcal{S}, p) \lesssim_\epsilon |\mathcal{S}|^\epsilon \tag{9.2}$$

holds true for each $\epsilon > 0$ and each family $\mathcal{S}$ under consideration.

Of course, (9.2) only makes sense if there is no upper bound on the cardinalities of the families $\mathcal{S}$. We allow small growth of the terms $\mathrm{Dec}(\mathcal{S}, p)$

as $|\mathcal{S}| \to \infty$ (we refer to these as $\epsilon$ losses), rather than seeking $O(1)$ upper bounds. This is enforced by the limitations of our methods, which rely at their heart on multiple iterations and induction on scales.

The leap from the triangle inequality to decoupling is superficially measured by replacing the $l^1$ norm with the smaller $l^2$ norm. Let us comment briefly on the significance of $l^2$. In the absence of any restriction on the Fourier transform, (9.1) may assume even stronger forms. For example, the inequality

$$\left\| \sum_{j=1}^{N} F_j \right\|_{L^p} \leq \left( \sum_{j=1}^{N} \|F_j\|_{L^p}^p \right)^{1/p}$$

holds whenever $F_j$ are disjointly supported. However, Exercise 9.7 shows (9.2) cannot hold if $l^2$ is replaced with $l^q$ ($q > 2$) in the definition of $\mathrm{Dec}(\mathcal{S}, p)$.

It turns out that we have already encountered a few nontrivial instances of $l^2$ decoupling in this book, though they were not presented this way. The underlying principle behind these examples is that $l^2$ decoupling holds whenever one has a reverse square function estimate. More precisely, let us take note of the following immediate consequence of Minkowski's inequality, valid for $p \geq 2$ and arbitrary functions $F_j$:

$$\left\| \left( \sum_{j=1}^{N} |F_j|^2 \right)^{1/2} \right\|_{L^p(\mathbb{R}^n)} \leq \left( \sum_{j=1}^{N} \|F_j\|_{L^p(\mathbb{R}^n)}^2 \right)^{1/2} . \tag{9.3}$$

Our first example is a consequence of (9.3) and of the Littlewood–Paley theorem 5.15. It states that $\mathrm{Dec}(\mathcal{R}_{LP}, p) < \infty$ for $2 \leq p < \infty$.

**Example 9.2** Let $\mathcal{R}_{LP}$ be the collection of all boxes $R = I_1 \times \cdots \times I_n$ in $\mathbb{R}^n$, with each $I_i$ of the form $[2^{k_i}, 2^{k_i+1}]$ or $[-2^{k_i+1}, -2^{k_i}]$ for some $k_i \in \mathbb{Z}$. Then for each $2 \leq p < \infty$ and each $F \in L^p(\mathbb{R}^n)$, we have

$$\|F\|_{L^p(\mathbb{R}^n)} \lesssim \left( \sum_{R \in \mathcal{R}_{LP}} \|\mathcal{P}_R F\|_{L^p(\mathbb{R}^n)}^2 \right)^{1/2} .$$

The next example will be the object of a thorough investigation in Chapter 10. In the terminology we have just introduced, it states that there is a genuine $l^2(L^p)$ decoupling for the families of sets $\Theta_{\mathbb{P}^{n-1}}(\delta)$ defined in the following example (as $\delta \to 0$), in the range $2 \leq p \leq 2n/(n-1)$.

**Example 9.3** For $\delta \leq 1$, let $\Theta_{\mathbb{P}^{n-1}}(\delta)$ be a partition of the $\delta$-neighborhood $\mathcal{N}_{\mathbb{P}^{n-1}}(\delta)$ of $\mathbb{P}^{n-1}$ into almost rectangular boxes $\theta$ of the form

$$\theta = (\tau \times \mathbb{R}) \cap \mathcal{N}_{\mathbb{P}^{n-1}}(\delta),$$

where $\tau$ is an almost cube with diameter $\sim \delta^{1/2}$.

Assume that Conjecture 5.19 holds. Then whenever $\widehat{F}$ is supported on $\mathcal{N}_{\mathbb{P}^{n-1}}(\delta)$ and $2 \leq p \leq 2n/(n-1)$, we have

$$\|F\|_{L^p(\mathbb{R}^n)} \lesssim_\epsilon \delta^{-\epsilon} \left( \sum_{\theta \in \Theta_{\mathbb{P}^{n-1}}(\delta)} \|\mathcal{P}_\theta F\|^2_{L^p(\mathbb{R}^n)} \right)^{1/2}. \tag{9.4}$$

This is a consequence of (9.3) and Corollary 5.21. Let us recall that Conjecture 5.19, while known for $n = 2$, is a very deep open problem when $n \geq 3$. However, in Chapter 10 we will prove that (9.4) is true unconditionally. In fact, and this is one of the main strengths of decoupling, we will see that (9.4) holds in the larger range $2 \leq p \leq 2(n + 1)/(n - 1)$.

The paraboloid in Example 9.3 is just one among a large host of manifolds for which the decoupling question is worth asking. The problem we propose to investigate in the remainder of this book can be (somewhat vaguely) formulated as follows. For a manifold $\mathcal{M}$ as in (1.3), we continue to write

$$\mathcal{N}_{\mathcal{M}}(\delta) = \{(\xi, \psi(\xi) + \eta) : \xi \in U, \ \eta \in B_{n-d}(0, \delta)\}.$$

**Problem 9.4** ($l^2$ decoupling for manifolds) *Let $\mathcal{M}$ be a $d$-dimensional manifold in $\mathbb{R}^n$, as in (1.3). For each $\delta \in (0, 1)$, let $\tilde{\mathcal{N}}_{\mathcal{M}}(\delta) \subset \mathbb{R}^n$ be a set containing the $\delta$-neighborhood $\mathcal{N}_{\mathcal{M}}(\delta)$ of $\mathcal{M}$. Let also $\Theta_{\mathcal{M}}(\delta)$ be a partition of $\tilde{\mathcal{N}}_{\mathcal{M}}(\delta)$ into sets of the form*

$$\theta = (\tau \times \mathbb{R}^{n-d}) \cap \tilde{\mathcal{N}}_{\mathcal{M}}(\delta),$$

*where $\tau$ ranges over a collection $\mathcal{P}_{\mathcal{M}}(\delta)$ of pairwise disjoint, almost rectangular boxes in $\mathbb{R}^d$.*

*Find the range $2 \leq p \leq p_c$ such that*

$$\mathrm{Dec}(\Theta_{\mathcal{M}}(\delta), p) \lesssim_\epsilon \delta^{-\epsilon}. \tag{9.5}$$

We will refer to $\delta$ as the *thickness parameter*, and to $p_c$ as the *critical exponent*. In general, we will work with sets $\theta$ that are almost rectangular boxes, though sometimes it is more convenient (and at the same time, sufficient for applications) to allow for more curved sets. See Section 12.5.

When $\mathcal{M}$ is a hypersurface, we will always take $\tilde{\mathcal{N}}_{\mathcal{M}}(\delta) = \mathcal{N}_{\mathcal{M}}(\delta)$ and each $\theta$ will be an almost rectangular box. On the other hand, when $\mathcal{M}$ is the moment curve in $\mathbb{R}^n$ we will find it more convenient to work with the larger neighborhoods $\Gamma_n(\delta^{1/n})$ that will be introduced in Chapter 11. Most of our investigation will focus on the case when all $\tau \in \mathcal{P}_{\mathcal{M}}(\delta)$ have comparable dimensions. This is a feature that distinguishes the new decouplings from the Littlewood–Paley-type decoupling in Example 9.2. While the latter relies on lacunary growth, the main engine that powers the new decouplings is curvature. Sometimes the two methods can be combined, as in Section 12.6.

In general, given a manifold $\mathcal{M}$, the problem of identifying good candidates for the neighborhoods $\tilde{\mathcal{N}}_{\mathcal{M}}(\delta)$ and their partitions $\Theta_{\mathcal{M}}(\delta)$ for which an interesting decoupling holds is itself interesting. Different choices for $\tilde{\mathcal{N}}_{\mathcal{M}}(\delta)$ and $\Theta_{\mathcal{M}}(\delta)$ will lead to different critical exponents $p_c$, and quite often to qualitatively different types of applications.

It is clear that (9.5) holds with $p = 2$ for each $\mathcal{M}$ and each family of partitions. On the other hand, the next result severely restricts the geometry of the partitions $\Theta_{\mathcal{M}}(\delta)$ for which a genuine decoupling holds for $p > 2$. The guiding principle is that one cannot decouple in a direction where the manifold is flat.

**Proposition 9.5** *Let $T$ be a tube in $\mathbb{R}^n$. Consider a partition $\mathcal{T}_N$ of $T$ into $N$ shorter, essentially congruent tubes. Then*

$$\mathrm{Dec}(\mathcal{T}_N, p) \sim N^{\frac{1}{2} - \frac{1}{p}},$$

*with implicit similarity constants independent of $N$.*

*Proof* The upper bound on $\mathrm{Dec}(\mathcal{T}_N, p)$ follows by interpolating between $L^2$ and $L^\infty$, as in Lemma 4.5. The lower bound is achieved by testing with a function $F$ whose Fourier transform is a smooth approximation of $1_T$. The details are left to the reader. □

Let us identify two consequences of this proposition regarding decoupling for a hypersurface $\mathcal{M}$ given by the graph of a smooth function $\psi$. Consider a partition $\Theta_{\mathcal{M}}(\delta)$ of $\mathcal{N}_{\mathcal{M}}(\delta)$ into almost rectangular boxes of the form

$$\theta = \mathcal{N}_\tau(\delta) = \{(\xi, \psi(\xi) + s) \colon \xi \in \tau, \ |s| < \delta\},$$

and assume that (9.5) holds for some $p > 2$.

Let us first assume that all $\tau$ are cubes of diameter $\sim \Delta$. Then Proposition 9.5 forces the lower bound $\Delta \gtrsim_\epsilon \delta^{1/2+\epsilon}$, for each $\epsilon > 0$. Indeed, since $\mathcal{N}_q(\delta)$ is an almost rectangular box whenever $q$ is a cube with side length $\delta^{1/2}$, the inequality $\Delta \lesssim \delta^{1/2+\epsilon_0}$ for some $\epsilon_0 > 0$ would force the existence (for some $\delta_* \sim \delta$) of a $(\delta_*, \delta_*^{1/2})$-tube $T$ inside $\mathcal{N}_{\mathcal{M}}(\delta)$, such that its partition

$$\mathcal{T}_N := \{T \cap \theta \colon \theta \in \Theta_{\mathcal{M}}(\delta)\}$$

with $N \sim \delta^{1/2}/\Delta \gtrsim \delta^{-\epsilon_0}$ is like that in Proposition 9.5. For hypersurfaces with nonzero Gaussian curvature, we are left with essentially only one choice for the diameter of $\tau$, namely $\Delta \approx \delta^{1/2}$ (if $\Delta$ is much larger than $\delta^{1/2}$, then $\theta$ is no longer almost rectangular). We will deal with this case in Chapters 10 and 12.

Another consequence of Proposition 9.5 is that each line segment contained in $\mathcal{M}$ would have to intersect $O_\epsilon(\delta^{-\epsilon})$ elements of the partition $\Theta_{\mathcal{M}}(\delta)$. Examples of manifolds containing lines through every point include both singly ruled and doubly ruled hypersurfaces. From the first category, we will analyze the cone. To accommodate the aforementioned restriction, in Section 12.2 we will decouple the cone using partitions into planks having length $\sim 1$ in the direction of zero principal curvature. An example of surface from the second category that will be investigated in Section 12.3 is the hyperbolic paraboloid. Its doubly ruled nature will in fact rule out any genuine $l^2(L^p)$ decoupling for $p > 2$, and will force us to introduce $l^q$ decouplings.

We can also formulate multilinear analogs of decoupling. The lack of curvature is less of an enemy in this setting. In particular, in the range $2 \leq p \leq 2n/d$ transversality suffices. Moreover, the diameter of the sets in the partition can be taken to be as small as the thickness parameter $\delta$.

**Theorem 9.6** (Multilinear decoupling) *Let* $\mathbb{M} = (\mathcal{M}_1, \ldots, \mathcal{M}_m)$ *with each* $\mathcal{M}_j$ *a $d$-dimensional manifold in $\mathbb{R}^n$ as in (6.21). Assume that $\mathbb{M}$ satisfies the BL-transversality condition (6.22).*

*Given* $\delta \leq \delta' < 1$, *let* $\Theta_j$ *be a partition of* $\mathcal{N}_{\mathcal{M}_j}(\delta)$ *into sets* $\theta$ *of the form* $(q \times \mathbb{R}^{n-d}) \cap \mathcal{N}_{\mathcal{M}_j}(\delta)$, *with each $q$ a cube in $\mathbb{R}^d$ with diameter $\sim \delta'$.*

*Then for each* $2 \leq p \leq 2n/d$ *and each $F_j$ with the Fourier transform supported on $\mathcal{N}_{\mathcal{M}_j}(\delta)$, we have*

$$\left\| \left( \prod_{j=1}^{m} F_j \right)^{\frac{1}{m}} \right\|_{L^p(\mathbb{R}^n)} \lesssim_{\epsilon} \delta^{-\epsilon} \prod_{j=1}^{m} \left( \sum_{\theta \in \Theta_j} \| \mathcal{P}_\theta F_j \|_{L^p(\mathbb{R}^n)}^2 \right)^{\frac{1}{2m}}.$$

*Proof* Since $\mathcal{N}_{\mathcal{M}_j}(\delta) \subset \mathcal{N}_{\mathcal{M}_j}(\delta')$, we may assume that $\delta' = \delta$. In this case, the sets $\theta$ are almost cubes with diameter $\sim \delta$. The inequality is easily seen to hold when $p = 2$. Invoking interpolation (a multilinear variant of Exercise 9.21) reduces the problem to the case $p = 2n/d$. On the other hand, this case follows via a standard local-to-global argument that we sketch in the following.

Indeed, for each ball $B$ in $\mathbb{R}^n$ with radius $\delta^{-1}$, let $w_B$ be a smooth approximation of $1_B$. The local inequality

$$\left\| \left( \prod_{j=1}^{m} F_j \right)^{\frac{1}{m}} \right\|_{L^{\frac{2n}{d}}(B)} \lesssim_{\epsilon} \delta^{-\epsilon + \frac{n-d}{2}} \prod_{j=1}^{m} \left( \sum_{\theta \in \Theta_j} \| \mathcal{P}_\theta F_j \|_{L^2(w_B)}^2 \right)^{\frac{1}{2m}}$$

is a consequence of the multilinear restriction Theorem 6.24 and $L^2$ orthogonality. Combining this with Hölder's inequality leads to

$$\left\| \left( \prod_{j=1}^{m} F_j \right)^{\frac{1}{m}} \right\|_{L^{\frac{2n}{d}}(B)}^{\frac{2n}{d}} \lesssim_{\epsilon} \delta^{-\epsilon} \prod_{j=1}^{m} \left( \sum_{\theta \in \Theta_j} \| \mathcal{P}_\theta F_j \|_{L^{\frac{2n}{d}}(w_B)}^2 \right)^{\frac{n}{dm}}.$$

To close the argument, we sum this inequality over balls $B$ in a finitely overlapping cover of $\mathbb{R}^n$. $\qquad \square$

**Exercise 9.7** Let $U_1, \ldots, U_N$ be arbitrary pairwise disjoint open sets in $\mathbb{R}^n$. Let $p, q \geq 2$. Show that there are smooth functions $f_j$ with Fourier transforms compactly supported on $U_j$ such that

$$\left\| \sum_{j=1}^{N} f_j \right\|_{L^p(\mathbb{R}^n)} \gtrsim N^{\frac{1}{2}-\frac{1}{q}} \left( \sum_{j=1}^{N} \|f_j\|_{L^p(\mathbb{R}^n)}^q \right)^{1/q},$$

with implicit constant independent of $N$ and $U_1, \ldots, U_N$.

Hint: Reduce matters to inequalities about exponential sums.

**Exercise 9.8** Let $U_1, \ldots, U_N$ be arbitrary pairwise disjoint open sets in $\mathbb{R}^n$. Let $p < 2$. Show that there are smooth functions $f_j$ with Fourier transforms compactly supported on $U_j$ such that

$$\left\| \sum_{j=1}^{N} f_j \right\|_{L^p(\mathbb{R}^n)} \gtrsim \left( \sum_{j=1}^{N} \|f_j\|_{L^p(\mathbb{R}^n)}^2 \right)^{1/2} N^{\frac{1}{p}-\frac{1}{2}},$$

with implicit constant independent of $N$ and $U_1, \ldots, U_N$.

Hint: Arrange that $f_j$ are spatially concentrated on pairwise disjoint sets.

**Exercise 9.9** Let $\Theta$ be a collection of sets in $\mathbb{R}^n$ and let $T : \mathbb{R}^n \to \mathbb{R}^n$ be a nonsingular affine map. Denote by $\Theta'$ the collection consisting of the sets $T(\theta), \theta \in \Theta$.

Prove that $\mathrm{Dec}(\Theta, p) = \mathrm{Dec}(\Theta', p)$.

Hint: Use (4.1).

**Exercise 9.10** Let $\mathcal{R}$ be a collection of rectangular boxes in $\mathbb{R}^n$ such that the boxes $\{2R : R \in \mathcal{R}\}$ are pairwise disjoint. Let $2 \leq p < \infty$. Prove that for each $F \in L^p(\mathbb{R}^n)$ with the Fourier transform supported on $\cup_{R \in \mathcal{R}} R$, we have

$$\left( \sum_{R \in \mathcal{R}} \|\mathcal{P}_R F\|_p^p \right)^{\frac{1}{p}} \lesssim \|F\|_p.$$

**Exercise 9.11** Let $\mathcal{R}_1, \mathcal{R}_2$ be two partitions of a set $S$, consisting of rectangular boxes. Assume that each $R_1 \in \mathcal{R}_1$ intersects at most $C$ sets from $\mathcal{R}_2$ and that each $R_2 \in \mathcal{R}_2$ intersects at most $C$ sets from $\mathcal{R}_1$. Prove that $\mathrm{Dec}(\mathcal{R}_1, p) \sim_C \mathrm{Dec}(\mathcal{R}_2, p)$.

## 9.2 Local and Global Decoupling

Given a ball (or a cube) $B \subset \mathbb{R}^n$ centered at $c$ and of radius (side length) $R$, we consider the weight

$$w_B(x) = \frac{1}{\left(1 + \frac{|x-c|}{R}\right)^{50n}}. \tag{9.6}$$

We will occasionally use the notation

$$\|F\|_{L^p_\sharp(w_B)} = \left(\frac{1}{|B|}\int |F|^p w_B\right)^{1/p}.$$

It is immediate that $w_B$ decays slower than any Schwartz function. The only special fact about our choice for the exponent $50n$ is that it is large enough to guarantee various (implicit) integrability requirements. All the inequalities we will prove hold equally well if a larger exponent is chosen instead.

A key, easy-to-check property of the weights $w_B$ that will be used a few times is the following inequality

$$1_B \lesssim \sum_{B'\in\mathcal{B}} w_{B'} \lesssim w_B, \tag{9.7}$$

valid for all balls $B$ with radius $R$ and all finitely overlapping covers $\mathcal{B}$ of $B$ with balls $B'$ of (fixed) radius $1 \le R' \le R$. The implicit constants in (9.7) will be independent of $R, R'$.

We will find extremely useful the following simple result. Sometimes we will be able to check a stronger assumption than (W1), with $O_2(w_B)$ replaced with $O_2(\eta_B)$, for some Schwartz function $\eta_B$.

**Lemma 9.12** *Let $\mathcal{W}$ be the collection of all positive integrable functions on $\mathbb{R}^n$. Fix $R > 0$. Let $O_1, O_2 \colon \mathcal{W} \to [0,\infty]$ have the following four properties:*

*(W1) $O_1(1_B) \lesssim O_2(w_B)$ for all balls $B \subset \mathbb{R}^n$ of radius $R$.*
*(W2) $O_1(\alpha u + \beta v) \le \alpha O_1(u) + \beta O_1(v)$, for each $u, v \in \mathcal{W}$ and $\alpha, \beta > 0$.*
*(W3) $O_2(\alpha u + \beta v) \ge \alpha O_2(u) + \beta O_2(v)$, for each $u, v \in \mathcal{W}$ and $\alpha, \beta > 0$.*
*(W4) If $u \le v$, then $O_1(u) \le O_1(v)$ and $O_2(u) \le O_2(v)$.*

*Then*

$$O_1(w_B) \lesssim O_2(w_B)$$

*for each ball $B$ with radius $R$. The implicit constant is independent of $R, B$ and only depends on the implicit constant from (W1).*

*Proof* Let $\mathcal{B}$ be a finitely overlapping cover of $\mathbb{R}^n$ with balls $B' = B'(c_{B'}, R)$. Let us recall the two inequalities in (7.5)

$$w_B(x) \lesssim \sum_{B'\in\mathcal{B}} 1_{B'}(x) w_B(c_{B'})$$

and

$$\sum_{B'\in\mathcal{B}} w_{B'}(x) w_B(c_{B'}) \lesssim w_B(x).$$

We combine these inequalities with (W1) through (W4) to write

$$O_1(w_B) \lesssim O_1 \left( \sum_{B' \in \mathcal{B}} 1_{B'} w_B(c_{B'}) \right)$$

$$\leq \sum_{B' \in \mathcal{B}} O_1(1_{B'}) w_B(c_{B'})$$

$$\leq \sum_{B' \in \mathcal{B}} O_2(w_{B'}) w_B(c_{B'})$$

$$\lesssim O_2(w_B). \qquad \square$$

**Remark 9.13** It is rather immediate that for each function $F$, the quantity

$$O_1(v) := \|F\|_{L^p(v)}^p$$

satisfies (W2) and (W4). Also, for each $p \geq 2$ and arbitrary functions $F_i$, Minkowski's inequality in $l_{2/p}$ shows that

$$O_2(v) := \left( \sum_i \|F_i\|_{L^p(v)}^2 \right)^{p/2}$$

satisfies (W3) and (W4).

Given $\mathcal{M}$, $\Theta_{\mathcal{M}}(\delta)$ and $\tilde{\mathcal{N}}_{\mathcal{M}}(\delta)$ as in Problem 9.4, we introduce three useful ways to measure the associated decoupling constant. The first one is a restatement of Definition 9.1.

**Definition 9.14** (Decoupling constants) Let $\mathrm{Dec}_{\mathrm{global}}(\delta, p)$ be the smallest constant such that

$$\|F\|_{L^p(\mathbb{R}^n)} \leq \mathrm{Dec}_{\mathrm{global}}(\delta, p) \left( \sum_{\theta \in \Theta_{\mathcal{M}}(\delta)} \|\mathcal{P}_\theta F\|_{L^p(\mathbb{R}^n)}^2 \right)^{1/2}$$

holds for each $F$ with the Fourier transform supported on $\tilde{\mathcal{N}}_{\mathcal{M}}(\delta)$.

Let $\mathrm{Dec}_{\mathrm{local}}(\delta, p)$ be the smallest constant such that

$$\|F\|_{L^p(B_R)} \leq \mathrm{Dec}_{\mathrm{local}}(\delta, p) \left( \sum_{\theta \in \Theta_{\mathcal{M}}(\delta)} \|\mathcal{P}_\theta F\|_{L^p(w_{B_R})}^2 \right)^{1/2}$$

holds for each $F$ with the Fourier transform supported on $\tilde{\mathcal{N}}_{\mathcal{M}}(\delta)$ and each ball $B_R$ of radius $R = \delta^{-1}$.

Let $\mathrm{Dec}_{\mathrm{weighted}}(\delta, p)$ be the smallest constant such that

$$\|F\|_{L^p(w_{B_R})} \leq \mathrm{Dec}_{\mathrm{weighted}}(\delta, p) \left( \sum_{\theta \in \Theta_{\mathcal{M}}(\delta)} \|\mathcal{P}_\theta F\|_{L^p(w_{B_R})}^2 \right)^{1/2}$$

holds for each $F$ with Fourier transform supported on $\tilde{\mathcal{N}}_{\mathcal{M}}(\delta)$ and each ball $B_R$ of radius $R = \delta^{-1}$.

The reader will find it rather pleasing that the three decoupling constants are in fact comparable to each other in all cases of interest. As a consequence, in the future we will simply write $\mathrm{Dec}(\delta, p)$ to refer to either $\mathrm{Dec}_{\mathrm{global}}(\delta, p)$ or $\mathrm{Dec}_{\mathrm{weighted}}(\delta, p)$ or $\mathrm{Dec}_{\mathrm{local}}(\delta, p)$.

To keep things simple, we illustrate this principle under the following assumption: for each $\theta \in \Theta_{\mathcal{M}}(\delta)$, there is a rectangular box $R_\theta$ such that

$$R_\theta \subset \theta \quad \text{and} \quad \theta + B(0, \delta) \subset R_\theta + T, \tag{9.8}$$

for a finite set of points $T$ independent of $\theta$ and whose cardinality is $O(1)$ (independent of $\delta$).

It is easy to check that our partition $\Theta_{\mathbb{P}^1}(\delta)$ for the parabola – this manifold will be thoroughly investigated in the next chapter – satisfies this assumption. Small modifications of this assumption will lead to the verification of the principle for other manifolds that we will encounter.

**Proposition 9.15** *The following equivalence holds with implicit constants independent of $\delta$*

$$\mathrm{Dec}_{\mathrm{global}}(\delta, p) \sim \mathrm{Dec}_{\mathrm{local}}(\delta, p) \sim \mathrm{Dec}_{\mathrm{weighted}}(\delta, p).$$

*Proof* Throughout the whole argument, we write $R = \delta^{-1}$.

Assume that $\mathrm{supp}\,(\widehat{F}) \subset \tilde{\mathcal{N}}_{\mathcal{M}}(\delta)$. Let $\mathcal{B}_R$ be a finitely overlapping cover of $\mathbb{R}^n$ with balls of radius $R$. To prove that $\mathrm{Dec}_{\mathrm{global}}(\delta, p) \lesssim \mathrm{Dec}_{\mathrm{local}}(\delta, p)$, we rely on (9.7) and Minkowski's inequality:

$$\|F\|_{L^p(\mathbb{R}^n)} \leq \left( \sum_{B_R \in \mathcal{B}_R} \|F\|_{L^p(B_R)}^p \right)^{1/p}$$

$$\leq \mathrm{Dec}_{\mathrm{local}}(\delta, p) \left( \sum_{B_R \in \mathcal{B}_R} \left( \sum_{\theta \in \Theta_{\mathcal{M}}(\delta)} \|\mathcal{P}_\theta F\|_{L^p(w_{B_R})}^2 \right)^{p/2} \right)^{1/p}$$

$$\lesssim \mathrm{Dec}_{\mathrm{local}}(\delta, p) \left( \sum_{\theta \in \Theta_{\mathcal{M}}(\delta)} \|\mathcal{P}_\theta F\|_{L^p(\mathbb{R}^n)}^2 \right)^{1/2}.$$

To prove that $\mathrm{Dec}_{\mathrm{local}}(\delta, p) \lesssim \mathrm{Dec}_{\mathrm{global}}(\delta, p)$, fix $B_R$ and let $\phi_R$ be a positive Schwartz function such that $1_{B_R} \leq \phi_R \lesssim w_{B_R}$ and $\mathrm{supp}\,(\widehat{\phi_R}) \subset B(0, \delta)$. Assume that $\mathrm{supp}\,(\widehat{F}) \subset \tilde{\mathcal{N}}_{\mathcal{M}}(\delta)$. Note that $(\mathcal{P}_\theta F)\phi_R$ has the Fourier transform supported on $\theta + B(0, \delta)$. Let $R_\theta, T$ be as in (9.8). Write

$$(\mathcal{P}_\theta F)\phi_R = \sum_{t \in T} F_{\theta, t},$$

where

$$\widehat{F_{\theta,t}} = \frac{1_{R_\theta+t}}{\sum_{t'\in T} 1_{R_\theta+t'}}(\widehat{\mathcal{P}_\theta F})\phi_R.$$

Note that there is a partition $\mathcal{R}$ of $R_\theta + t$ into $O(1)$ rectangular boxes, and there are coefficients $a_R \in [0, 1]$ such that

$$\frac{1_{R_\theta+t}}{\sum_{t'\in T} 1_{R_\theta+t'}} = \sum_{R\in\mathcal{R}} a_R 1_R.$$

Since the Fourier projection onto a rectangular box has $L^p$ norm independent of the box, it follows that

$$\|F_{\theta,t}\|_{L^p(\mathbb{R}^n)} \lesssim \|(\mathcal{P}_\theta F)\phi_R\|_{L^p(\mathbb{R}^n)},$$

with the implicit constant independent of $\theta, \delta$.

Since $R_\theta \subset \theta$, we trivially have $\mathrm{Dec}(\{R_\theta : \theta \in \Theta_\mathcal{M}(\delta)\}, p) \le \mathrm{Dec}_{\mathrm{global}}(\delta, p)$. Exercise 9.9 shows that for each $t \in T$, we have

$$\left\| \sum_{\theta\in\Theta_\mathcal{M}(\delta)} F_{\theta,t} \right\|_{L^p(\mathbb{R}^n)} \le \mathrm{Dec}_{\mathrm{global}}(\delta, p)\left( \sum_{\theta\in\Theta_\mathcal{M}(\delta)} \|F_{\theta,t}\|^2_{L^p(\mathbb{R}^n)} \right)^{1/2}.$$

Combining the last two inequalities with the triangle inequality, we may complete the proof of the inequality $\mathrm{Dec}_{\mathrm{local}}(\delta, p) \lesssim \mathrm{Dec}_{\mathrm{global}}(\delta, p)$ as follows:

$$\|F\|_{L^p(B_R)} \le \left\| \sum_{\theta\in\Theta_\mathcal{M}(\delta)} (\mathcal{P}_\theta F)\phi_R \right\|_{L^p(\mathbb{R}^n)}$$

$$\le \sum_{t\in T}\left\| \sum_{\theta\in\Theta_\mathcal{M}(\delta)} F_{\theta,t} \right\|_{L^p(\mathbb{R}^n)}$$

$$\le \mathrm{Dec}_{\mathrm{global}}(\delta, p)\sum_{t\in T}\left( \sum_{\theta\in\Theta_\mathcal{M}(\delta)} \|F_{\theta,t}\|^2_{L^p(\mathbb{R}^n)} \right)^{1/2}$$

$$\lesssim \mathrm{Dec}_{\mathrm{global}}(\delta, p)\left( \sum_{\theta\in\Theta_\mathcal{M}(\delta)} \|(\mathcal{P}_\theta F)\phi_R\|^2_{L^p(\mathbb{R}^n)} \right)^{1/2}$$

$$\lesssim \mathrm{Dec}_{\mathrm{global}}(\delta, p)\left( \sum_{\theta\in\Theta_\mathcal{M}(\delta)} \|\mathcal{P}_\theta F\|^2_{L^p(w_{B_R})} \right)^{1/2}.$$

Note that the inequality $\mathrm{Dec}_{\mathrm{local}}(\delta, p) \lesssim \mathrm{Dec}_{\mathrm{weighted}}(\delta, p)$ is trivial. Also, the reverse inequality $\mathrm{Dec}_{\mathrm{weighted}}(\delta, p) \lesssim \mathrm{Dec}_{\mathrm{local}}(\delta, p)$ is a consequence of Lemma 9.12 and Remark 9.13. $\qquad\square$

**Remark 9.16** A similar equivalence exists and will be exploited later in the multilinear setting. This is due to the fact that if $O_2^{(i)}$ satisfies (W3) and (W4) for each $1 \le i \le N$, then so does the geometric average $(\prod_{i=1}^N O_2^{(i)})^{1/N}$.

There is an essentially equivalent way to formulate $l^2$ decoupling, using the extension operator. For example, it is rather immediate that for each ball $B$ of radius $\delta^{-1}$ we have

$$\|E^{\mathcal{M}} f\|_{L^p(w_B)} \lesssim \mathrm{Dec}_{\mathrm{weighted}}(\delta, p) \left( \sum_{\tau \in \mathcal{P}_{\mathcal{M}}(\delta)} \|E_\tau^{\mathcal{M}} f\|_{L^p(w_B)}^2 \right)^{\frac{1}{2}}. \quad (9.9)$$

We will occasionally prefer this formulation.

## 9.3 A Few Basic Tools

This section explores (partly through exercises) some of the general principles and techniques that will be used frequently throughout the remaining chapters.

The first result shows that decouplings can be iterated. This is a key feature that distinguishes them from the reverse square function estimates that have been analyzed earlier in the book. The proof is immediate.

**Proposition 9.17** (Iteration) *Let $\Theta$ be a collection consisting of pairwise disjoint sets $\theta$. Assume that each $\theta$ is partitioned into sets $\theta_1$. Call $\Theta_1$ the collection of all these sets $\theta_1$. Assume that for each $\theta$ and each $F$ we have*

$$\|\mathcal{P}_\theta F\|_p \le D_1 \left( \sum_{\theta_1 \subset \theta} \|\mathcal{P}_{\theta_1} F\|_p^2 \right)^{1/2}.$$

*Assume also that*

$$\|F\|_p \le D_2 \left( \sum_{\theta \in \Theta} \|\mathcal{P}_\theta F\|_p^2 \right)^{1/2},$$

*whenever $\widehat{F}$ is supported on $\cup_{\theta \in \Theta} \theta$.*

*Then, for each such $F$ we also have*

$$\|F\|_p \le D_1 D_2 \left( \sum_{\theta_1 \in \Theta_1} \|\mathcal{P}_{\theta_1} F\|_p^2 \right)^{1/2}.$$

An application of $L^2$ (global) orthogonality and Lemma 9.12 leads to the following useful local version.

**Proposition 9.18** (Local $L^2$ decoupling) *Let $S_1, \dots, S_N$ be sets in $\mathbb{R}^n$ such that their $\delta$-neighborhoods are pairwise disjoint. Then for each $F$ with $\mathrm{supp}\,(\widehat{F}) \subset \cup S_i$ and each ball $B$ with radius $\delta^{-1}$, we have*

$$\|F\|_{L^2(w_B)} \lesssim \left( \sum_i \|\mathcal{P}_{S_i} F\|_{L^2(w_B)}^2 \right)^{1/2}.$$

The following inequality will serve as a rigorous quantification of the fact that the absolute value of a function with the Fourier transform supported on a set of diameter $O(R^{-1})$ is essentially constant at spatial scale $R$. This is a close relative of Proposition 7.17.

**Lemma 9.19** (Reverse Hölder's inequality) *Assume that $F: \mathbb{R}^n \to \mathbb{C}$ has a Fourier transform supported on a set with diameter $\lesssim 1/R$.*

*Then for each $q \geq p \geq 1$ and each ball $B$ in $\mathbb{R}^n$ with radius $R$, we have*

$$\|F\|_{L^q_\sharp(B)} \lesssim \|F\|_{L^p_\sharp(w_B)},$$

*with the implicit constant independent of $R$, $B$, and $F$.*

*Proof* Let $\eta$ be a Schwartz function on $\mathbb{R}^n$ satisfying $1_{[-1,1]^n} \leq \eta$ and such that the Fourier transform of $\eta$ is supported on $[-1,1]^n$. Letting $\eta_B(x) = \eta((x - c_B)/R)$, we may write

$$\|F\|_{L^q(B)} \leq \|\eta_B F\|_{L^q(\mathbb{R}^n)}.$$

Since the Fourier transform of $\eta_B F$ is supported on a set with diameter $\lesssim 1/R$, invoking Young's inequality for convolutions, we have

$$\|\eta_B F\|_{L^q(\mathbb{R}^n)} \lesssim R^{-n/r} \|\eta_B F\|_{L^p(\mathbb{R}^n)},$$

with

$$\frac{1}{q} = \frac{1}{p} - \frac{1}{r}.$$

The desired inequality is now immediate since $(\eta_B)^p \lesssim w_B$. $\qquad \square$

**Exercise 9.20** Given functions $F, F_i: \mathbb{R}^n \to \mathbb{C}$, assume that the inequality

$$\|F\|_{L^p(w_B)} \leq D \left( \sum_i \|F_i\|_{L^p(w_B)}^2 \right)^{1/2}$$

holds for all balls $B$ in $\mathbb{R}^n$ with some fixed radius $R$. Prove that

$$\|F\|_{L^p(w_B)} \lesssim \left( \sum_i \|F_i\|_{L^p(w_B)}^2 \right)^{1/2}$$

holds for all balls of radius at least $R$, with an implicit constant that is independent of $R$. Conclude that in the Definition 9.14, the requirement $R = \delta^{-1}$ can be replaced with $R \geq \delta^{-1}$.

The next exercise shows that decouplings can be interpolated. This will conveniently allow us to restrict attention to proving decouplings for certain critical exponents.

**Exercise 9.21** (Interpolation) Let $\Theta$ be a collection of congruent rectangular boxes $\theta$ in $\mathbb{R}^n$ with dimensions in the interval $[\delta, 1]$. We assume that the boxes $4\theta$ are pairwise disjoint. Let $\Theta'$ be the collection consisting of the boxes $4\theta$.

(a) Assume that the following mixed local–global decoupling holds true

$$\|F\|_{L^p(w_B)} \le C \left( \sum_{\theta \in \Theta} \|\mathcal{P}_{2\theta} F\|_{L^p(\mathbb{R}^n)}^2 \right)^{\frac{1}{2}}$$

uniformly over all balls $B \subset \mathbb{R}^n$ of radius $\delta^{-1}$ and all $F$ with $\widehat{F}$ supported on the union of $2\theta$. Prove that

$$\mathrm{Dec}(\Theta, p) \lesssim C.$$

(b) Prove that for each $p_1 < p < p_2$ with $1/p = \alpha/p_1 + (1-\alpha)/p_2$, we have

$$\mathrm{Dec}(\Theta, p) \lesssim_\epsilon \delta^{-\epsilon} |\Theta|^\epsilon \mathrm{Dec}(\Theta', p_1)^\alpha \mathrm{Dec}(\Theta', p_2)^{1-\alpha}.$$

Hint for (b): Using (a), it suffices to prove that if $\mathrm{supp}\,(\widehat{F}) \subset \cup 2\theta$, then

$$\|F\|_{L^p(w_B)} \lesssim_\epsilon \delta^{-\epsilon} |\Theta|^\epsilon \mathrm{Dec}(\Theta', p_1)^\alpha \mathrm{Dec}(\Theta', p_2)^{1-\alpha} \left( \sum_{\theta \in \Theta} \|\mathcal{P}_{2\theta} F\|_{L^p(\mathbb{R}^n)}^2 \right)^{\frac{1}{2}}.$$

Consider a wave packet decomposition (Exercise 2.7)

$$\mathcal{P}_{2\theta} F = \sum_{T \in \mathcal{T}_\theta} w_T W_T$$

with $\mathrm{supp}\,(\widehat{W_T}) \subset 4\theta$. Split the collection $\cup \mathcal{T}_\theta$ into subcollections $\mathcal{T}_i$, such that each $\mathcal{T}_i$ has the following two properties. First, all $|w_T|$ are roughly the same for $T \in \mathcal{T}_i$. Second, there is a number $N_i \ge 1$ such that for each $\theta$ we either have $|\mathcal{T}_\theta \cap \mathcal{T}_i| \sim N_i$ or $\mathcal{T}_\theta \cap \mathcal{T}_i = \emptyset$.

Analyze each $F_i = \sum_{T \in \mathcal{T}_i} w_T W_T$ separately using Exercise 2.7. In particular, prove that

$$\left( \sum_{\theta \in \Theta} \|\mathcal{P}_{4\theta} F_i\|_{L^p(\mathbb{R}^n)}^2 \right)^{\frac{1}{2}} \sim \left( \sum_{\theta \in \Theta} \|\mathcal{P}_{4\theta} F_i\|_{L^{p_1}(\mathbb{R}^n)}^2 \right)^{\frac{\alpha}{2}} \left( \sum_{\theta \in \Theta} \|\mathcal{P}_{4\theta} F_i\|_{L^{p_2}(\mathbb{R}^n)}^2 \right)^{\frac{1-\alpha}{2}}.$$

Also, argue that only $O(\delta^{-\epsilon} |\Theta|^\epsilon)$ values of $i$ contribute significantly to the inequality

$$\|F\|_{L^p(w_B)} \le \sum_i \|F_i\|_{L^p(w_B)}.$$

**Exercise 9.22** (Cylindrical decoupling) Let $\mathcal{S}_n$ be a collection of sets in $\mathbb{R}^n$. Let $\mathcal{S}_{n+1}$ be the collection of all vertical sets $S \times \mathbb{R} \subset \mathbb{R}^{n+1}$ with $S \in \mathcal{S}_n$. Prove that $\mathrm{Dec}(\mathcal{S}_n, p) = \mathrm{Dec}(\mathcal{S}_{n+1}, p)$.

# 10

# Decoupling for the Elliptic Paraboloid

We will devote this chapter to investigating decoupling for $\mathbb{P}^{n-1}$. The parabola will serve as our main case study. We will start with a wave packet perspective on the significance and sharpness of decoupling. In Sections 10.2 and 10.3, we present an argument due to Bourgain and the author. We will later see that this type of argument is robust enough to serve as a template for proving decouplings for other hypersurfaces as well as for curves. Section 10.4 explores a slightly different, perhaps more intuitive, argument due to Guth.

Let $\Theta(\delta)$ be a partition of the $\delta$-neighborhood $\mathcal{N}(\delta)$ of $\mathbb{P}^{n-1}$ into almost rectangular boxes

$$\mathcal{N}_\tau(\delta) = \{(\xi, |\xi|^2 + s) : \xi \in \tau, \ |s| < \delta\},$$

where each $\tau$ is an almost cube with diameter $\sim \delta^{1/2}$. For $p \geq 2$, we denote by $\mathrm{Dec}(\delta, p)$ the decoupling constant associated with $\Theta(\delta)$. If $\delta^{-1/2} \in 2^{\mathbb{N}}$, each $\tau$ can be chosen to be a genuine cube with side length exactly $\delta^{1/2}$. Otherwise, we will have to content ourselves with almost cubes $\tau$, and there are infinitely many choices for $\Theta(\delta)$. This minor issue will in general be ignored, since the quantity $\mathrm{Dec}(\delta, p)$ will in fact not change in a significant way if different partitions $\Theta(\delta)$ are used at scale $\delta$. See Exercise 9.11.

The following result is the first sharp decoupling in the literature that was proved in the full range. It will serve us as a reference point throughout the rest of the book.

**Theorem 10.1** ([29]) *Assume that $\widehat{F}$ is supported on $\mathcal{N}(\delta)$. In the range $2 \leq p \leq 2(n+1)/(n-1)$ we have*

$$\|F\|_{L^p(\mathbb{R}^n)} \lesssim_\epsilon \delta^{-\epsilon} \left( \sum_{\theta \in \Theta(\delta)} \|\mathcal{P}_\theta F\|_{L^p(\mathbb{R}^n)}^2 \right)^{1/2},$$

*while if $2(n+1)/(n-1) < p \leq \infty$, we have*

197

$$\|F\|_{L^p(\mathbb{R}^n)} \lesssim_\epsilon \delta^{-\epsilon + \frac{n+1}{2p} - \frac{n-1}{4}} \left( \sum_{\theta \in \Theta(\delta)} \|\mathcal{P}_\theta F\|^2_{L^p(\mathbb{R}^n)} \right)^{1/2}. \tag{10.1}$$

The estimates for $p = 2$ and $p = \infty$ are rather immediate, using orthogonality and the Cauchy–Schwarz inequality, respectively. The only $p$ that needs a careful examination is the critical exponent $p = 2(n+1)/(n-1)$. Indeed, the other values of $p$ are covered using interpolation. See Exercise 10.4.

The exponent of $\delta$ in (10.1) is sharp, apart from the $\epsilon$ loss. To see this, let us consider

$$\widehat{F} = \sum_{\theta \in \Theta(\delta)} \psi_\theta, \tag{10.2}$$

with $\psi_\theta$ a positive, smooth approximation of $1_\theta$ supported on $\theta$. It is clear that if $|x| \ll 1$, then

$$|F(x)| \geq \int \widehat{F}(\xi)\, d\xi - \int \widehat{F}(\xi)|e(\xi \cdot x) - 1|\, d\xi \gtrsim \delta.$$

In particular, we have $\|F\|_p \gtrsim \delta$. Note also that

$$\left( \sum_{\theta \in \Theta(\delta)} \|\mathcal{P}_\theta F\|^2_{L^p(\mathbb{R}^n)} \right)^{1/2} = \left( \sum_{\theta \in \Theta(\delta)} \|\widehat{\psi_\theta}\|^2_{L^p(\mathbb{R}^n)} \right)^{1/2}$$

$$\sim \delta^{\frac{1-n}{4}} \delta^{\frac{n+1}{2p'}}.$$

There are two fundamental features that distinguish $\mathbb{P}^{n-1}$ among all hypersurfaces in $\mathbb{R}^n$, and that ultimately make its decoupling theory slightly easier. These features have already been exploited in the restriction theory of $\mathbb{P}^{n-1}$ discussed in Section 7.3. One is the fact that the intersection of $\mathbb{P}^{n-1}$ with a vertical hyperplane is an affine copy of $\mathbb{P}^{n-2}$. This simplifies the multilinear-to-linear reduction and will allow us to prove Theorem 10.1 using induction on $n$.

The other important feature is the fact that small caps on $\mathbb{P}^{n-1}$ can be stretched to the whole $\mathbb{P}^{n-1}$ using affine transformations. This will enable us to rescale decouplings as follows.

**Proposition 10.2** *Assume* $\delta \leq \sigma \leq 1$. *Let* $Q = c + [-\sigma^{1/2}, \sigma^{1/2}]^{n-1}$ *be a cube with center* $c = (c_1, \ldots, c_{n-1})$. *Let* $\Theta_Q(\delta)$ *be a partition of* $\mathcal{N}_Q(\delta)$ *into a subfamily of sets* $\theta \in \Theta(\delta)$. *Then*

$$\|F\|_{L^p(\mathbb{R}^n)} \leq \mathrm{Dec}\left( \frac{\delta}{\sigma}, p \right) \left( \sum_{\theta \in \Theta_Q(\delta)} \|\mathcal{P}_\theta F\|^2_{L^p(\mathbb{R}^n)} \right)^{1/2}, \tag{10.3}$$

*whenever* $\widehat{F}$ *is supported on* $\mathcal{N}_Q(\delta)$.

*Proof* The proof is a standard application of parabolic rescaling. It follows from the observation that the affine function

$$(\bar{\xi}, \xi_n) \mapsto \left( \frac{\bar{\xi} - c}{\sigma^{1/2}}, \frac{\xi_n - 2\bar{\xi} \cdot c + |c|^2}{\sigma} \right)$$

maps the partition $\Theta_Q(\delta)$ to a partition $\Theta(\delta/\sigma)$ of $\mathcal{N}(\delta/\sigma)$. □

Inequality (10.3) is trivially true with $\mathrm{Dec}(\delta/\sigma, p)$ replaced with $\mathrm{Dec}(\delta, p)$, but the first bound is morally stronger than the second one.

A rather immediate consequence of this and of Proposition 9.17 is the following submultiplicative structure of the $l^2$ decoupling constants for the paraboloid.

**Corollary 10.3** *For each* $0 < \delta_1, \delta_2 < 1$,

$$\mathrm{Dec}(\delta_1 \delta_2, p) \lesssim \mathrm{Dec}(\delta_1, p)\mathrm{Dec}(\delta_2, p). \qquad (10.4)$$

**Exercise 10.4** Prove that if $1/p = \alpha/p_1 + 1 - \alpha/p_2$ then

$$\mathrm{Dec}(\delta, p) \lesssim_\epsilon \delta^{-\epsilon}\mathrm{Dec}(O(\delta), p_1)^\alpha \mathrm{Dec}(O(\delta), p_2)^{1-\alpha}.$$

Hint: Use Exercise 9.21. Split $\Theta(\delta)$ into $O(1)$ collections $\Theta_i(\delta)$, each with the following property: for each $\theta \in \Theta_i(\delta)$ there is a rectangle $R_\theta$ containing it such that the sets $4R_\theta$ are pairwise disjoint, and each $4R_\theta$ is covered by $O(1)$ sets in $\Theta(C\delta)$, for some $C = O(1)$.

## 10.1 Almost Extremizers

We will consider a few examples to illustrate the sharpness of decoupling at the critical exponent $p = 2(n + 1)/(n - 1)$. In preparation for the proofs in the next sections, we restrict attention to the case of the parabola. The examples presented here have natural analogs in higher dimensions.

We will use the notation $R = \delta^{-1}$. Let $F : \mathbb{R}^2 \to \mathbb{C}$ be such that $\mathrm{supp}\,(\widehat{F}) \subset \mathcal{N}_{\mathbb{P}^1}(\delta)$ and with wave packet decomposition (Exercise 2.7)

$$F(x) = \sum_{T \in \mathcal{T}} w_T W_T(x).$$

We assume all $(R^{1/2}, R)$-rectangles (tubes) $T$ are inside $[-R, R]^2$. The wave packet $W_T$ ($L^\infty$ normalized this time) is spatially concentrated on $T$. It has the Fourier transform supported on some $\theta_T \in \Theta_{\mathbb{P}^1}(\delta)$, and we will write $T \sim \theta_T$. The following approximation holds

$$W_T(x) \approx 1_T(x)e(\xi_T \cdot x),$$

where $\xi_T$ can be taken to be the center of $\theta_T$.

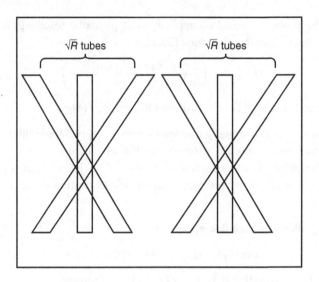

√R tubes  √R tubes

Figure 10.1  Two complete bushes.

Theorem 10.1 shows that

$$\|F\|_{L^6(\mathbb{R}^2)} \lesssim_\epsilon R^{\frac{1}{4}+\epsilon} \left( \sum_{\theta\in\Theta_{\mathfrak{p}1}(\delta)} \left( \sum_{T\sim\theta} |w_T|^6 \right)^{\frac{1}{3}} \right)^{\frac{1}{2}}.$$

Assume now that $\mathcal{T}$ contains $N \lesssim R^{1/2}$ rectangles for each direction in the interval $[\pi/4, 3\pi/4]$. Assume also that there is a collection $\mathcal{S}$ of roughly $N$ squares $S$ with unit side length inside $[-R, R]^2$, so that each $T \in \mathcal{T}$ intersects exactly one $S$. We will refer to this configuration as "$N$ complete bushes" (see Figure 10.1). One way to enforce this scenario is to place all squares $S$ along a horizontal line segment at distance $\sim R/N$ from each other, and to place a tube in each direction through each $S$. Call $\mathcal{T}_S$ the collection of those $T \in \mathcal{T}$ that intersect $S$.

We choose all weights $w_T$ with magnitude one. Moreover, for each $S \in \mathcal{S}$ with center $c_S$ we can align the phases of $w_T$ so that we have constructive interference

$$\left| \sum_{T\in\mathcal{T}_S} w_T W_T(c_S) \right| \gtrsim R^{1/2}.$$

This can be achieved due to our assumption that the collections $\mathcal{T}_S$ are pairwise disjoint. The same inequality will in fact hold with $c_S$ replaced with $x$ satisfying $|x - c_S| \ll 1$. Also, for each such $x$, the contribution from those $T$ outside $\mathcal{T}_S$ will be negligible. To conclude, we can arrange that

$$\left| \sum_{T \in \mathcal{T}} w_T W_T(x) \right| \gtrsim R^{1/2}$$

holds on a set of area $\sim N$. A simple computation shows that

$$\|F\|_{L^6(\mathbb{R}^2)} \gtrsim N^{1/6} R^{1/2}$$

and

$$\left( \sum_{\theta \in \Theta(\delta)} \|\mathcal{P}_\theta F\|^2_{L^6(\mathbb{R}^2)} \right)^{1/2} \sim N^{1/6} R^{1/2}.$$

We are learning two things from these estimates. First, ignoring $\delta^{-\epsilon}$ losses, this example provides an (almost) extremizer for the $l^2(L^6)$ decoupling. Second, this is achieved by only considering the mass of $F$ concentrated in the rather small area covered by the $N$ unit squares $S$. Theorem 10.1 in fact reassures us that by doing so we are not missing any important contribution to $\|F\|_{L^6(\mathbb{R}^2)}$.

It is worth dissecting this example even further, by considering two special cases.

**Example 10.5** (One complete bush) When $N = 1$, we recover the example in (10.2). Indeed each $\widehat{\psi_\theta}$ is essentially one wave packet spatially concentrated in a tube containing the origin.

**Example 10.6** ($R^{1/2}$ complete bushes) Let $B_\theta = B(\xi_\theta, \delta)$ be a ball of radius $\delta = R^{-1}$ inside $\theta$, with $\xi_\theta \in \mathbb{P}^1$. Let us assume that

$$\widehat{F} = \sum_\theta \phi_\theta,$$

with $\phi_\theta$ a positive smooth approximation of $1_{B_\theta}$ supported on $B_\theta$. Then $\widehat{\phi_\theta}(x)$ is a smooth approximation of $R^{-2} 1_{[-R, R]^2}(x) e(-\xi_\theta \cdot x)$. Thus, $F$ has essentially $N \sim R^{1/2}$ wave packets for each direction, each with weight $w_T \sim R^{-2}$.

Let us also explore another related example.

**Example 10.7** (One incomplete bush) Assume that $\mathcal{T}$ consists of $M \lesssim R^{1/2}$ rectangles $T_1, \ldots, T_M$ centered at the origin with $\xi_{T_j} = (j/R^{1/2}, j^2/R)$. See Figure 10.2.

Assume that each $W_T$ has weight $w_T = 1$, so that

$$F(x) \approx \sum_{j=1}^M 1_{T_j}(x) e(\xi_{T_j} \cdot x).$$

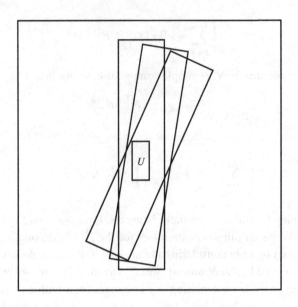

Figure 10.2 One incomplete bush.

Note that the rectangle $U$ with dimensions $R^{1/2}/(100M) \times R/(100M^2)$ centered at the origin is inside each $T_j$ and moreover, we have constructive interference

$$\left| \sum_{j=1}^{M} e\left( \frac{j}{R^{1/2}} x_1 + \frac{j^2}{R} x_2 \right) \right| \sim M$$

for $(x_1, x_2) \in U$. A simple computation reveals that

$$\|F\|_{L^6(\mathbb{R}^2)} \geq \|F\|_{L^6(U)} \gtrsim M^{1/2} R^{1/4}$$

and

$$\left( \sum_{\theta \in \Theta(\delta)} \|\mathcal{P}_\theta F\|_{L^6(\mathbb{R}^2)}^2 \right)^{1/2} \sim M^{1/2} R^{1/4}.$$

The next exercises show that decoupling does not always provide the strongest possible estimate.

**Exercise 10.8** Assume $F = \sum w_T W_T$ with $|w_T| \sim 1$ and one wave packet $W_T$ for each direction. Prove that for each $p \geq 4$,

$$\|F\|_{L^p} \lesssim R^{\frac{1}{2}}$$

and

$$\left( \sum_{\theta \in \Theta_{\mathbb{P}^1}(\delta)} \|\mathcal{P}_\theta F\|_{L^p}^2 \right)^{1/2} \sim R^{\frac{1}{4} + \frac{3}{2p}}.$$

Conclude that decoupling is not sharp in the range $4 \le p < 6$.

Hint: For the first inequality, combine Kakeya-type estimates with Exercise 7.25.

**Exercise 10.9** Let $F = \sum w_T W_T$ with $|w_T| \sim 1$. Assume that for each direction there are $M$ wave packets, and they are spatially concentrated inside a fat $(MR^{1/2}, R)$- tube. For $4 \le p \le 6$, compare the upper bounds for $\|F\|_{L^p}$ found using decoupling on the one hand, and combining Kakeya-type estimates with Exercise 7.25 on the other hand.

**Exercise 10.10** Assume $F = \sum w_T W_T$ with $|w_T| \sim 1$ and one wave packet $W_T$ for each direction. Assume also that the tubes $T$ are pairwise disjoint. Prove that decoupling is not sharp for $F$ when $2 < p \le 6$.

## 10.2 A Detailed Proof of the Decoupling for the Parabola

In this section, we present a rigorous proof of Theorem 10.1 for $\mathbb{P}^1$. We may and will restrict attention to the part of the parabola above $[0, 1]$, and will denote by $\mathcal{N}(\delta)$ its (vertical) $\delta$-neighborhood.

In the attempt to keep the argument as rigorous as possible, we will work with dyadic scales. For $m \ge 0$, let $\mathbb{I}_m$ be the collection of the $2^m$ dyadic subintervals of $[0, 1]$ of length $2^{-m}$. Thus $\mathbb{I}_0$ consists of only $[0, 1]$. Note that each $I \in \mathbb{I}_{m+1}$ is inside some $I' \in \mathbb{I}_m$ and each $I' \in \mathbb{I}_m$ has two "children" in $\mathbb{I}_{m+1}$, adjacent to each other.

It will be helpful to simplify notation as follows. For an interval $I \subset \mathbb{R}$ and for $F: \mathbb{R}^2 \to \mathbb{C}$, we let $\mathcal{P}_I F$ be the Fourier projection of $F$ onto the infinite vertical strip $I \times \mathbb{R}$.

For $p \ge 2$, let $\mathrm{Dec}(n, p)$ be the smallest constant such that the inequality

$$\|F\|_{L^p(\mathbb{R}^2)} \le \mathrm{Dec}(n, p) \left( \sum_{I \in \mathbb{I}_n} \|\mathcal{P}_I F\|_{L^p(\mathbb{R}^2)}^2 \right)^{1/2}$$

holds true for each $F: \mathbb{R}^2 \to \mathbb{C}$ with the Fourier transform supported on $\mathcal{N}(4^{-n})$.

As mentioned earlier, due to interpolation it will suffice to prove decoupling for the critical exponent $p = 6$.

**Theorem 10.11** *We have*

$$\text{Dec}(n,6) \lesssim_\epsilon 2^{n\epsilon}. \tag{10.5}$$

We will first seek to understand why the most naive approach to proving (10.5) fails. Assume that $F: \mathbb{R}^2 \to \mathbb{C}$ has the Fourier transform supported on $\mathcal{N}(4^{-n})$. Parabolic rescaling (Proposition 10.2) shows that for each $I \in \mathbb{I}_m$ with children $I_1, I_2$, we have

$$\|\mathcal{P}_I F\|_{L^6(\mathbb{R}^2)} \le \text{Dec}(1,6) \left( \|\mathcal{P}_{I_1} F\|^2_{L^6(\mathbb{R}^2)} + \|\mathcal{P}_{I_2} F\|^2_{L^6(\mathbb{R}^2)} \right)^{1/2}.$$

By rescaling back so that $I$ becomes $[0,1]$, this inequality is seen to be sharp for an appropriate choice of $F$, due to the definition of $\text{Dec}(1,6)$. We may thus write

$$\left( \sum_{I \in \mathbb{I}_m} \|\mathcal{P}_I F\|^2_{L^6(\mathbb{R}^2)} \right)^{1/2} \le \text{Dec}(1,6) \left( \sum_{I \in \mathbb{I}_{m+1}} \|\mathcal{P}_I F\|^2_{L^6(\mathbb{R}^2)} \right)^{1/2}. \tag{10.6}$$

If we iterate this $n$ times, we arrive at the inequality

$$\|F\|_{L^6(\mathbb{R}^2)} \le \text{Dec}(1,6)^n \left( \sum_{I \in \mathbb{I}_n} \|\mathcal{P}_I F\|^2_{L^6(\mathbb{R}^2)} \right)^{1/2},$$

which in turn implies that

$$\text{Dec}(n,6) \le \text{Dec}(1,6)^n.$$

Exercise 10.12 proves that $\text{Dec}(1,6) > 1$. This forces $\text{Dec}(1,6)^n \gg 2^{n\epsilon_0}$, for some $\epsilon_0 > 0$, leading to an unfavorable estimate for $\text{Dec}(n,6)$. The failure of this strategy can be accounted for by the fact that, while (10.6) is sharp for some function $F_m$ depending on $m$, there is no function that makes all these inequalities simultaneously sharp.

The moral of this computation is that we cannot afford to lose a multiplicative factor equal to $\text{Dec}(1,6)$ each time we move between consecutive levels $m$ and $m + 1$. Instead, we are going to take huge leaps, and will jump from $m$ to $2m$. The choice for the size of this leap is motivated by the fact that intervals in $\mathbb{I}_{2m}$ have length equal to those in $\mathbb{I}_m$ squared. We have a very efficient mechanism to decouple from scale $\delta$ to $\delta^2$, relying on the bilinear Kakeya inequality and $L^2$ orthogonality. The loss for each application of the bilinear Kakeya is rather tiny, at most $n^C$ (compare this with the accumulated loss $\text{Dec}(1,6)^m$, if instead we went from level $m$ to $2m$ in $m$ steps). From $\mathbb{I}_0$ to $\mathbb{I}_n$, we need $\log n$ such leaps, so the overall loss from the repeated use of bilinear Kakeya amounts to $n^{O(\log n)}$. This is easily seen to be $O(2^{n\epsilon})$ for each $\epsilon > 0$, as desired. A small caveat to our approach is that in each leap we

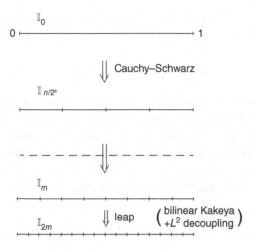

Figure 10.3 Illustration of the proof.

only decouple a certain fraction of the operator. See Proposition 10.23 for a precise statement.

Here is a brief account of how we put things together (see Figure 10.3). First, we will do a trivial decoupling (using the Cauchy–Schwarz inequality) to get from $\mathbb{I}_0$ to $\mathbb{I}_{n/2^s}$, at the cost of a small multiplicative constant $2^{O(n/2^s)}$. We will be able to choose $s$ as large as we wish, so this loss will end up being of the form $O(2^{n\epsilon})$. The transition from $\mathbb{I}_{n/2^s}$ to $\mathbb{I}_n$ will then be done in $s$ leaps, each time applying Proposition 10.23. Collecting all contributions, an a priori bound of the form

$$\mathrm{Dec}(n,6) \lesssim 2^{nA}, \text{ for some } A > 0$$

will get upgraded to a stronger estimate (assuming $s$ is large enough)

$$\mathrm{Dec}(n,6) \lesssim 2^{n\left(A\left(1-\frac{s+2}{2^s+1}\right)+\frac{1}{2^s-1}\right)}.$$

This bootstrapping argument will force $A$ to get smaller and smaller, arbitrarily close to 0.

The leaps are operated in a bilinear format, in order to take advantage of the bilinear Kakeya phenomenon. The fact that there is no significant loss in bilinearization is proved in Proposition 10.14.

Unless specified otherwise, we will present most of the forthcoming computations for $4 \le p \le 6$, and will specialize to $p = 6$ at the very end of the argument.

We will perform a local analysis. It will be slightly more convenient to work with squares rather than balls, since we can partition big squares into smaller ones. $Q_R$ will denote a typical square in $\mathbb{R}^2$ with side length $R$. Given such a square $Q_R$ centered at $c$, we will write

$$w_Q(x) = \left(1 + \frac{|x - c|}{R}\right)^{-100}.$$

The exponent 100 will not change throughout the argument in this section.

The results from the previous chapter involving weights and balls continue to hold true for squares. In particular, Proposition 9.15 shows that $\text{Dec}(n, p)$ is comparable to the smallest constants $\text{Dec}_{\text{weighted}}(n, p)$, $\text{Dec}_{\text{local}}(n, p)$ such that the following inequalities hold true for each $F: \mathbb{R}^2 \to \mathbb{C}$ with the Fourier transform supported on $\mathcal{N}(4^{-n})$, and for arbitrary squares $Q_{4^n}$

$$\|F\|_{L^p(w_{Q_{4^n}})} \leq \text{Dec}_{\text{weighted}}(n, p) \left(\sum_{I \in \mathbb{I}_n} \|\mathcal{P}_I F\|^2_{L^p(w_{Q_{4^n}})}\right)^{1/2},$$

$$\|F\|_{L^p(Q_{4^n})} \leq \text{Dec}_{\text{local}}(n, p) \left(\sum_{I \in \mathbb{I}_n} \|\mathcal{P}_I F\|^2_{L^p(w_{Q_{4^n}})}\right)^{1/2}.$$

**Exercise 10.12** Prove that for each pair $\mathcal{S} = \{S_1, S_2\}$ of open sets in $\mathbb{R}^2$ we have

$$\text{Dec}(\mathcal{S}, 6) > 1.$$

Conclude that $\text{Dec}(1, 6) > 1$.
Hint: Let $\psi$ be smooth, compactly supported near the origin. Let $\xi_1, \xi_2 \in S_1$ and $\xi_3 \in S_2$. Test decoupling with $F_\epsilon$ satisfying

$$\widehat{F_\epsilon}(\xi) = \psi\left(\frac{\xi - \xi_1}{\epsilon}\right) + \psi\left(\frac{\xi - \xi_2}{\epsilon}\right) + \psi\left(\frac{\xi - \xi_3}{\epsilon}\right)$$

and let $\epsilon \to 0$.

### 10.2.1 Linear vs. Bilinear Decoupling

Let us start by restating the content of Proposition 10.2 for the parabola. The reader is referred to Proposition 9.15 for two equivalent local/weighted versions of this inequality that will be used interchangeably in the future.

Given $I = [t, t + 2^{-l}] \subset \mathbb{R}$ with $l < n$, let the collection $\mathbb{I}_n(I)$ consist of all subintervals of $I$ of the form $[t + j2^{-n}, t + (j + 1)2^{-n}]$, with $j \in \mathbb{N}$.

**Proposition 10.13** *Let $I = [t, t + 2^{-l}] \subset \mathbb{R}$ with $l < n$. Then for each $F$ with the Fourier transform supported on $\mathcal{N}_I(4^{-n})$, we have*

$$\|\mathcal{P}_I F\|_{L^p(\mathbb{R}^2)} \leq \text{Dec}(n - l, p) \left(\sum_{J \in \mathbb{I}_n(I)} \|\mathcal{P}_J F\|^2_{L^p(\mathbb{R}^2)}\right)^{1/2}.$$

Throughout the rest of the argument in this section, we will fix the intervals $I_1 = [0, 1/4]$, $I_2 = [1/2, 1]$ and will use them to introduce a bilinear version of the decoupling inequality. The most important feature of $I_1, I_2$ is that they are disjoint. The fact that they are dyadic intervals leads to technical simplifications.

Define $\text{BilDec}(n, p)$ to be the smallest constant such that the inequality

$$\| |F_1 F_2|^{1/2} \|_{L^p(\mathbb{R}^2)}$$

$$\leq \text{BilDec}(n, p) \left( \sum_{I \in \mathbb{I}_n(I_1)} \|\mathcal{P}_I F_1\|_{L^p(\mathbb{R}^2)}^2 \sum_{I \in \mathbb{I}_n(I_2)} \|\mathcal{P}_I F_2\|_{L^p(\mathbb{R}^2)}^2 \right)^{1/4}$$

holds true for all $F_1$, $F_2$ with Fourier transforms supported on $\mathcal{N}_{I_1}(4^{-n})$ and $\mathcal{N}_{I_2}(4^{-n})$, respectively.

Remark 9.16 shows that if we replace $\mathbb{R}^2$ with $w_{Q_{4^n}}$, the value of $\text{BilDec}(n, p)$ does not change significantly.

From Hölder's inequality, it is immediate that

$$\text{BilDec}(n, p) \leq \text{Dec}(n, p).$$

The next result proves a satisfactory converse.

**Proposition 10.14** *For each $\epsilon > 0$,*

$$\text{Dec}(n, p) \lesssim_\epsilon 2^{n\epsilon} \left( 1 + \max_{m \leq n} \text{BilDec}(m, p) \right).$$

*Proof*  It will suffice to prove that for each $k < n$,

$$\text{Dec}(n, p) \lesssim C^{\frac{n}{k}} \left( 1 + C_k n \max_{m \leq n} \text{BilDec}(m, p) \right).$$

This will instead follow by iterating the inequality

$$\text{Dec}(n, p) \leq C\text{Dec}(n - k, p) + C_k \max_{m \leq n} \text{BilDec}(m, p), \qquad (10.7)$$

with $C$ independent of $n, k$. Let us next prove this inequality.

Fix $k$ and assume $F$ has the Fourier transform supported on $\mathcal{N}(4^{-n})$. Since

$$F(x) = \sum_{I \in \mathbb{I}_k} \mathcal{P}_I F(x),$$

it is not difficult to see that

$$|F(x)| \leq 4 \max_{I \in \mathbb{I}_k} |\mathcal{P}_I F(x)| + 2^{O(k)} \sum_{\substack{J_1, J_2 \in \mathbb{I}_k \\ 2J_1 \cap 2J_2 = \emptyset}} |\mathcal{P}_{J_1} F(x) \mathcal{P}_{J_2} F(x)|^{1/2}, \qquad (10.8)$$

where the sum on the right is taken over all pairs of intervals $J_1, J_2 \in \mathbb{I}_k$ that are not neighbors. See Lemma 7.2 for a proof.

Fix such a pair $J_1 = [a, a + 2^{-k}]$, $J_2 = [b, b + 2^{-k}]$, and let $m$ be a positive integer satisfying $2^{-m} \leq b - a < 2^{-m+1}$. Since $J_1, J_2$ are not adjacent to each other, we must have $m \leq k - 1$. It follows that the affine function $T(\xi) = (\xi - a)/2^{-m+1}$ maps $J_1$ to a dyadic subinterval of $[0, 1/4]$ and $J_2$ to a dyadic subinterval of $[1/2, 1]$. Thus, parabolic rescaling shows that

$$\||\mathcal{P}_{J_1} F \, \mathcal{P}_{J_2} F|^{1/2}\|_{L^p(\mathbb{R}^2)}$$

$$\lesssim \mathrm{BilDec}(n - m + 1, p) \left( \sum_{I \in \mathbb{I}_n(J_1)} \|\mathcal{P}_I F\|^2_{L^p(\mathbb{R}^2)} \sum_{I \in \mathbb{I}_n(J_2)} \|\mathcal{P}_I F\|^2_{L^p(\mathbb{R}^2)} \right)^{1/4}$$

$$\leq \mathrm{BilDec}(n - m + 1, p) \left( \sum_{I \in \mathbb{I}_n} \|\mathcal{P}_I F\|^2_{L^p(\mathbb{R}^2)} \right)^{1/2}. \qquad (10.9)$$

Finally, invoking again Proposition 10.13, we get

$$\| \max_{I \in \mathbb{I}_k} |\mathcal{P}_I F| \|_{L^p(\mathbb{R}^2)} \leq \left( \sum_{I \in \mathbb{I}_k} \|\mathcal{P}_I F\|^2_{L^p(\mathbb{R}^2)} \right)^{\frac{1}{2}}$$

$$\leq \mathrm{Dec}(n - k, p) \left( \sum_{I \in \mathbb{I}_k} \sum_{I' \in \mathbb{I}_n(I)} \|\mathcal{P}_{I'} F\|^2_{L^p(\mathbb{R}^2)} \right)^{\frac{1}{2}}$$

$$= \mathrm{Dec}(n - k, p) \left( \sum_{I' \in \mathbb{I}_n} \|\mathcal{P}_{I'} F\|^2_{L^p(\mathbb{R}^2)} \right)^{\frac{1}{2}}. \qquad (10.10)$$

Now (10.7) follows by combining (10.8), (10.9), and (10.10). $\qquad \square$

## 10.2.2 A Multiscale Inequality

To simplify notation, we will denote by $Q^r$ an arbitrary square in $\mathbb{R}^2$ with side length $2^r$.

Given $m, r \in \mathbb{N}$, $t \geq 2$ and $F : \mathbb{R}^2 \to \mathbb{C}$, we let

$$D_t(m, Q^r, F) = \left[ \left( \sum_{I \in \mathbb{I}_m(I_1)} \|\mathcal{P}_I F\|^2_{L^t_\sharp(w_{Q^r})} \right) \left( \sum_{I \in \mathbb{I}_m(I_2)} \|\mathcal{P}_I F\|^2_{L^t_\sharp(w_{Q^r})} \right) \right]^{\frac{1}{4}}.$$

The two values of $t$ that are important for our argument are $t = 2$ and $t = 6$. In light of the similarity $\mathrm{Dec}(n, 6) \approx \mathrm{BilDec}(n, 6)$ established earlier, inequality (10.5) is equivalent to the fact that for each $F$ with the Fourier transform supported on $\mathcal{N}_{I_1 \cup I_2}(4^{-n})$,

$$\|(\mathcal{P}_{I_1} F \, \mathcal{P}_{I_2} F)^{1/2}\|_{L^6_\sharp(w_{Q^{2n}})} \lesssim_\epsilon 2^{n\epsilon} D_6(n, Q^{2n}, F).$$

For $m \leq r$, we will denote by $\mathcal{Q}_m(Q^r)$ the partition of $Q^r$ into squares $Q^m$. Define

$$A_p(m, Q^r, F) = \left( \frac{1}{|\mathcal{Q}_m(Q^r)|} \sum_{Q^m \in \mathcal{Q}_m(Q^r)} D_2(m, Q^m, F)^p \right)^{\frac{1}{p}}. \quad (10.11)$$

In our applications, the term $A_p(m, Q^r, F)$ will always have the entry $F$ a function with the Fourier transform supported on $\mathcal{N}(2^{-m})$. Thus, each $|\mathcal{P}_I F|$ with $I \in \mathbb{I}_m$ can be thought of as being essentially constant on $Q^m$, cf. Proposition 7.16. This leads to the more palpable representation

$$A_p(m, Q^r, F) \approx \left\| \left( \sum_{I \in \mathbb{I}_m(I_1)} |\mathcal{P}_I F|^2 \right)^{\frac{1}{4}} \left( \sum_{I \in \mathbb{I}_m(I_2)} |\mathcal{P}_I F|^2 \right)^{\frac{1}{4}} \right\|_{L^p_\sharp(w_{Q^r})}.$$

The use of the slightly more complicated definition (10.11) is for purely technical reasons. We have argued earlier that there is a close connection between square functions and Kakeya-type inequalities in the multilinear setting. See, for example, Exercise 6.30. It should thus come as no surprise that the terms $A_p$ encode Kakeya-type information. They will serve as auxiliary tools throughout the forthcoming argument, which will allow us to bring to bear the bilinear Kakeya inequality. The $A_p$ terms will be replaced with $D_p$ terms at the end of the argument, by essentially using the following elementary inequality.

**Lemma 10.15** *For each $p \geq 2$, we have*

$$A_p(m, Q^r, F) \lesssim D_p(m, Q^r, F).$$

*Proof* Write first using Hölder's and the Cauchy–Schwarz inequalities

$$|\mathcal{Q}_m(Q^r)|^{\frac{1}{p}} A_p(m, Q^r, F) \lesssim \left( \sum_{Q^m \in \mathcal{Q}_m(Q^r)} D_p(m, Q^m, F)^p \right)^{\frac{1}{p}}$$

$$\leq \left[ \sum_{Q^m \in \mathcal{Q}_m(Q^r)} \left( \sum_{I \in \mathbb{I}_m(I_1)} \|\mathcal{P}_I F\|^2_{L^p_\sharp(w_{Q^m})} \right)^{\frac{p}{2}} \right]^{\frac{1}{2p}}$$

$$\times \left[ \sum_{Q^m \in \mathcal{Q}_m(Q^r)} \left( \sum_{I \in \mathbb{I}_m(I_2)} \|\mathcal{P}_I F\|^2_{L^p_\sharp(w_{Q^m})} \right)^{\frac{p}{2}} \right]^{\frac{1}{2p}}.$$

The desired inequality now follows immediately by applying twice the following Minkowski's inequality in $l^{2/p}$ with $X(Q^m, I) = \|\mathcal{P}_I F\|^p_{L^p_\sharp(w_{Q^m})}$

$$\sum_{Q^m} \left( \sum_I X(Q^m, I)^{\frac{2}{p}} \right)^{\frac{p}{2}} \le \left( \sum_I \left( \sum_{Q^m} X(Q^m, I) \right)^{\frac{p}{2}} \right)^{\frac{p}{2}},$$

and using (9.7). $\qquad\qquad\qquad\qquad\qquad\qquad\qquad\qquad\qquad\qquad\square$

For $p \ge 4$, let $0 \le \kappa_p \le 1$ satisfy

$$\frac{2}{p} = \frac{1 - \kappa_p}{2} + \frac{\kappa_p}{p},$$

that is,

$$\kappa_p = \frac{p - 4}{p - 2}.$$

The proof of the following multiscale inequality will be the focus of the remaining subsections. Note that it takes a particularly nice form when $p = 4$.

**Theorem 10.16** *Assume that $n = 2^u$, $u \in \mathbb{N}$. Assume also that $F$ has the Fourier transform supported on $\mathcal{N}_{I_1 \cup I_2}(4^{-n})$. Then for each $s \le u$, we have*

$$A_p\left(\frac{n}{2^s}, Q^{2n}, F\right) \lesssim_{s,\epsilon} 2^{\epsilon s n} D_p(n, Q^{2n}, F) \prod_{l=1}^{s} \mathrm{Dec}\left(n - \frac{n}{2^l}, p\right)^{\kappa_p(1 - \kappa_p)^{s-l}}.$$

Before proving this theorem, we will next show how it implies inequality (10.5).

### 10.2.3 Proof of Theorem 10.11

Let us prove

$$\mathrm{Dec}(n, 6) \lesssim_\epsilon 2^{n\epsilon}.$$

Assume that $F$ has the Fourier transform supported on $\mathcal{N}_{I_1 \cup I_2}(4^{-n})$. Let $F_1 = \mathcal{P}_{I_1} F$, $F_2 = \mathcal{P}_{I_2} F$, and $n = 2^u$.

Applying the Cauchy–Schwarz inequality and Minkowski's inequality (9.3), we write for each $s \le u$,

$$\||F_1 F_2|^{1/2}\|_{L^p_\sharp(Q^{2n})} = \left( \frac{1}{|\mathcal{Q}_{\frac{n}{2^s}}(Q^{2n})|} \sum_{Q \in \mathcal{Q}_{\frac{n}{2^s}}(Q^{2n})} \||F_1 F_2|^{1/2}\|_{L^p_\sharp(Q)}^p \right)^{1/p}$$

$$\le 2^{\frac{n}{2^s + 1}} \left[ \frac{1}{|\mathcal{Q}_{\frac{n}{2^s}}(Q^{2n})|} \sum_{Q \in \mathcal{Q}_{\frac{n}{2^s}}(Q^{2n})} \right.$$

$$\left. \times \left( \sum_{I \in \mathbb{I}_{\frac{n}{2^s}}(I_1)} \|\mathcal{P}_I F\|_{L^p_\sharp(Q)}^2 \sum_{I \in \mathbb{I}_{\frac{n}{2^s}}(I_2)} \|\mathcal{P}_I F\|_{L^p_\sharp(Q)}^2 \right)^{\frac{p}{4}} \right]^{\frac{1}{p}}.$$

Let us recall the inequality from Lemma 9.19:

$$\|\mathcal{P}_I F\|_{L^p_\sharp(Q)} \lesssim \|\mathcal{P}_I F\|_{L^2_\sharp(w_Q)}. \tag{10.12}$$

Combining this with the previous inequality, our find can be summarized as follows:

$$\||F_1 F_2|^{1/2}\|_{L^p_\sharp(Q^{2n})} \lesssim 2^{\frac{n}{2^s+1}} A_p\left(\frac{n}{2^s}, Q^{2n}, F\right).$$

This inequality represents the top part of the diagram in Figure 10.3.

Combining this with Theorem 10.16 gives the following estimate.

**Theorem 10.17** *For each $s \leq u$,*

$$\||F_1 F_2|^{1/2}\|_{L^p_\sharp(Q^{2n})} \lesssim_{s,\epsilon} 2^{\epsilon s n} 2^{\frac{n}{2^s+1}} D_p(n, Q^{2n}, F)$$

$$\times \prod_{l=1}^{s} \mathrm{Dec}\left(n - \frac{n}{2^l}, p\right)^{\kappa_p(1-\kappa_p)^{s-l}}.$$

Taking the supremum over $F$, we get the following inequality.

**Corollary 10.18** *For each $s \leq u$,*

$$\mathrm{BilDec}(n, p) \lesssim_{s,\epsilon} 2^{\epsilon s n} 2^{\frac{n}{2^s+1}} \prod_{l=1}^{s} \mathrm{Dec}\left(n - \frac{n}{2^l}, p\right)^{\kappa_p(1-\kappa_p)^{s-l}}.$$

Let us now specialize this to $p = 6$. Note that $\kappa_6 = 1/2$. We will finish the proof with a bootstrapping argument. Assume $\mathrm{Dec}(n, 6) \lesssim 2^{nA}$ holds for some $A$ and all $n$. It is easy to see that $A = 1/2$ works. We will first show that a smaller value of $A$ always works, too. Corollary 10.18 implies that for each $s$ and each $n$

$$\mathrm{BilDec}(n, 6) \lesssim_s 2^{n\left(Ac_s + \frac{1}{2^s}\right)},$$

where

$$c_s = \sum_{l=1}^{s}\left(1 - \frac{1}{2^l}\right)\frac{1}{2^{s-l+1}} = 1 - \frac{s+2}{2^{s+1}}.$$

Combining this with Proposition 10.14 leads to

$$\mathrm{Dec}(n, 6) \lesssim_s 2^{n\left(Ac_s + \frac{1}{2^s-1}\right)}. \tag{10.13}$$

Define now

$$\mathcal{A} := \{A > 0 : \mathrm{Dec}(n, 6) \lesssim 2^{nA}\}$$

and let $A_0 = \inf \mathcal{A}$. Note that $\mathcal{A}$ is either $(A_0, \infty)$ or $[A_0, \infty)$. We claim that $A_0$ must be zero, which will finish the proof of our theorem. Indeed, if $A_0 > 0$, then

$$Ac_s + \frac{1}{2^{s-1}} < A_0$$

for some $A \in \mathcal{A}$ sufficiently close to $A_0$ and $s$ sufficiently large. Combining this with (10.13) contradicts the definition of $A_0$. In conclusion, we must have $A_0 = 0$.

**Remark 10.19** It is worth realizing that the bootstrapping argument presented here works because $2(1 - \kappa_6) \geq 1$. The relevance of the coefficient 2 is that it measures the size of the leap (from $m$ to $2m$ in Proposition 10.23). See also Exercise 10.24.

In the remaining subsections, we will prove Theorem 10.16.

### 10.2.4 $L^2$ Decoupling

We will use Proposition 9.18 to decouple in $L^2$ into frequency intervals of smallest possible scale, equal to the inverse of the spatial scale. We restate the result for the reader's convenience.

**Proposition 10.20** *For each $F: \mathbb{R}^2 \to \mathbb{C}$, each dyadic interval $I$ of length greater than $2^{-m}$, and each square $Q^m \subset \mathbb{R}^2$, we have*

$$\|\mathcal{P}_I F\|_{L^2(w_{Q^m})} \lesssim \left( \sum_{J \in \mathbb{I}_m(I)} \|\mathcal{P}_J F\|_{L^2(w_{Q^m})}^2 \right)^{1/2}.$$

### 10.2.5 A Consequence of the Bilinear Kakeya Inequality

For the convenience of the reader, we recall the bilinear Kakeya inequality from Chapter 6.

**Theorem 10.21** *Consider two families $\mathbb{T}_1, \mathbb{T}_2$ consisting of rectangles $T$ in $\mathbb{R}^2$ having the following properties:*

(i) *Each $T$ has the short side of length $R^{1/2}$ and the long side of length $R$ pointing in the direction of the unit vector $v_T$.*

(ii) *$|v_{T_1} \wedge v_{T_2}| \gtrsim 1$ for each $T_1 \in \mathbb{T}_1$ and $T_2 \in \mathbb{T}_2$.*

*Then the following inequality holds:*

$$\int_{\mathbb{R}^2} \prod_{i=1}^{2} g_i \lesssim \frac{1}{R^2} \prod_{i=1}^{2} \int_{\mathbb{R}^2} g_i \tag{10.14}$$

*for all functions $g_i$ of the form*

$$g_i = \sum_{T \in \mathbb{T}_i} c_T 1_T, \quad c_T \in [0, \infty).$$

*The implicit constant in $\lesssim$ does not depend on $R, c_T, \mathbb{T}_i$.*

If $I \subset \mathbb{R}$ is an interval of length $2^{-l}$ and $\delta = 2^{-k}$ with $k \geq l$, we will denote by $\text{Part}_\delta(I)$ the partition of $I$ into intervals of length $\delta$.

The following result is part of a two-stage process. Note that, strictly speaking, this inequality is not a decoupling, since the size of the frequency intervals $J_i$ remains unchanged. However, the side length of the spatial squares increases from $\delta^{-1}$ to $\delta^{-2}$. This will facilitate a subsequent decoupling, as we shall later see in Proposition 10.23.

**Theorem 10.22** *Let $p \geq 2$ and $\delta < 1$. Let $Q$ be an arbitrary square in $\mathbb{R}^2$ with side length $\delta^{-2}$, and let $\mathcal{Q}$ be the partition of $Q$ into squares $\Delta$ with side length $\delta^{-1}$. Then for each $F : \mathbb{R}^2 \to \mathbb{C}$ with the Fourier transform supported on $\mathcal{N}(\delta^2)$, we have*

$$
\frac{1}{|\mathcal{Q}|} \sum_{\Delta \in \mathcal{Q}} \left[ \prod_{i=1}^{2} \left( \sum_{J_i \in \text{Part}_\delta(I_i)} \|\mathcal{P}_{J_i} F\|^2_{L^p_\sharp(w_\Delta)} \right)^{\frac{1}{2}} \right]^p
$$

$$
\lesssim \left[ \prod_{i=1}^{2} \left( \sum_{J_i \in \text{Part}_\delta(I_i)} \|\mathcal{P}_{J_i} F\|^2_{L^p_\sharp(w_Q)} \right)^{\frac{1}{2}} \right]^p . \tag{10.15}
$$

*Moreover, the implicit constant is independent of $F, \delta, Q$.*

*Proof* We will reduce the proof to an application of Theorem 10.21. Indeed, for each interval $J$ of length $\delta$, the Fourier transform of $\mathcal{P}_J F$, is supported inside a rectangle with dimensions comparable to $\delta$ and $\delta^2$. This in turn forces $|\mathcal{P}_J F|$ to be essentially constant on rectangles dual to this rectangle. Note that due to the separation of $I_1$ and $I_2$, the rectangles corresponding to intervals $J_1 \subset I_1, J_2 \subset I_2$ satisfy the requirements in Theorem 10.21, with $R = \delta^{-2}$.

Since we can afford logarithmic losses in $\delta$, it suffices to prove the inequality with the summation on both sides restricted to families of intervals $J_i$ for which $\|\mathcal{P}_{J_i} F\|_{L^p_\sharp(w_Q)}$ have comparable size (within a multiplicative factor of 2), for each $i$. Indeed, the intervals $J'_i$ satisfying (for some large enough $C = O(1)$)

$$
\|\mathcal{P}_{J'_i} F\|_{L^p_\sharp(w_Q)} \leq \delta^C \max_{J_i \in \text{Part}_\delta(I_i)} \|\mathcal{P}_{J_i} F\|_{L^p_\sharp(w_Q)}
$$

can be easily dealt with by using the triangle inequality, since we automatically have

$$
\max_{\Delta \in \mathcal{Q}} \|\mathcal{P}_{J'_i} F\|_{L^p_\sharp(w_\Delta)} \leq \delta^{O(1)} \max_{J_i \in \text{Part}_\delta(I_i)} \|\mathcal{P}_{J_i} F\|_{L^p_\sharp(w_Q)} .
$$

This leaves only $\log(\delta^{-O(1)})$ sizes to consider.

Let us now assume that we have $N_i$ intervals $J_i$, with $\|\mathcal{P}_{J_i}F\|_{L^p_\sharp(w_Q)}$ of comparable size. Since $p \geq 2$, by Hölder's inequality, (10.15) is at most

$$\left(\prod_{i=1}^{2} N_i^{\frac{1}{2}-\frac{1}{p}}\right)^p \frac{1}{|Q|} \sum_{\Delta \in Q} \prod_{i=1}^{2} \left(\sum_{J_i} \|\mathcal{P}_{J_i}F\|_{L^p_\sharp(w_\Delta)}^p\right). \tag{10.16}$$

For each $I = J_i$ centered at $c_I$, consider a family $\mathcal{F}_I$ of pairwise disjoint, mutually parallel rectangles $T$ covering $Q$. They have the short side of length $\delta^{-1}$ and the longer side of length $\delta^{-2}$, pointing in the direction $(2c_I, -1)$.

The function $|\mathcal{P}_I F|$, and thus also the function

$$g_I(x) := \|\mathcal{P}_I F\|_{L^p_\sharp\left(w_{Q(x,\delta^{-1})}\right)}^p$$

can be thought of as being essentially constant on each rectangle from $\mathcal{F}_I$. We will write

$$g_I \approx \sum_{T \in \mathcal{F}_I} c_T 1_T.$$

This can be made precise, but we will sacrifice a bit of the rigor for the sake of keeping the argument simple enough (the interested reader may check the proof of Theorem 11.10 for a rigorous argument). Thus we may write

$$\frac{1}{|Q|} \sum_{\Delta \in Q} \prod_{i=1}^{2} \left(\sum_{J_i} \|\mathcal{P}_{J_i}F\|_{L^p_\sharp(w_\Delta)}^p\right) \approx \frac{1}{|Q|} \int \prod_{i=1}^{2} g_i,$$

with $g_i(x) = \sum_{J_i} g_{J_i}(x)$.

Applying Theorem 10.21, we can dominate the term on the right by

$$\frac{1}{|Q|^2} \prod_{i=1}^{2} \int g_i.$$

Note also that

$$\frac{1}{|Q|} \int g_i \approx \sum_{J_i} \|\mathcal{P}_{J_i}F\|_{L^p_\sharp(w_Q)}^p.$$

It follows that (10.16) is dominated by

$$\left(\prod_{i=1}^{2} N_i^{\frac{1}{2}-\frac{1}{p}}\right)^p \prod_{i=1}^{2} \left(\sum_{J_i} \|\mathcal{P}_{J_i}F\|_{L^p_\sharp(w_Q)}^p\right). \tag{10.17}$$

Recalling the restriction we have made on $J_i$, (10.17) is comparable to

$$\left[\prod_{i=1}^{2} \left(\sum_{J_i} \|\mathcal{P}_{J_i}F\|_{L^p_\sharp(w_Q)}^2\right)^{1/2}\right]^p,$$

as desired.                                                                    $\square$

### 10.2.6 A Two-Scale Inequality

Recall that $\kappa_p = (p-4)/(p-2)$.

The following result shows how to (partially) decouple from scale $\delta = 2^{-m}$ to scale $\delta^2$. It corresponds to what we have previously referred to as *leap*.

**Proposition 10.23** *We have for each $p \geq 4$, $n \geq r \geq m$ and each $F$ with the Fourier transform supported on $\mathcal{N}_{I_1 \cup I_2}(4^{-n})$*

$$A_p(m, Q^{2r}, F) \lesssim_\epsilon 2^{m\epsilon} A_p(2m, Q^{2r}, F)^{1-\kappa_p} D_p(m, Q^{2r}, F)^{\kappa_p}.$$

*Proof* Using elementary inequalities, it suffices to prove the proposition for $r = m$. By Hölder's inequality,

$$\|\mathcal{P}_I F\|_{L^2_\sharp(w_{Q^m})} \lesssim \|\mathcal{P}_I F\|_{L^{\frac{p}{2}}_\sharp(w_{Q^m})}.$$

Combining this with Theorem 10.22 we may write

$$A_p(m, Q^{2m}, F) \lesssim_\epsilon 2^{m\epsilon} \Bigg[ \Bigg( \sum_{I \in \mathbb{I}_m(I_1)} \|\mathcal{P}_I F\|^2_{L^{\frac{p}{2}}_\sharp(w_{Q^{2m}})} \Bigg)$$
$$\times \Bigg( \sum_{I \in \mathbb{I}_m(I_2)} \|\mathcal{P}_I F\|^2_{L^{\frac{p}{2}}_\sharp(w_{Q^{2m}})} \Bigg) \Bigg]^{\frac{1}{4}}.$$

Using Hölder's inequality again, we can dominate this by

$$2^{m\epsilon} \Bigg[ \Bigg( \sum_{I \in \mathbb{I}_m(I_1)} \|\mathcal{P}_I F\|^2_{L^2_\sharp(w_{Q^{2m}})} \Bigg) \Bigg( \sum_{I \in \mathbb{I}_m(I_2)} \|\mathcal{P}_I F\|^2_{L^2_\sharp(w_{Q^{2m}})} \Bigg) \Bigg]^{\frac{1-\kappa_p}{4}}$$
$$\times \Bigg[ \Bigg( \sum_{I \in \mathbb{I}_m(I_1)} \|\mathcal{P}_I F\|^2_{L^p_\sharp(w_{Q^{2m}})} \Bigg) \Bigg( \sum_{I \in \mathbb{I}_m(I_2)} \|\mathcal{P}_I F\|^2_{L^p_\sharp(w_{Q^{2m}})} \Bigg) \Bigg]^{\frac{\kappa_p}{4}}.$$

Finally, we can further process the first term by invoking $L^2$ decoupling (Proposition 10.20) for each $I \in \mathbb{I}_m$:

$$\|\mathcal{P}_I F\|^2_{L^2_\sharp(w_{Q^{2m}})} \lesssim \sum_{J \in \mathbb{I}_{2m}(I)} \|\mathcal{P}_J F\|^2_{L^2_\sharp(w_{Q^{2m}})}. \qquad \square$$

### 10.2.7 Iteration

We now prove Theorem 10.16 using iteration. Recall that $n = 2^u$. Take $F$ with the Fourier transform supported on $\mathcal{N}_{I_1 \cup I_2}(4^{-n})$.

Iterating Proposition 10.23, $s \leq u$ times leads to the following inequality:

$$A_p \left( \frac{n}{2^s}, Q^{2n}, F \right) \lesssim_{s,\epsilon} 2^{\epsilon sn} A_p(n, Q^{2n}, F)^{(1-\kappa_p)^s}$$

$$\times \prod_{l=1}^{s} D_p \left( \frac{n}{2^l}, Q^{2n}, F \right)^{\kappa_p(1-\kappa_p)^{s-l}}.$$

We combine this with two elementary inequalities. The first one is a consequence of Proposition 10.13:

$$D_p \left( \frac{n}{2^l}, Q^{2n}, F \right) \leq \mathrm{Dec} \left( n - \frac{n}{2^l}, p \right) D_p(n, Q^{2n}, F).$$

The second one is a particular case of Lemma 10.15:

$$A_p(n, Q^{2n}, F) \lesssim D_p(n, Q^{2n}, F).$$

This completes the argument.

**Exercise 10.24** Let $l$ be an integer greater than 1, and let $p \geq 2$ and $\kappa \in (0, 1)$. Assume we have the following inequality for each $m \in \mathbb{N}$:

$$A_p(m, Q^{2r}, F) \lesssim_{\epsilon} \delta^{-\epsilon} A_p(lm, Q^{2r}, F)^{1-\kappa} D_p(m, Q^{2r}, F)^{\kappa}.$$

Prove that if $l(1 - \kappa) \geq 1$, then for each $\epsilon > 0$ we have $\mathrm{Dec}(n, p) \lesssim_{\epsilon} 2^{n\epsilon}$. Hint: Repeat the previous arguments with $n = l^u$.

## 10.3 Decoupling for $\mathbb{P}^{n-1}$

Throughout this section, we will denote by $\mathrm{Dec}_n(\delta, p)$ the $l^2$ decoupling constant $\mathrm{Dec}(\Theta_{\mathbb{P}^{n-1}}(\delta), p)$ for $\mathbb{P}^{n-1}$. Our goal here is to give a brief sketch of how the argument from the previous section can be generalized to show that

$$\mathrm{Dec}_n \left( \delta, \frac{2(n+1)}{n-1} \right) \lesssim_{\epsilon} \delta^{-\epsilon}$$

for each $n \geq 3$.

There is only one noteworthy difference between $n = 2$ and higher values of $n$. This is manifested when proving the equivalence between linear and multilinear decoupling, and has to do with estimating lower-dimensional contributions. The case $n = 3$ is entirely representative for this nuance and will be discussed in this section.

Some of the notation used earlier for $\mathbb{P}^1$ will now refer to $\mathbb{P}^2$. For example, given $S \subset \mathbb{R}^2$ we will denote by $\mathcal{P}_S$ the Fourier projection onto $S \times \mathbb{R}$. Also, $\mathcal{N}_S(\delta)$ will refer to the $\delta$-neighborhood of the cap above $S$ on $\mathbb{P}^2$.

For a square $S \subset [-1, 1]^2$, let $\mathcal{C}_\mu(S)$ denote its partition into squares with side length $2\mu^{-1}$. We will simply write $\mathcal{C}_\mu$ if $S = [-1, 1]^2$.

Let $K \in 2^{\mathbb{N}}$. The collection of all triples $(\alpha_1, \alpha_2, \alpha_3) \in \mathcal{C}_K \times \mathcal{C}_K \times \mathcal{C}_K$ that are $K^{-2}$-transverse (Definition 7.12) will be denoted by $\mathcal{I}_K$.

We introduce the trilinear $l^2$ decoupling constant $\mathrm{TriDec}_3(\delta, p, K)$ to be the smallest number for which the following inequality holds for each $(\alpha_1, \alpha_2, \alpha_3) \in \mathcal{I}_K$ and for each $F$ with the Fourier transform supported on $\mathcal{N}_{\alpha_1}(\delta) \cup \mathcal{N}_{\alpha_2}(\delta) \cup \mathcal{N}_{\alpha_3}(\delta)$:

$$\left\| \left( \prod_{i=1}^{3} \mathcal{P}_{\alpha_i} F \right)^{1/3} \right\|_{L^p(\mathbb{R}^3)} \le \mathrm{TriDec}_3(\delta, p, K) \left( \prod_{i=1}^{3} \sum_{\tau \in \mathcal{C}_{\frac{1}{\delta^{1/2}}}(\alpha_i)} \| \mathcal{P}_\tau F \|^2_{L^p(\mathbb{R}^3)} \right)^{\frac{1}{6}}.$$

It is immediate that $\mathrm{TriDec}_3(\delta, p, K) \le \mathrm{Dec}_3(\delta, p)$. The reverse inequality is also essentially true, with some negligible losses.

**Proposition 10.25** *There is* $\epsilon_p(K) > 0$ *with* $\lim_{K \to \infty} \epsilon_p(K) = 0$ *such that*

$$\mathrm{Dec}_3(\delta, p) \lesssim_K \delta^{-\epsilon_p(K)} \left( \sup_{\delta \le \delta' \le 1} \mathrm{Dec}_2(\delta', p) + \sup_{\delta \le \delta' \le 1} \mathrm{TriDec}_3(\delta', p, K) \right).$$

The proof of this proposition is in the spirit of the analysis from Section 7.3. Let us just briefly sketch it. Assume $\mathrm{supp}\,(\widehat{F}) \subset \mathcal{N}_{[-1, 1]^2}(\delta)$. Using the local–global equivalence of decoupling constants, it suffices to estimate $\|F\|_{L^p(B_R)}$ on a ball $B_R$ of radius $R = \delta^{-1}$.

Cover $B_R$ with balls $B_K$. Write

$$F = \sum_{\alpha \in \mathcal{C}_K} \mathcal{P}_\alpha F.$$

On each $B_K$, a significant contribution to $F$ comes from either those $\alpha$ inside a $K^{-1/2}$-wide strip, or from three $K^{-2}$-transverse squares in $\mathcal{C}_K$. The latter scenario is covered using trilinear decoupling. The first scenario is covered by the following result, a close relative of Lemma 7.19.

**Lemma 10.26** (Lower-dimensional decoupling) *Let $L$ be a line in $\mathbb{R}^2$. Let $\mathcal{F}$ be a collection of squares $\beta \subset \mathcal{C}_{K^{1/2}}$ with* $\mathrm{dist}\,(\beta, L) \lesssim K^{-1/2}$. *For each $F$ with the Fourier transform supported on $\cup_{\beta \in \mathcal{F}} \mathcal{N}_\beta(1/K)$ and each ball $B_K \subset \mathbb{R}^3$ with radius $K$, we have*

$$\|F\|_{L^p(w_{B_K})} \lesssim \mathrm{Dec}_2(K^{-1}, p) \left( \sum_{\beta \in \mathcal{F}} \| \mathcal{P}_\beta F \|^2_{L^p(w_{B_K})} \right)^{1/2}.$$

*Proof* We sketch the proof in the following paragraph. An alternative argument is explored in Exercise 10.29.

Assume the equation of the line is $A\xi_1 + B\xi_2 + C = 0$ with $A^2 + B^2 = 1$. The Fourier transform of $F$ is supported inside the $O(K^{-1})$-neighborhood $\mathcal{N}$ of the strip on the paraboloid

$$\{(\xi_1, \xi_2, \xi_1^2 + \xi_2^2) : \operatorname{dist}((\xi_1, \xi_2), L) \lesssim K^{-1/2}\}.$$

As remarked in the proof of Lemma 7.19, $\mathcal{N}$ is a subset of the $O(K^{-1})$-neighborhood of the cylinder with directrix given by the parabola

$$\{(\xi_1, \xi_2, \xi_1^2 + \xi_2^2) : (\xi_1, \xi_2) \in L \cap [-1, 1]^2\}$$

and generatrix parallel to $\mathbf{v} = (A, B, -2C)$. This parabola is an affine image of $\mathbb{P}^1$. The result will follow by combining cylindrical decoupling (Exercise 9.22) with Exercise 9.9. $\qquad\square$

Theorem 9.6 shows that for $2 \le p \le 3$

$$\operatorname{TriDec}_3(\delta, p, K) \lesssim_{K, \epsilon} \delta^{-\epsilon}.$$

Combining this with Proposition 10.25 leads to the decoupling

$$\operatorname{Dec}_3(\delta, p) \lesssim_\epsilon \delta^{-\epsilon}.$$

This inequality, as well as its higher-dimensional counterpart explored in Exercise 10.28, was proved by Bourgain in [20].

To extend this to the full range $2 \le p \le 4$, we will essentially repeat the earlier argument for the parabola.

Fix $(\alpha_1, \alpha_2, \alpha_3) \in \mathcal{I}_K$ and let $\delta \ll_K 1$. Given a function $F : \mathbb{R}^3 \to \mathbb{C}$, a ball $B^r$ with radius $\delta^{-r}$ in $\mathbb{R}^3$, $t \ge 2$ and $m \le r$, we let

$$D_t(m, B^r, F, \delta) = \left[ \prod_{i=1}^3 \left( \sum_{\tau_{i,m} \in \mathcal{C}_{\delta-m}(\alpha_i)} \|\mathcal{P}_{\tau_{i,m}} F\|_{L_\sharp^t(w_{B^r})}^2 \right)^{\frac{1}{2}} \right]^{\frac{1}{3}}.$$

Also, given a finitely overlapping cover $\mathcal{B}_m(B^r)$ of $B^r$ with balls $B^m$ of radius $\delta^{-m}$, we let

$$A_p(m, B^r, F, \delta) = \left( \frac{1}{|\mathcal{B}_m(B^r)|} \sum_{B^m \in \mathcal{B}_m(B^r)} D_2(m, B^m, F, \delta)^p \right)^{\frac{1}{p}}.$$

Note that the quantities $D_t(m, B^r, F, 1/2)$ and $A_p(m, B^r, F, 1/2)$ are the analogs of the quantities $D_t(m, Q^r, F)$ and $A_p(m, Q^r, F)$ introduced earlier for $\mathbb{P}^1$. Let us define $\kappa_{p,3}$ using the relation

$$\frac{3}{2p} = \frac{1 - \kappa_{p,3}}{2} + \frac{\kappa_{p,3}}{p}.$$

The proof of the following two-scale inequality is very similar to the one of Proposition 10.23. The only difference is that it relies on the trilinear (rather than the bilinear) Kakeya inequality for tubes, Theorem 6.16.

**Proposition 10.27** *Assume* $\widehat{F}$ *is supported on* $\mathcal{N}_{\mathbb{P}^2}(\delta^2)$. *We have for each* $B^2$ *and* $p \geq 3$

$$A_p(1, B^2, F, \delta) \lesssim_{\epsilon, K} \delta^{-\epsilon} A_p(2, B^2, F, \delta)^{1-\kappa_{p,3}} D_p(1, B^2, F, \delta)^{\kappa_{p,3}}. \quad (10.18)$$

This inequality can be iterated as we did in the case of $\mathbb{P}^1$. Note that $2(1 - \kappa_{p,3}) \geq 1$ precisely when $p \leq 4$. The relevance of such a condition was discussed in Exercise 10.24. In particular, it implies that $\mathrm{Dec}_3(\delta, 4) \lesssim_\epsilon \delta^{-\epsilon}$.

**Exercise 10.28** Prove the analog of inequality (10.18) for $\mathbb{P}^{n-1}$, with

$$\frac{n}{p(n-1)} = \frac{1 - \kappa_{p,n}}{2} + \frac{\kappa_{p,n}}{p}.$$

Note that since $\kappa_{2n/(n-1),n} = 0$, the two-scale inequality can be easily iterated. Use this to give a more direct proof of the following:

$$\mathrm{Dec}_n\left(\delta, \frac{2n}{n-1}\right) \lesssim_\epsilon \delta^{-\epsilon}.$$

**Exercise 10.29** For $1 \leq k \leq n-1$, let $V$ be a $k$-dimensional affine space in $\mathbb{R}^{n-1}$ and let

$$S_V = \{\xi \in [-1, 1]^{n-1} : \mathrm{dist}\,(\xi, V) \lesssim K^{-1/2}\}.$$

Cover $S_V$ with roughly $K^{k/2}$ pairwise disjoint cubes $\beta$ with side length $\sim K^{-1/2}$.

Prove that for each $2 \leq p \leq 2(k+2)/k$, each $f : S_V \to \mathbb{C}$, and each ball $B_K$ of radius $\sim K$ in $\mathbb{R}^n$ we have

$$\|Ef\|_{L^p(w_{B_K})} \lesssim_\epsilon K^\epsilon \left(\sum_\beta \|E_\beta f\|_{L^p(w_{B_K})}^2\right)^{1/2}.$$

Hint: Assume for simplicity that

$$V = \{(\xi_1, \ldots, \xi_{n-1}) : \xi_1 = \cdots = \xi_{n-k-1} = 0\}.$$

Prove that for each fixed $x_1, \ldots, x_{n-k-1}$, the function $F$ on $\mathbb{R}^{k+1}$ given by

$$F(x_{n-k}, \ldots, x_n) = Ef(x_1, \ldots, x_n)$$

has the Fourier transform supported on the $\mathcal{N}_{\mathbb{P}^k}(O(K^{-1}))$. Combine decoupling for $\mathbb{P}^k$ with Fubini.

## 10.4 Another Look at the Decoupling for the Parabola

In this section, we present a different angle on the proof of Theorem 10.1. We again use the parabola as a case study. Though the argument is structured a bit differently here than in Section 10.2, the key ingredients are the same: $L^2$ local orthogonality, bilinear Kakeya, the bilinear-to-linear reduction, multiscale iteration, and bootstrapping. We hope that the reader will benefit from comparing the two proofs.

Our presentation follows an argument due to Guth. Unlike in the proof from Section 10.2, our emphasis here will be less on details, and more on the main ideas. Most technicalities will be swept under the rug. In particular, we will ignore weights and will assume perfect spatial localization for wave packets.

Recall that for an interval $I \subset [-1, 1]$,

$$\mathcal{N}_I(\sigma) = \{(\xi, \xi^2 + t) : \xi \in I, |t| < \sigma\}$$

is the $\sigma$-neighborhood of an arc on the parabola. We will simply write $\mathcal{N}(\sigma)$ to denote $\mathcal{N}_{[-1, 1]}(\sigma)$.

Fix the spatial scale $R = 2^{2^s}$, with $s \in \mathbb{N}$. For $1 \le i \le s$, we denote by $\Theta_i$ the partition of $\mathcal{N}(R^{-2^{-i+1}})$ into almost rectangular boxes $\theta_i$ with dimensions $\sim R^{-2^{-i}}$ and $\sim R^{-2^{-i+1}}$. As before, we let $\mathcal{P}_\theta F$ be the Fourier projection of $F$ onto $\theta$.

We define (somewhat casually) $\mathrm{Dec}(R)$ to be the smallest constant such that for each $F \colon \mathbb{R}^2 \to \mathbb{C}$ with the Fourier transform supported on $\mathcal{N}(R^{-1})$, we have

$$\|F\|_{L^6([-R, R]^2)} \le \mathrm{Dec}(R) \left( \sum_{\theta_1 \in \Theta_1} \|\mathcal{P}_{\theta_1} F\|^2_{L^6([-R, R]^2)} \right)^{\frac{1}{2}}.$$

Our goal is to prove that $\mathrm{Dec}(R) \lesssim_\epsilon R^\epsilon$.

Fix two disjoint intervals $I_1, I_2 \subset [-1, 1]$, and denote by $\Theta_i(I_j)$ those $\theta_i \in \Theta_i$ that lie inside $\mathcal{N}_{I_j}(R^{-2^{-i+1}})$. We also define a bilinear version of $\mathrm{Dec}(R)$. More precisely, let $\mathrm{BilDec}(R)$ be the smallest constant such that for each $F_j \colon \mathbb{R}^2 \to \mathbb{C}$ with the Fourier transform supported on $\mathcal{N}_{I_j}(R^{-1})$, we have

$$\|(F_1 F_2)^{\frac{1}{2}}\|_{L^6([-R, R]^2)} \le \mathrm{BilDec}(R) \left( \sum_{\theta_1 \in \Theta_1(I_1)} \|\mathcal{P}_{\theta_1} F_1\|^2_{L^6([-R, R]^2)} \right)^{\frac{1}{4}}$$

$$\times \left( \sum_{\theta_1 \in \Theta_1(I_2)} \|\mathcal{P}_{\theta_1} F_2\|^2_{L^6([-R, R]^2)} \right)^{\frac{1}{4}}.$$

The advantage of bilinearization is that it will give us access to the bilinear Kakeya inequality. This in turn will allow us to obtain more favorable estimates on the number of squares at various scales where lots of wave packets contribute substantially.

As in the previous argument from Section 10.2, we will make use of the approximate equivalence

$$\text{Dec}(R) \approx \text{BilDec}(R). \tag{10.19}$$

For the rest of the argument, we fix $F_1, F_2$ with Fourier transforms supported on $\mathcal{N}_{I_1}(R^{-1})$, $\mathcal{N}_{I_2}(R^{-1})$, and we write $F = F_1 + F_2$. Let us introduce the quantity $I_{1,R}(F_1, F_2)$ defined as follows:

$$\frac{\|(F_1 F_2)^{\frac{1}{2}}\|_{L^6([-R,R]^2)}}{\left(\sum_{\theta_1 \in \Theta_1(I_1)} \|\mathcal{P}_{\theta_1} F_1\|^2_{L^6([-R,R]^2)}\right)^{\frac{1}{4}} \left(\sum_{\theta_1 \in \Theta_1(I_2)} \|\mathcal{P}_{\theta_1} F_2\|^2_{L^6([-R,R]^2)}\right)^{\frac{1}{4}}}.$$

The equivalence (10.19) allows us to reformulate our goal in the form of the following inequality:

$$I_{1,R}(F_1, F_2) \lesssim_\epsilon R^\epsilon.$$

We will seek various estimates for $I_{1,R}(F_1, F_2)$ via wave packet decompositions at multiple scales and repeated pigeonholing. There will be many parameters at each scale that reflect the local properties of the two functions. However, our final estimates will be independent of $F_1, F_2$.

Let us denote $[-R, R]^2$ by $S_1$ and let $\mathcal{S}_2 = \mathcal{S}_2(S_1)$ be the partition of $S_1$ into squares $S_2$ of side length $R^{1/2}$. For $j \in \{1, 2\}$ we use a wave packet representation of $F_j$ at scale $R$, (Exercise 2.7)

$$F_j = \sum_{T_1 \in \mathcal{T}_j} w_{T_1} W_{T_1}.$$

To recap, each $T_1$ is a rectangle with dimensions $R, R^{1/2}$, which is roughly dual to some $\theta_1(T_1) \in \Theta_1(I_j)$. Also, $W_{T_1}$ has the Fourier transform supported on $\theta_1(T_1)$, and it decays rapidly outside of $T_1$.

We let $\mathcal{T}_{j,S_1}$ consist of the rectangles in $\mathcal{T}_j$ intersecting $S_1$ and define

$$F_{j,S_1} = \sum_{T_1 \in \mathcal{T}_{j,S_1}} w_{T_1} W_{T_1}.$$

We will assume various approximations meant to simplify the technical level of the presentation. For example, we will write

$$\|(F_1 F_2)^{\frac{1}{2}}\|_{L^6(S_1)} \approx \|(F_{1,S_1} F_{2,S_1})^{\frac{1}{2}}\|_{L^6(S_1)}.$$

Also, we will assume

$$W_{T_1}(x) \approx 1_{T_1}(x)e(\xi_{T_1} \cdot x)$$

for some $\xi_{T_1} \in \theta_1(T_1) \cap \mathbb{P}^1$. These approximations leave out error terms that are negligible. We will use $\lesssim$ to denote $(\log R)^{O(1)}$ losses.

Let us give a brief overview of the argument. When evaluating $F_1(x)$ and $F_2(x)$, a typical $x$ receives contributions from many wave packets $W_{T_1}$, each with a different oscillatory phase $e(\xi_{T_1} \cdot x)$. We expect lots of cancellations for a generic $x$, but at the earliest stage of the argument the only available estimate is the triangle inequality

$$\left| \sum_{T_1} w_{T_1} e(\xi_{T_1} \cdot x) \right| \leq \sum_{T_1} |w_{T_1}|.$$

This estimate can only be thought of as being essentially sharp if the sum contains $O(1)$ elements. We will eventually achieve this scenario by performing a wave packet decomposition at scale $\sim 1$. The transition between the initial scale $R$ and the final scale $\sim 1$ will be rather methodical, visiting each of the intermediate scales $R^{1/2^i}$, $1 \leq i \lesssim \log \log R$, collecting and processing the relevant parameters at each scale. The role of pigeonholing in the forthcoming argument is merely to simplify notation. It will help us focus on just one genealogy of significant squares, whose last generation will have the desired unit scale.

### 10.4.1 Pigeonholing at Scale $R$

The next proposition makes the point that when analyzing the term $\|(F_1 F_2)^{\frac{1}{2}}\|_{L^6(S_1)}$, we may restrict attention to uniform families of wave packets.

**Proposition 10.30** (Pigeonholing at scale $R$) *There are positive integers* $M_1, N_1, U_1, V_1, \beta_1, \gamma_1$ *and real numbers* $g_1, h_1 > 0$, *a collection* $\mathcal{S}_2^* \subset \mathcal{S}_2$ *of squares* $S_2$ *with side length* $R^{1/2}$, *and families of rectangles* $\mathcal{T}_{1,S_1}^* \subset \mathcal{T}_{1,S_1}$, $\mathcal{T}_{2,S_1}^* \subset \mathcal{T}_{2,S_1}$ *such that*

- *(uniform weight) For each* $T_1 \in \mathcal{T}_{1,S_1}^*$ *we have* $|w_{T_1}| \sim g_1$ *and for each* $T_1 \in \mathcal{T}_{2,S_1}^*$ *we have* $|w_{T_1}| \sim h_1$.
- *(uniform number of rectangles per direction) There is a family of* $\sim M_1$ *boxes in* $\Theta_1(I_1)$ *such that each* $T_1 \in \mathcal{T}_{1,S_1}^*$ *is dual to some* $\theta_1$ *in this family, with* $\sim U_1$ *rectangles for each such* $\theta_1$. *In particular, the size of* $\mathcal{T}_{1,S_1}^*$ *is* $\sim M_1 U_1$.

*Similarly, there is a family of $\sim N_1$ boxes in $\Theta_1(I_2)$ such that each $T_1 \in \mathcal{T}^*_{2,S_1}$ is dual to some $\theta_1$ in this family, with $\sim V_1$ rectangles for each such $\theta_1$. In particular, the size of $\mathcal{T}^*_{2,S_1}$ is $\sim N_1 V_1$.*

- *(uniform number of rectangles per square) Each $S_2 \in \mathcal{S}^*_2$ intersects $\sim M_1/\beta_1$ rectangles from $\mathcal{T}^*_{1,S_1}$ and $\sim N_1/\gamma_1$ rectangles from $\mathcal{T}^*_{2,S_1}$. Moreover,*

$$\left\| (F_1 F_2)^{\frac{1}{2}} \right\|_{L^6(S_1)} \lesssim \left\| \left( F_1^{(1)} F_2^{(1)} \right)^{\frac{1}{2}} \right\|_{L^6\left( \cup_{S_2 \in \mathcal{S}^*_2} S_2 \right)},$$

*where for $1 \le j \le 2$*

$$F_j^{(1)} = \sum_{T_1 \in \mathcal{T}^*_{j,S_1}} w_{T_1} W_{T_1}.$$

*Proof*  The proof involves three rounds of pigeonholing and simple applications of the triangle inequality.

First, it is easy to see that the contribution to the integral $\| (F_1 F_2)^{1/2} \|_{L^6(S_1)}$ coming from the wave packets $W_{T'_1}$, $T'_1 \in \mathcal{T}_{1,S_1}$ with weights smaller than $R^{-100} \max_{T_1 \in \mathcal{T}_{1,S_1}} |w_{T_1}|$ is negligible. Similarly, the contribution to $\| (F_1 F_2)^{1/2} \|_{L^6(S_1)}$ coming from the wave packets $W_{T'_1}$, $T'_1 \in \mathcal{T}_{2,S_1}$ with weights smaller than $R^{-100} \max_{T_1 \in \mathcal{T}_{2,S_1}} |w_{T_1}|$ is negligible. This leaves us with $\lesssim 1$ dyadic blocks of significant weights. Two of these blocks (one for $F_1$ and one for $F_2$) will contribute a significant (logarithmic) fraction to the integral, and for the next two rounds of pigeonholing we will restrict attention to the wave packets corresponding to these blocks. At this point, we have determined the values of $g_1$ and $h_1$.

Second, note that the range for each of the parameters $M_1, N_1, U_1, V_1$ is between 1 and $O(R^{1/2})$. There are $\lesssim 1$ dyadic blocks in this range. Again, the contribution to the integral coming from wave packets associated with one such block will be significant, and we fix such a block. Call $\mathcal{T}^*_{1,S_1}$ the rectangles corresponding to the selected parameters $M_1, U_1$ and call $\mathcal{T}^*_{2,S_1}$ the rectangles corresponding to the selected parameters $N_1, V_1$.

Third, let us see how to select the parameters $\beta_1, \gamma_1$. Each square in $\mathcal{S}_2$ is intersected by $O(M_1)$ rectangles from $\mathcal{T}^*_{1,S_1}$ and by $O(N_1)$ rectangles from $\mathcal{T}^*_{2,S_1}$. This observation allows us to partition $\mathcal{S}_2$ into $\lesssim 1$ collections of squares, with each square in a given collection being intersected by roughly $M_1/\beta_1$ rectangles from $\mathcal{T}^*_{1,S_1}$ and by roughly $N_1/\gamma_1$ rectangles from $\mathcal{T}^*_{2,S_1}$, for some $1 \le \beta_1, \gamma_1 \lesssim R^{1/2}$. We choose $\mathcal{S}^*_2$ to be the collection contributing the most to the integral $\| (F_1^{(1)} F_2^{(1)})^{1/2} \|_{L^6(S_1)}$. $\qquad\square$

At this point, let us observe that we have the following lower bound for the denominator of $I_{1,R}(F_1, F_2)$.

**Proposition 10.31** *We have*

$$\left( \sum_{\theta_1 \in \Theta_1(I_1)} \|\mathcal{P}_{\theta_1} F_1\|^2_{L^6([-R,R]^2)} \right)^{\frac{1}{4}} \left( \sum_{\theta_1 \in \Theta_1(I_2)} \|\mathcal{P}_{\theta_1} F_2\|^2_{L^6([-R,R]^2)} \right)^{\frac{1}{4}}$$
$$\gtrsim R^{1/4}(M_1 N_1)^{1/4}(g_1 h_1)^{1/2}(U_1 V_1)^{1/12}.$$

*Proof* We know that there are roughly $M_1$ boxes $\theta_1 \in \Theta_1(I_1)$, and that each of them contributes with $U_1$ wave packets of magnitude $\sim g_1$. Thus, for each such $\theta_1$, due to spatial almost orthogonality, we may write

$$\|\mathcal{P}_{\theta_1} F_1\|_{L^6([-R,R]^2)} \gtrsim R^{1/4} g_1 U_1^{1/6}.$$

Hence

$$\sum_{\theta_1 \in \Theta_1(I_1)} \|\mathcal{P}_{\theta_1} F_1\|^2_{L^6([-R,R]^2)} \gtrsim R^{1/2} M_1 g_1^2 U_1^{1/3}.$$

Reasoning similarly for $F_2$, we get

$$\sum_{\theta_1 \in \Theta_1(I_2)} \|\mathcal{P}_{\theta_1} F_2\|^2_{L^6([-R,R]^2)} \gtrsim R^{1/2} N_1 h_1^2 V_1^{1/3}. \qquad \square$$

It is worth observing that Proposition 10.30 sheds very little light on the size of the numerator $\|(F_1 F_2)^{1/2}\|_{L^6([-R,R]^2)}$ of $I_{1,R}(F_1, F_2)$. This is because there are many wave packets $W_{T_1}$ contributing to a typical point $x \in S_1$, and the cancellations between them is very hard to quantify. All we can say at this point is the fact that most of the mass of $F_1 F_2$ is concentrated inside the smaller region covered by the squares from $\mathcal{S}_2^*$. Getting a favorable upper bound on the size of $\mathcal{S}_2^*$ will be one of the key ingredients in our proof.

**Proposition 10.32** (Counting squares using bilinear Kakeya) *We have*

$$|\mathcal{S}_2^*| \lesssim \beta_1 \gamma_1 U_1 V_1.$$

*Proof* We apply the bilinear Kakeya inequality (Theorem 10.21). More precisely, we have the following estimate for the fat rectangles $10T_1$:

$$\int_{\mathbb{R}^2} \left( \sum_{T_1 \in \mathcal{T}^*_{1,S_1}} 1_{10T_1} \right) \left( \sum_{T_1 \in \mathcal{T}^*_{2,S_1}} 1_{10T_1} \right) \lesssim R M_1 N_1 U_1 V_1.$$

Recall also that each $S_2 \in \mathcal{S}_2^*$ is contained in $\sim M_1/\beta_1$ fat rectangles $10T_1$ with $T_1 \in \mathcal{T}_{1,S_1}^*$ and in $\sim N_1/\gamma_1$ fat rectangles $10T_1$ with $T_1 \in \mathcal{T}_{2,S_1}^*$. With this observation, the desired upper bound on $|\mathcal{S}_2^*|$ follows immediately. $\square$

To get a better grasp on the size of $\|(F_1 F_2)^{1/2}\|_{L^6([-R,R]^2)}$, we need to use wave packets of smaller scale. In fact, we will aim to eventually work with wave packets of scale roughly one. At this tiny scale, there are only $O(1)$ rectangles intersecting each point $x$, and $|F_1(x)F_2(x)|$ can be sharply dominated using the triangle inequality.

### 10.4.2 Pigeonholing at Scale $R^{1/2}$

To eventually achieve a decomposition at scale $\sim 1$, we will repeat the previous pigeonholing at scales smaller than $R$. In this subsection, we discuss the scale $R^{1/2}$.

Recall that for $1 \le j \le 2$, the function $F_j^{(1)}$ is a sum of wave packets at scale $R$. Note also that $F_j^{(1)}$ has its Fourier transform supported on $\mathcal{N}(R^{-1})$ and thus also in $\mathcal{N}(R^{-1/2})$. Let us consider the wave packet decomposition of $F_j^{(1)}$ at scale $R^{1/2}$:

$$F_j^{(1)} = \sum_{T_2 \in \mathcal{T}_{j,2}} w_{T_2} W_{T_2}.$$

Each $T_2 \in \mathcal{T}_{j,2}$ is a rectangle with dimensions $\sim R^{1/2}, R^{1/4}$, which is essentially dual to some $\theta_2 \in \Theta_2(I_j)$. For each $S_2 \in \mathcal{S}_2^*$, we let $\mathcal{T}_{j,S_2}$ consist of those $T_2 \in \mathcal{T}_{j,2}$ intersecting $S_2$. We may write

$$F_j^{(1)}(x) \approx \sum_{T_2 \in \mathcal{T}_{j,S_2}} w_{T_2} W_{T_2}(x), \quad x \in S_2.$$

**Proposition 10.33** (Pigeonholing at scale $R^{1/2}$) *We may refine the collection $\mathcal{S}_2^*$ to get a smaller collection $\mathcal{S}_2^{**}$, so that for each $S_2 \in \mathcal{S}_2^{**}$ the following hold.*

*There are positive integers $M_2, N_2, U_2, V_2, \beta_2, \gamma_2$ and real numbers $g_2, h_2 > 0$ (all independent of $S_2$), a collection $\mathcal{S}_3^*(S_2)$ of squares $S_3$ with side length $R^{1/4}$ inside $S_2$, and two families of rectangles $\mathcal{T}_{1,S_2}^* \subset \mathcal{T}_{1,S_2}$, $\mathcal{T}_{2,S_2}^* \subset \mathcal{T}_{2,S_2}$ such that*

- *(uniform weight) For each $T_2 \in \mathcal{T}_{1,S_2}^*$ we have $|w_{T_2}| \sim g_2$ and for each $T_2 \in \mathcal{T}_{2,S_2}^*$ we have $|w_{T_2}| \sim h_2$.*
- *(uniform number of rectangles per direction) There is a family of $\sim M_2$ boxes $\theta_2$ in $\Theta_2(I_1)$ such that each $T_2 \in \mathcal{T}_{1,S_2}^*$ is dual to such a $\theta_2$, with $\sim U_2$ rectangles for each such $\theta_2$. In particular, the size of $\mathcal{T}_{1,S_2}^*$ is $\sim M_2 U_2$. Also,*

there is a family of $\sim N_2$ boxes $\theta_2 \in \Theta_2(I_2)$ such that each $T_2 \in \mathcal{T}^*_{2, S_2}$ is dual to such a $\theta_2$, with $\sim V_2$ rectangles for each such $\theta_2$. In particular, the size of $\mathcal{T}^*_{2, S_2}$ is $\sim N_2 V_2$.

- *(uniform number of rectangles per square) Each $S_3 \in \mathcal{S}^*_3(S_2)$ intersects $\sim M_2/\beta_2$ rectangles from $\mathcal{T}^*_{1, S_2}$ and $\sim N_2/\gamma_2$ rectangles from $\mathcal{T}^*_{2, S_2}$.*

*Moreover,*

$$\left\| (F_1 F_2)^{\frac{1}{2}} \right\|_{L^6(S_1)} \lesssim \left\| (F_1^{(2)} F_2^{(2)})^{\frac{1}{2}} \right\|_{L^6(\cup_{S_3 \in \mathcal{S}^*_3} S_3)},$$

*where for $1 \leq j \leq 2$,*

$$F_j^{(2)} = \sum_{S_2 \in \mathcal{S}^{**}_2} \sum_{T_2 \in \mathcal{T}^*_{j, S_2}} w_{T_2} W_{T_2}$$

*and*

$$\mathcal{S}^*_3 = \{ S_3 \in \mathcal{S}^*_3(S_2) \colon S_2 \in \mathcal{S}^{**}_2 \}.$$

*Proof* The three uniformity estimates are first achieved for each individual $S_2 \in \mathcal{S}^*_2$, exactly as in the previous round of pigeonholing at scale $R$. More precisely, the choice of these parameters is made in such a way that the corresponding families of rectangles $T_2$ and squares $S_3$ produce a significant contribution:

$$\left\| (F_1^{(1)} F_2^{(1)})^{\frac{1}{2}} \right\|_{L^6(S_2)} \lesssim \left\| \left( \sum_{T_2 \in \mathcal{T}^*_{1, S_2}} w_{T_2} W_{T_2} \sum_{T_2 \in \mathcal{T}^*_{2, S_2}} w_{T_2} W_{T_2} \right)^{\frac{1}{2}} \right\|_{L^6(\cup_{S_3 \in \mathcal{S}^*_3(S_2)} S_3)}.$$

At the end of this stage, the parameters $M_2, \ldots, h_2$ might be different for each $S_2$, but for each of the eight parameters there will be at most $\lesssim 1$ significant dyadic blocks in the range of all possible values. By using an additional round of pigeonholing, we may restrict attention to the family of squares $S_2$ corresponding to a fixed dyadic block for each parameter. More precisely, we select the parameters whose corresponding family of squares $S_2$ contribute most significantly to the integral. This will be the family $\mathcal{S}^{**}_2$. See Figure 10.4. □

Before iterating this procedure at even smaller scales, let us pause to collect three estimates. The first two are analogous to the ones in Propositions 10.31 and 10.32.

**Proposition 10.34** *We have*

$$\left( \sum_{\theta_2 \in \Theta_2(I_1)} \| \mathcal{P}_{\theta_2} F_1 \|^2_{L^6([-R,R]^2)} \right)^{\frac{1}{4}} \left( \sum_{\theta_2 \in \Theta_2(I_2)} \| \mathcal{P}_{\theta_2} F_2 \|^2_{L^6([-R,R]^2)} \right)^{\frac{1}{4}}$$
$$\gtrsim R^{1/8} |\mathcal{S}^{**}_2|^{1/6} (M_2 N_2)^{1/4} (g_2 h_2)^{1/2} (U_2 V_2)^{1/12}.$$

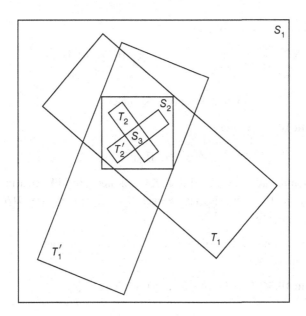

Figure 10.4 Transverse wave packets at scales $R$ and $R^{1/2}$.

*Proof*  Let us write $\theta_2 \sim S_2$ if $\theta_2$ is one of the roughly $M_2$ boxes contributing (with roughly $U_2$ wave packets) to $S_2$. For each $1 \leq 2^m \leq |\mathcal{S}_2^{**}|$, we let $\Theta_2(I_1, m)$ consist of those $\theta_2$ such that

$$2^m \leq |\{S_2 \in \mathcal{S}_2^{**} : \theta_2 \sim S_2\}| < 2^{m+1}.$$

We reason as in the proof of Proposition 10.31 and write for each $S_2 \in \mathcal{S}_2^{**}$ and each $\theta_2 \sim S_2$

$$\|\mathcal{P}_{\theta_2} F_1\|_{L^6(S_2)}^6 \gtrsim R^{3/4}(g_2)^6 U_2.$$

Thus, if $\theta_2 \in \Theta_2(I_1, m)$, then

$$\|\mathcal{P}_{\theta_2} F_1\|_{L^6([-R, R]^2)}^6 \gtrsim 2^m R^{3/4}(g_2)^6 U_2.$$

It follows that

$$\sum_{\theta_2 \in \Theta_2(I_1)} \|\mathcal{P}_{\theta_2} F_1\|_{L^6([-R, R]^2)}^2$$

$$\gtrsim R^{1/4}(g_2)^2 (U_2)^{1/3} \sum_{1 \leq 2^m \leq |\mathcal{S}_2^{**}|} 2^{m/3} |\Theta_2(I_1, m)|$$

$$\geq |\mathcal{S}_2^{**}|^{-2/3} R^{1/4}(g_2)^2 (U_2)^{1/3} \sum_{1 \leq 2^m \leq |\mathcal{S}_2^{**}|} 2^m |\Theta_2(I_1, m)|$$

$$\sim |\mathcal{S}_2^{**}|^{1/3} M_2 R^{1/4}(g_2)^2 (U_2)^{1/3}.$$

A similar computation shows that

$$\sum_{\theta_2 \in \Theta_2(I_2)} \|\mathcal{P}_{\theta_2} F_2\|^2_{L^6([-R,R]^2)} \gtrsim |\mathcal{S}_2^{**}|^{1/3} N_2 R^{1/4} (h_2)^2 (V_2)^{1/3}.$$

It now suffices to combine the last two estimates.     □

**Proposition 10.35** *We have*

$$|\mathcal{S}_3^*| \lesssim |\mathcal{S}_2^{**}| \beta_2 \gamma_2 U_2 V_2.$$

*Proof* An application of the bilinear Kakeya inequality like that in the proof of Proposition 10.32 reveals that for each $S_2 \in \mathcal{S}_2^{**}$, there are $O(\beta_2 \gamma_2 U_2 V_2)$ squares $S_3$ in $\mathcal{S}_3^*(S_2)$.     □

The third estimate relates the parameters from scales $R$ and $R^{1/2}$, by means of $L^2$ orthogonality.

**Proposition 10.36** (Local orthogonality) *We have*

$$U_2 g_2^2 R^{\frac{3}{4}} M_2 \lesssim R g_1^2 \frac{M_1}{\beta_1},$$

$$V_2 h_2^2 R^{\frac{3}{4}} N_2 \lesssim R h_1^2 \frac{N_1}{\gamma_1}$$

*and thus*

$$\frac{g_2 h_2}{g_1 h_1} \lesssim R^{1/4} \frac{1}{(\beta_1 \gamma_1)^{1/2}} \frac{1}{(U_2 V_2)^{1/2}} \left( \frac{M_1 N_1}{M_2 N_2} \right)^{1/2}.$$

*Proof* Let us prove the first inequality, as the second one has an identical proof. Recall that on each $S_2 \in \mathcal{S}_2^{**}$, we have two ways of representing $F_1^{(1)}$. The first one uses wave packets at scale $R$

$$F_1^{(1)}(x) \approx \sum_{\substack{T_1 \in \mathcal{T}_{1,S_1}^* \\ T_1 \cap S_2 \neq \emptyset}} w_{T_1} W_{T_1}(x), \quad x \in S_2,$$

while the second one uses wave packets at scale $R^{1/2}$

$$F_1^{(1)}(x) \approx \sum_{T_2 \in \mathcal{T}_{1,S_2}} w_{T_2} W_{T_2}(x), \quad x \in S_2.$$

Note that the wave packets $W_{T_1}$ in the first sum are almost orthogonal on $S_2$. Indeed, let $\eta$ be a smooth approximation of $1_{S_2}$ with its Fourier transform supported on $B(0, 1/10R^{1/2})$. Then the support of $\widehat{\eta W_{T_1}}$ is only slightly larger than the support of $\widehat{W_{T_1}}$. This observation allows us to write

$$\|F_1^{(1)}\|_{L^2(S_2)} \approx R^{1/2} \left( \sum_{\substack{T_1 \in \mathcal{T}_{1,S_1}^* \\ T_1 \cap S_2 \neq \emptyset}} |w_{T_1}|^2 \right)^{1/2} \approx R^{1/2} g_1 \left( \frac{M_1}{\beta_1} \right)^{1/2}.$$

The wave packets $W_{T_2}$ in the second sum are also almost orthogonal on $S_2$, so we also have

$$\|F_1^{(1)}\|_{L^2(S_2)} \approx R^{3/8} \left( \sum_{T_2 \in \mathcal{T}_{1,S_2}} |w_{T_2}|^2 \right)^{1/2}$$

$$\geq R^{3/8} \left( \sum_{T_2 \in \mathcal{T}_{1,S_2}^*} |w_{T_2}|^2 \right)^{1/2}$$

$$\approx R^{3/8} g_2 (M_2 U_2)^{1/2}.$$

The desired inequality follows by comparing the two $L^2$ estimates. $\qquad\square$

### 10.4.3 The Multiscale Decomposition

It should now be clear how to iterate the wave packet decomposition and pigeonholing at smaller scales. The following result summarizes the process and collects the relevant estimates at each step.

**Proposition 10.37** *For each* $1 \leq k \leq s$, *there are numbers* $g_k, h_k > 0$ *and* $M_k, N_k, U_k, V_k, \beta_k, \gamma_k \geq 1$, *two collections* $\mathcal{S}_k^{**} \subset \mathcal{S}_k^*$ *of squares* $S_k \subset [-R, R]^2$ *with side length* $\sim R^{2^{-k+1}}$ *(we let* $\mathcal{S}_1^{**} = \mathcal{S}_1^* = \{S_1\}$*) and also include a collection* $\mathcal{S}_{s+1}^*$ *for convenience), and two families* $\mathcal{T}_1^{(k)}$, $\mathcal{T}_2^{(k)}$ *of rectangles with dimensions* $R^{2^{-k+1}}$, $R^{2^{-k}}$ *and dual to boxes* $\theta_k \in \Theta_k$ *such that*

$$\left( \sum_{\theta_k \in \Theta_k(I_1)} \|\mathcal{P}_{\theta_k} F_1\|_{L^6([-R,R]^2)}^2 \right)^{\frac{1}{4}} \left( \sum_{\theta_k \in \Theta_k(I_2)} \|\mathcal{P}_{\theta_k} F_2\|_{L^6([-R,R]^2)}^2 \right)^{\frac{1}{4}}$$

$$\gtrsim R^{1/2^{k+1}} |\mathcal{S}_k^{**}|^{1/6} (M_k N_k)^{1/4} (g_k h_k)^{1/2} (U_k V_k)^{1/12}, \tag{10.20}$$

$$|\mathcal{S}_{k+1}^*| \lesssim |\mathcal{S}_k^{**}| \beta_k \gamma_k U_k V_k, \tag{10.21}$$

$$\frac{g_k h_k}{g_{k-1} h_{k-1}} \lesssim R^{2^{-k}} \frac{1}{(\beta_{k-1} \gamma_{k-1})^{1/2}} \frac{1}{(U_k V_k)^{1/2}} \left( \frac{M_{k-1} N_{k-1}}{M_k N_k} \right)^{1/2}, \tag{10.22}$$

$$\|(F_1 F_2)^{\frac{1}{2}}\|_{L^6(S_1)} \lesssim (\log R)^{O(\log \log R)} \|(F_1^{(k)} F_2^{(k)})^{\frac{1}{2}}\|_{L^6\left(\cup_{S_{k+1} \in \mathcal{S}_{k+1}^*} S_{k+1}\right)}. \tag{10.23}$$

*Also, on each $S_{k+1} \in \mathcal{S}^*_{k+1}$, we have*

$$F_1^{(k)} \approx \sum_{T_k} w_{T_k} W_{T_k},$$

*with the sum containing $\sim M_k/\beta_k$ rectangles $T_k \in \mathcal{T}_1^{(k)}$ with weights $|w_{T_k}| \sim g_k$, and*

$$F_2^{(k)} \approx \sum_{T_k} w_{T_k} W_{T_k},$$

*with the sum containing $\sim N_k/\gamma_k$ rectangles $T_k \in \mathcal{T}_2^{(k)}$ with weights $|w_{T_k}| \sim h_k$.*

Recall that $R = 2^{2^s}$. Thus, when $k = s$, the rectangles $T_s$ are in fact almost squares with diameter $\sim 1$. Moreover, for $j \in \{1, 2\}$, we have $|\Theta_s(I_j)| \lesssim 1$. It follows that

$$\|F_1^{(s)}\|_\infty \lesssim g_s \quad \text{and} \quad \|F_2^{(s)}\|_\infty \lesssim h_s. \qquad (10.24)$$

These estimates are essentially sharp. Note that the initial difficulty of quantifying cancellations for large-scale wave packets is no longer present at scale $\sim 1$.

We are finally in position to produce a good upper bound for $\|(F_1 F_2)^{1/2}\|_{L^6([-R,R]^2)}$, and thus also for $I_{1,R}(F_1, F_2)$. In fact, let us extend our earlier definition and denote by $I_{k,R}(F_1, F_2)$ the quantity

$$\frac{\|(F_1 F_2)^{\frac{1}{2}}\|_{L^6([-R,R]^2)}}{\left(\sum_{\theta_k \in \Theta_k(I_1)} \|\mathcal{P}_{\theta_k} F_1\|^2_{L^6([-R,R]^2)}\right)^{\frac{1}{4}} \left(\sum_{\theta_k \in \Theta_k(I_2)} \|\mathcal{P}_{\theta_k} F_2\|^2_{L^6([-R,R]^2)}\right)^{\frac{1}{4}}}.$$

**Corollary 10.38** *We have*

$$\|(F_1 F_2)^{\frac{1}{2}}\|_{L^6([-R,R]^2)} \lesssim (\log R)^{O(\log \log R)} (g_s h_s)^{1/2} |\mathcal{S}^*_{s+1}|^{1/6},$$

*and for each $1 \leq k \leq s - 1$,*

$$I_{k,R}(F_1, F_2) \lesssim (\log R)^{O(\log \log R)} \left(\frac{U_k V_k}{\prod_{l=k+1}^s (U_l V_l)}\right)^{\frac{1}{12}}. \qquad (10.25)$$

*Proof* Recall that $s = O(\log \log R)$. Thus, given $C = O(1)$, we have $C^s \lesssim 1$.

The first inequality in the corollary follows from (10.23) with $k = s$ and from (10.24). Combining this with (10.20), we get

$$I_{k,R}(F_1, F_2) \lesssim (\log R)^{O(\log \log R)} R^{-1/2^{k+1}}$$

$$\times \left(\frac{g_s h_s}{g_k h_k}\right)^{1/2} \left(\frac{|\mathcal{S}^*_{s+1}|}{|\mathcal{S}^{**}_k|}\right)^{1/6} (M_k N_k)^{-1/4} (U_k V_k)^{-1/12}.$$

On the other hand, from repeated applications of (10.22), we get

$$\left(\frac{g_s h_s}{g_k h_k}\right)^{1/2} \lesssim \prod_{l=k+1}^{s} R^{1/2^{l+1}} \frac{1}{(\beta_{l-1}\gamma_{l-1})^{1/4}} \frac{1}{(U_l V_l)^{1/4}} \left(\frac{M_{l-1}N_{l-1}}{M_l N_l}\right)^{1/4}$$

$$\lesssim R^{1/2^{k+1}} (M_k N_k)^{1/4} \prod_{l=k}^{s-1} \frac{1}{(\beta_l \gamma_l)^{1/4}} \prod_{l=k+1}^{s} \frac{1}{(U_l V_l)^{1/4}}.$$

Also, applying (10.21) many times, we get

$$\left(\frac{|\mathcal{S}^*_{s+1}|}{|\mathcal{S}^{**}_k|}\right)^{1/6} \lesssim \prod_{l=k}^{s} (\beta_l \gamma_l V_l U_l)^{1/6}.$$

Combining the last three inequalities, we may write

$$I_{k,R}(F_1, F_2) \lesssim (\log R)^{O(\log \log R)} \left(\frac{U_k V_k}{\prod_{l=k+1}^{s}(U_l V_l)}\right)^{\frac{1}{12}} \frac{(\beta_s \gamma_s)^{1/6}}{\prod_{l=k}^{s-1}(\beta_l \gamma_l)^{1/12}}.$$

Finally, recall that $\beta_l, \gamma_l \geq 1$ for each $l \geq 1$, and also $\beta_s, \gamma_s \lesssim 1$. $\qquad \square$

In the next subsection, we will combine these estimates for $I_{k,R}(F_1, F_2)$ with parabolic rescaling to produce a range of upper bounds for $I_{1,R}(F_1, F_2)$, one of which will necessarily prove to be favorable.

### 10.4.4 Bootstrapping

The argument here mirrors the one in Section 10.2.3. We are interested in getting an $O(R^\epsilon)$ upper bound for $I_{1,R}(F_1, F_2)$. Inequality (10.25) with $k = 1$ would provide such a bound, if it were the case that

$$U_1 V_1 \lesssim_\epsilon R^\epsilon \prod_{l=2}^{s}(U_l V_l).$$

While this inequality may fail, a very elementary analysis will reveal that at least one of the terms $I_{k,R}(F_1, F_2)$ has to be sufficiently small. This will turn out to be enough for our purpose, as $I_{1,R}(F_1, F_2)$ can be easily related to $I_{k,R}(F_1, F_2)$ via parabolic rescaling.

**Proposition 10.39** *For each $1 \leq k \leq s - 1$, we have*

$$I_{1,R}(F_1, F_2) \lesssim (\log R)^{O(\log \log R)} \text{Dec}(R^{1-2^{1-k}}) \left(\frac{U_k V_k}{\prod_{l=k+1}^{s}(U_l V_l)}\right)^{\frac{1}{12}}.$$

$$(10.26)$$

*Proof*

$$I_{1,R}(F_1, F_2) \le I_{k,R}(F_1, F_2)$$

$$\times \left( \frac{\sum_{\theta_k \in \Theta_k(I_1)} \|P_{\theta_k} F_1\|_{L^6([-R,R]^2)}^2}{\sum_{\theta_1 \in \Theta_1(I_1)} \|P_{\theta_1} F_1\|_{L^6([-R,R]^2)}^2} \right)^{\frac{1}{4}} \left( \frac{\sum_{\theta_k \in \Theta_k(I_2)} \|P_{\theta_k} F_2\|_{L^6([-R,R]^2)}^2}{\sum_{\theta_1 \in \Theta_1(I_2)} \|P_{\theta_1} F_2\|_{L^6([-R,R]^2)}^2} \right)^{\frac{1}{4}}.$$

Each of the last two terms is $O\left( \sqrt{\mathrm{Dec}(R^{1-2^{1-k}})} \right)$ due to parabolic rescaling. See Proposition 10.13.     □

**Proposition 10.40** *Let $\epsilon > 0$. For each $s > 100\epsilon^{-1}$ and $R = 2^{2^s}$ there is $k \le 100\epsilon^{-1}$ such that*

$$I_{1,R}(F_1, F_2) \lesssim (\log R)^{O(\log \log R)} R^{\epsilon 2^{-k}} \mathrm{Dec}(R^{1-2^{1-k}}).$$

*The implicit constant is independent of $F_1, F_2$.*
    *In particular, for each $\epsilon, \delta > 0$,*

$$\mathrm{Dec}(R) \lesssim_{\epsilon,\delta} R^\delta \max_{k \lesssim \epsilon^{-1}} R^{\epsilon 2^{-k}} \mathrm{Dec}(R^{1-2^{1-k}}). \tag{10.27}$$

*Proof* The first inequality is a consequence of (10.26) and Exercise 10.43. The second inequality then follows by combining (10.19) with the fact that

$$\mathrm{BilDec}(R) = \sup_{F_1, F_2} I_{1,R}(F_1, F_2).     \square$$

**Corollary 10.41** (Bootstrapping) *For each $\epsilon > 0$, we have*

$$\mathrm{Dec}(R) \lesssim_\epsilon R^\epsilon.$$

*Proof* Let $\Sigma$ be the set of all $\sigma > 0$ such that $\mathrm{Dec}(R) \lesssim R^\sigma$. It suffices to prove that $\sigma_0 := \inf \Sigma = 0$. Note that (10.27) shows that for each $\epsilon, \delta > 0$,

$$\sigma \in \Sigma \implies \max_{k \lesssim \epsilon^{-1}} (\sigma(1 - 2^{1-k}) + \epsilon 2^{-k} + \delta) \in \Sigma.$$

In particular, letting $\sigma \to \sigma_0$ and $\delta \to 0$, we get that

$$\sigma_0 \le \max_{k \lesssim \epsilon^{-1}} (\sigma_0(1 - 2^{1-k}) + \epsilon 2^{-k}).$$

This forces $\sigma_0 = 0$, once we test the preceding inequality with small enough $\epsilon$.     □

The reader is invited to rework the computations from this section in order to prove the $l^2(L^p)$ decoupling for $4 \le p \le 6$. It is worth realizing that Exercise 10.43 is needed when $p > 4$, but not needed when $p = 4$.

**Exercise 10.42** Using the notation from Proposition 10.37, prove that if $k \geq 2$,

$$g_{k+1} \lesssim g_k \frac{M_k}{\beta_k}.$$

**Exercise 10.43** (a) Consider a sequence $x_n$ with $x_1 \geq 1$ and satisfying

$$x_{n+1} \geq x_1 + \cdots + x_n + 2^n,$$

for $n \geq 1$. Prove that

$$x_n \geq n 2^{n-2}.$$

(b) Conclude that if $(y_k)_{1 \leq k \leq s}$ is a sequence with terms in the interval $[1, R]$, then for each $\epsilon > 8s^{-1}$ there is $1 \leq k \leq 8\epsilon^{-1}$ such that

$$\frac{y_k}{\prod_{l=k+1}^{s} y_l} \leq R^{\epsilon 2^{-k}}.$$

**Exercise 10.44** Adapt the argument from this section to prove Theorem 10.1 in all dimensions. Convince yourself that the proof simplifies when $p = 2n/(n-1)$, in the sense that Exercise 10.43 is not needed. This is in line with the observation from Exercise 10.28.

**Exercise 10.45** Let $I_1, \ldots, I_n$ be a partition of $[0, 1]$ into intervals of arbitrary length, and let $\delta := \min |I_i|$. Prove that for each $R \geq \delta^{-2}$,

$$\|E_{[0,1]} f\|_{L^6(B_R)} \lesssim_\epsilon \delta^{-\epsilon} \left( \sum_i \|E_{I_i} f\|_{L^6(w_{B_R})}^2 \right)^{1/2}.$$

**Exercise 10.46** Let $I_1, \ldots, I_n$ be a partition of $[0, 1]$ into intervals of arbitrary length, and let $\delta := \min |I_i|$. Prove that for each $R \geq \delta^{-2}$,

$$\|E_{[0,1]} f\|_{L^4(B_R)} \lesssim \left( \sum_i \|E_{I_i} f\|_{L^4(w_{B_R})}^2 \right)^{1/2},$$

with implicit constant independent of $\delta$.

Hint: Use biorthogonality to prove a stronger reverse square function estimate.

### 10.4.5 Final Remarks

Let us close with a few remarks concerning the argument we have just presented. First, there is a gentle transition between the initial scale $R$ and the final scale $\sim 1$, that involves all scales of the form $R^{2^{-k}}$ with $1 \leq k \lesssim \log \log R$. The idea of going from one scale to its square root follows the classical paradigm enforced by curvature. However, the induction on scales here is slightly more complicated than the one used in the proof of the multilinear restriction theorem from Chapter 6. Note for example that (10.27) does not guarantee that

$$\text{Dec}(R) \lesssim R^\epsilon \text{Dec}(R^{\frac{1}{2}}).$$

Thus, there is no certainty of a gain when passing from *each* scale to its square root. It is rather the case that for each $\epsilon$ there is *some* good scale $R^{1-\alpha}$ such that

$$\text{Dec}(R) \lesssim R^{\alpha\epsilon} \text{Dec}(R^{1-\alpha}).$$

This is the key improvement over the inequality (10.4):

$$\text{Dec}(R) \lesssim \text{Dec}(R^\alpha)\text{Dec}(R^{1-\alpha}).$$

Second, let us reflect on the role of the multiscale decomposition. It can be seen as a methodical way to approach the unit scale, where the issue of estimating cancellations collapses to a trivial application of the triangle inequality. Each time we jump to a lower scale, we collect and keep track of the local information about $F_1, F_2$, via repeated applications of bilinear Kakeya and local orthogonality. Could we have performed a unit scale wave packet decomposition from the beginning, without passing through the intermediate scales? Can one avoid iteration and induction on scales, tools that are strongly built into all known arguments for proving genuine decoupling at (or near) the critical exponent $p = 6$? There is no easy answer to these questions. In Section 13.2, we will see that the $l^2(L^6)$ decoupling constant $\text{Dec}(R)$ must in fact grow (albeit very mildly) with $R$. However, it seems likely that there is a uniform estimate for the $l^2(L^p)$ decoupling constants when $p < 6$ (see Exercise 10.46 for $p = 4$). One may fantasize that if this is indeed the case, the argument to prove it would either brilliantly optimize induction on scales, or avoid it altogether.

Third, we can now make a more informed judgment on the merit of the bilinear approach. Let us see what happens if we run the argument with just one family (rather than two families) $\mathcal{T}^{(k)}$ of $(R^{2^{-k}}, R^{2^{-k+1}})$-rectangles. Assume we have about $M_k U_k$ such rectangles, with roughly $U_k$ of them for each direction from among $M_k$ possible directions. We need to count the number of squares of side length $R^{2^{-k}}$ that intersect about $M_k/\beta_k$ rectangles. The linear Kakeya estimate in Proposition 5.6 leads to the upper bound

$$\frac{\beta_k^2 U_k^2}{M_k} R^{2^{-k}}.$$

If $M_k \sim R^{2^{-k}}$, this bound is as good as the bound we have obtained using bilinear Kakeya (see Exercise 5.14). But the linear Kakeya bound gets worse as $M_k$ gets closer to 1. The reader may check that in fact such a bound is not good enough to prove the desired decoupling.

# 11

# Decoupling for the Moment Curve

As we have seen earlier in the book, the Fourier restriction theory of regular curves has essentially been settled. In particular, doing so proved to be much easier than settling the theory of hypersurfaces, which are still under a lot of investigation. Perhaps the main reason for this discrepancy is that curves do not contain nontrivial lower-dimensional submanifolds. In particular, this makes the multilinear-to-linear reduction rather straightforward. It will thus come as a surprise that proving sharp decouplings for the moment curve turns out to be more difficult than for the paraboloid. In this chapter, we will seek to understand why this is the case.

The argument for the moment curve will rely at its core on induction on dimension. The case $n = 2$ of the parabola that we have discussed in the previous chapter will serve as the base of the induction. While the moment curve "escapes" every hyperplane, small-scale arcs stay "close" to affine subspaces of $\mathbb{R}^n$, and will resemble lower-dimensional moment curves. Decoupling for small arcs will thus be handled using lower-dimensional decoupling, essentially ignoring the unimportant coordinates.

The structure of the argument is somewhat similar to the one for the paraboloid in the previous chapter. There is a rather straightforward multilinear-to-linear reduction using the Bourgain–Guth method. We will derive a multiscale inequality that will be iterated many times. A key difference here is that this inequality involves more than two scales, a subtle consequence of the delicate geometry of curves. The proof of the multiscale inequality will rely on the whole array of multilinear Kakeya-type inequalities from Chapter 6.

We devote the bulk of this chapter to studying the twisted cubic, which is representative for most of the new difficulties compared to the parabola. A brief discussion of the general case appears in Section 11.10.

## 11.1 Decoupling for the Twisted Cubic

Let us consider the moment curve $\Gamma = \Gamma_3$ in $\mathbb{R}^3$ (also known as the "twisted cubic"):

$$\gamma(t) = (t, t^2, t^3), \ t \in [0, 1].$$

For each $\delta < 1$ and each interval $J \subset [0, 1]$, we introduce the anisotropic neighborhood of the arc $\Gamma_J$:

$$\Gamma_J(\delta) = \{(\xi_1, \xi_2, \xi_3) \colon \xi_1 \in J, \ |\xi_2 - \xi_1^2| \leq \delta^2, \ |\xi_3 - 3\xi_1\xi_2 + 2\xi_1^3| \leq \delta^3\}.$$

When $J = [0, 1]$, we will simply write $\Gamma(\delta)$. See Figure 11.1.

To understand better what $\Gamma(\delta)$ is, let us consider the Frenet frame at each point $\gamma(t)$, consisting of the tangent vector $\mathbf{T}(t)$, the normal vector $\mathbf{N}(t)$, and the binormal $\mathbf{B}(t)$. It is not hard to see that for each interval $J = [t, t + \delta]$, $\Gamma_J(\delta)$ is an almost rectangular box with dimensions $\sim \delta, \delta^2, \delta^3$ with respect to the axes $\mathbf{T}(t)$, $\mathbf{N}(t)$ and $\mathbf{B}(t)$.

The surface

$$\xi_3 = 3\xi_1\xi_2 - 2\xi_1^3 \tag{11.1}$$

contains $\Gamma$, and stays locally close to the osculating plane of $\gamma$. More precisely, for each $\xi_1 \in [0, 1]$, its intersection with the plane $\xi_1' = \xi_1$ is a line segment inside the osculating plane at $\gamma(\xi_1)$. In other words, each point $(\xi_1, \xi_2, \xi_3)$ on the surface satisfies the equation

$$\det[(\xi_1, \xi_2, \xi_3) - \gamma(\xi_1), \gamma'(\xi_1), \gamma''(\xi_1)] = 0.$$

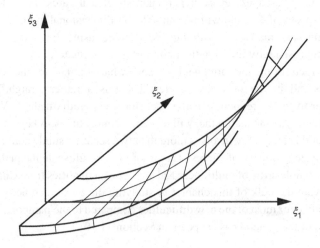

Figure 11.1  Partition of $\Gamma(\delta)$ into sets $\Gamma_J(\delta)$.

Thus, the intersection of $\Gamma(\delta)$ with each plane $\xi_1' = constant$ is a parallelogram with dimensions $2\delta^2, 2\delta^3$.

Throughout this chapter, various scales such as $\delta, \sigma, K$ will in general be implicitly assumed to be dyadic. The sums $\sum_{\substack{J \subset [0,1] \\ |J|=\delta}}$ will always be understood with $J$ ranging over the intervals $[j\delta, (j+1)\delta], 0 \le j \le \delta^{-1} - 1$. In an attempt to simplify notation, we will denote by $\mathcal{P}_J$ the Fourier projection operator onto the infinite strip $J \times \mathbb{R}^2$.

Here is the main result we will prove in this chapter.

**Theorem 11.1** ($l^2$ *decoupling for the twisted cubic*) *Assume* $g \colon \mathbb{R}^3 \to \mathbb{C}$ *has its Fourier transform supported on* $\Gamma(\delta)$. *Then*

$$
\|g\|_{L^{12}(\mathbb{R}^3)} \lesssim_\epsilon \delta^{-\epsilon} \left( \sum_{\substack{J \subset [0,1] \\ |J|=\delta}} \|\mathcal{P}_J g\|_{L^{12}(\mathbb{R}^3)}^2 \right)^{1/2}. \tag{11.2}
$$

The intersection of the strip $J \times \mathbb{R}^2$ with $\Gamma(\delta)$ is $\Gamma_J(\delta)$, so if $\text{supp}(\widehat{g}) \subset \Gamma(\delta)$, then $\mathcal{P}_J g$ coincides with $\mathcal{P}_{\Gamma_J(\delta)} g$.

The exponent $p = 12$ cannot be replaced with a larger one due to scaling considerations; see Exercise 11.18. Combining cylindrical decoupling with the decoupling for the parabola from the previous chapter proves (11.2), when 12 is replaced with any $2 \le p \le 6$. See also Exercise 11.19 for a different proof of this fact. To reach the critical exponent 12, we will take advantage of the nonplanarity of $\Gamma$.

Note that $\Gamma_J(\delta)$ contains, and in fact is substantially larger than, the more familiar isotropic neighborhood

$$
\mathcal{N}_J(\delta^3) = \{(t, t^2 + s_2, t^3 + s_3) \colon t \in J, \ (s_2^2 + s_3^2)^{1/2} \le \delta^3\}.
$$

Theorem 11.1 (and its higher-dimensional extension) was proved in [33] under the superficially stronger assumption that $\widehat{g}$ is supported on $\mathcal{N}_{[0,1]}(\delta^3)$. We will see that the modifications needed to prove Theorem 11.1 in this greater generality are rather small. In fact, as we will argue in the next section, the neighborhoods $\Gamma_J(\delta)$ are more natural to work with, as they scale rather canonically.

When $|J| = \delta$, it is easy to see that $\mathcal{N}_J(\delta^3)$ is a curved tube, while $\Gamma_J(\delta)$ is an almost rectangular box. Moreover, it is easy to check that $\widetilde{\mathcal{N}}_\Gamma(\delta) := \Gamma(\delta)$ and its partition into boxes $\Gamma_J(\delta)$ fall under the scope of Problem 9.4.

For $p \ge 2$, we will denote by $V_p(\delta)$ the smallest constant such that the inequality

$$\|g\|_{L^p(\mathbb{R}^3)} \leq V_p(\delta) \left( \sum_{\substack{J \subset [0,1] \\ |J| = \delta}} \|\mathcal{P}_J g\|^2_{L^p(\mathbb{R}^3)} \right)^{1/2}$$

holds whenever $\widehat{g}$ is supported on $\Gamma(\delta)$.

Let $w_B$ be the weight introduced in (9.6). A slight variant of Proposition 9.15 shows that $V_p(\delta)$ is comparable in size to its local counterpart $V_{p,loc}(\delta)$, defined as the smallest constant for which the inequality

$$\|g\|_{L^p(B)} \leq V_{p,loc}(\delta) \left( \sum_{\substack{J \subset [0,1] \\ |J| = \delta}} \|\mathcal{P}_J g\|^2_{L^p(w_B)} \right)^{1/2}$$

holds for arbitrary balls $B$ with radius $\delta^{-3}$, whenever $\text{supp}(\widehat{g}) \subset \Gamma(\delta)$. Moreover, $\|g\|_{L^p(B)}$ may be replaced with $\|g\|_{L^p(w_B)}$.

## 11.2 Rescaling the Neighborhoods $\Gamma(\delta)$

Let $\sigma < 1$ and $a \in [0,1]$. Consider the affine transformation

$$T_{\sigma,a}(\xi_1, \xi_2, \xi_3) = (\xi_1', \xi_2', \xi_3'),$$

where

$$\begin{cases} \xi_1' = \frac{\xi_1 - a}{\sigma} \\ \xi_2' = \frac{\xi_2 - 2a\xi_1 + a^2}{\sigma^2} \\ \xi_3' = \frac{\xi_3 - 3a\xi_2 + 3a^2\xi_1 - a^3}{\sigma^3}. \end{cases}$$

It is easy to see that $T_{\sigma,a}$ maps an arc on $\Gamma$ to another arc on $\Gamma$. However, more is true. What makes the neighborhoods $\Gamma_J(\delta)$ suitable for our analysis is the fact that they are also mapped to each other by the transformations $T_{\sigma,a}$. More precisely,

$$T_{\sigma_2,a}(\Gamma_{[b,b+\sigma_1]}(\delta)) = \Gamma_{\left[\frac{b-a}{\sigma_2}, \frac{b-a}{\sigma_2} + \frac{\sigma_1}{\sigma_2}\right]}\left(\frac{\delta}{\sigma_2}\right).$$

This is due to the fact that $T_{\sigma,a}$ also leaves the surface (11.1) invariant. Indeed, it is easy to check that

$$\xi_3 - 3\xi_1\xi_2 + 2\xi_1^3 = \xi_3' - 3\xi_1'\xi_2' + 2\xi_1'^3.$$

A consequence of rescaling that will be frequently used is the following decoupling for sets $\Gamma_I(\delta)$. Note that the inequality with $V_p(\delta)$ in place of $V_p(\delta/\sigma)$ is trivial. However, the latter is an essentially stronger bound.

**Proposition 11.2** *Assume $\widehat{g}$ is supported on $\Gamma_I(\delta)$, where $I = [a, a + \sigma]$ for some $\sigma > \delta$. Then*

$$\|g\|_{L^p(\mathbb{R}^3)} \le V_p\left(\frac{\delta}{\sigma}\right) \left( \sum_{\substack{J \subset I \\ |J| = \delta}} \|\mathcal{P}_J g\|^2_{L^p(\mathbb{R}^3)} \right)^{1/2}.$$

*Proof* Consider the linear map $A_{\sigma,a}$ on $\mathbb{R}^3$ given by $x' = A_{\sigma,a}(x)$ with

$$\begin{cases} x'_1 = \sigma(x_1 + 2ax_2 + 3a^2x_3) \\ x'_2 = \sigma^2(x_2 + 3ax_3) \\ x'_3 = \sigma^3 x_3. \end{cases}$$

Define a new function $h$ as follows:

$$g(x) = \sigma^6 h(A_{\sigma,a}x).$$

An easy computation shows that

$$|\widehat{g}| = |\widehat{h} \circ T_{\sigma,a}|,$$

thus $\widehat{h}$ is supported on $\Gamma(\delta/\sigma)$. Also

$$\mathcal{P}_J g(x) = \sigma^6 \mathcal{P}_H h(A_{\sigma,a}x),$$

where $H = \sigma^{-1}(J - a)$ is an interval of length $\delta/\sigma$. The intervals $H$ obtained this way partition $[0, 1]$, and consequently

$$\|h\|_{L^p(\mathbb{R}^3)} \le V_p\left(\frac{\delta}{\sigma}\right) \left( \sum_{\substack{H \subset [0,1] \\ |H| = \frac{\delta}{\sigma}}} \|\mathcal{P}_H h\|^2_{L^p(\mathbb{R}^3)} \right)^{1/2}.$$

The desired inequality now follows via a change of variables. $\qquad\square$

## 11.3 The Trilinear-to-Linear Reduction

Given $K \in 2^{\mathbb{N}}$, let $\mathcal{I}_K$ be the family of all triples of intervals $I_1, I_2, I_3 \subset [0, 1]$ of the form $[i/K, (i + 1)/K]$, that, in addition, are required to not be adjacent to each other. Note that $|\mathcal{I}_K| \lesssim_K 1$.

We define the trilinear $l^2$ decoupling constant $V_p(\delta, K)$ as the smallest number such that the inequality

$$\left\| \left( \prod_{i=1}^{3} \mathcal{P}_{I_i} g \right)^{1/3} \right\|_{L^p(\mathbb{R}^3)} \le V_p(\delta, K) \prod_{i=1}^{3} \left( \sum_{\substack{J \subset I_i \\ |J| = \delta}} \|\mathcal{P}_J g\|^2_{L^p(\mathbb{R}^3)} \right)^{\frac{1}{6}}$$

holds true for each $g : \mathbb{R}^3 \to \mathbb{C}$ with the Fourier transform supported on $\Gamma(\delta)$ and each 3-tuple $(I_1, I_2, I_3) \in \mathcal{I}_K$.

It is immediate that $V_p(\delta, K) \leq V_p(\delta)$. The reverse inequality is also essentially true, apart from negligible losses.

**Theorem 11.3** *Let* $p \geq 2$. *There exist constants* $\epsilon_p(K)$ *with* $\lim_{K \to \infty} \epsilon_p(K) = 0$ *and* $C_{K,p}$ *so that*

$$V_p(\delta) \leq C_{K,p} \delta^{-\epsilon_p(K)} \sup_{\delta \leq \delta' < 1} V_p(\delta', K), \qquad (11.3)$$

*for each* $0 < \delta \leq 1$.

*Proof* The values of $C$ in the following argument will only depend on $p$, and will be allowed to change from one line to the next one. Assume $\widehat{g}$ is supported on $\Gamma(\delta)$.

We start by observing the following elementary inequality:

$$|g(x)| \leq C \max_{|J_1| = K^{-1}} |\mathcal{P}_{J_1} g(x)| + K^C \max_{(I_1, I_2, I_3) \in \mathcal{I}_K} \left| \prod_{i=1}^{3} \mathcal{P}_{I_i} g(x) \right|^{1/3}.$$

Integration leads to

$$\|g\|_{L^p(\mathbb{R}^3)} \leq C \left( \sum_{|J_1| = K^{-1}} \|\mathcal{P}_{J_1} g\|_{L^p(\mathbb{R}^3)}^2 \right)^{1/2}$$

$$+ K^C \max_{(I_1, I_2, I_3) \in \mathcal{I}_K} \left\| \prod_{i=1}^{3} |\mathcal{P}_{I_i} g|^{1/3} \right\|_{L^p(\mathbb{R}^3)}$$

$$\leq C \left( \sum_{|J_1| = K^{-1}} \|\mathcal{P}_{J_1} g\|_{L^p(\mathbb{R}^3)}^2 \right)^{1/2}$$

$$+ K^C V_p(\delta, K) \left( \sum_{\substack{J \subset [0,1] \\ |J| = \delta}} \|\mathcal{P}_J g\|_{L^p(\mathbb{R}^3)}^2 \right)^{\frac{1}{2}}.$$

Rescaling this inequality as in Proposition 11.2, we may write for each $J_1$

$$\|\mathcal{P}_{J_1} g\|_{L^p(\mathbb{R}^3)} \leq C \left( \sum_{\substack{J_2 \subset J_1 \\ |J_2| = K^{-2}}} \|\mathcal{P}_{J_2} g\|_{L^p(\mathbb{R}^3)}^2 \right)^{1/2}$$

$$+ K^C V_p(\delta K, K) \left( \sum_{\substack{J \subset J_1 \\ |J| = \delta}} \|\mathcal{P}_J g\|_{L^p(\mathbb{R}^3)}^2 \right)^{\frac{1}{2}}.$$

Combining the last two inequalities leads to

$$\|g\|_{L^p(\mathbb{R}^3)} \leq C^2 \left( \sum_{|J_2|=K^{-2}} \|\mathcal{P}_{J_2} g\|_{L^p(\mathbb{R}^3)}^2 \right)^{1/2}$$

$$+ [CK^C V_p(\delta K, K) + K^C V_p(\delta, K)] \left( \sum_{\substack{J \subset [0,1] \\ |J|=\delta}} \|\mathcal{P}_J g\|_{L^p(\mathbb{R}^3)}^2 \right)^{\frac{1}{2}}.$$

If we repeat this process $l$ times, where $K^{-l} \sim \delta$, we get the following inequality:

$$\|g\|_{L^p(\mathbb{R}^3)} \leq C^l \left( \sum_{|J_l|=K^{-l}} \|\mathcal{P}_{J_l} g\|_{L^p(\mathbb{R}^3)}^2 \right)^{1/2}$$

$$+ l C^{l-1} K^C \sup_{\delta \leq \delta' < 1} V_p(\delta', K) \left( \sum_{\substack{J \subset [0,1] \\ |J|=\delta}} \|\mathcal{P}_J g\|_{L^p(\mathbb{R}^3)}^2 \right)^{\frac{1}{2}}$$

$$\lesssim l C^l K^C \sup_{\delta \leq \delta' < 1} V_p(\delta', K) \left( \sum_{\substack{J \subset [0,1] \\ |J|=\delta}} \|\mathcal{P}_J g\|_{L^p(\mathbb{R}^3)}^2 \right)^{\frac{1}{2}}.$$

The reader can now easily verify the conclusion of the theorem, since $C$ is independent of $K$. □

## 11.4 A Multiscale Inequality

Let us introduce some notation. Fix $K$, $(I_1, I_2, I_3) \in \mathcal{I}_K$ and a small enough $\delta > 0$. We will denote by $B^s$ a generic ball with radius $\delta^{-s}$ in $\mathbb{R}^3$. Recall the weight $w_B$ introduced in (9.6). Given a function $g : \mathbb{R}^3 \to \mathbb{C}$, numbers $t \geq 2$, $q \leq s$, and a ball $B^s$, we write

$$D_t(q, B^s, g) = \left[ \prod_{i=1}^{3} \left( \sum_{\substack{J_{i,q} \subset I_i \\ |J_{i,q}|=\delta^q}} \|\mathcal{P}_{J_{i,q}} g\|_{L_{\sharp}^t(w_{B^s})}^2 \right)^{1/2} \right]^{\frac{1}{3}}.$$

Also, given a finitely overlapping cover $\mathcal{B}_q(B^s)$ of $B^s$ with balls $B^q$, we let

$$A_p(q, B^s, g) = \left( \frac{1}{|\mathcal{B}_q(B^s)|} \sum_{B^q \in \mathcal{B}_q(B^s)} D_2(q, B^q, g)^p \right)^{\frac{1}{p}}.$$

The reader will recognize the similarity with quantities introduced earlier for the paraboloid. Note that $D_t, A_p, B^q$ also depend on the parameters $\delta$ and $K$. In the attempt to keep notation more compact, we choose to omit this dependence. The implicit constants in the inequalities that follow will be uniform over $\delta$.

In our applications, the term $A_p(q, B^s, g)$ will always be associated with a $g$ whose Fourier transform is supported on $\Gamma(\delta^q)$. Thus, each $|\mathcal{P}_{J_{i,q}} g|$ can be thought of as being essentially constant on $B^q$, and we may write

$$
A_p(q, B^s, g) \approx \left\| \prod_{i=1}^{3} \left( \sum_{\substack{J_{i,q} \subset I_i \\ |J_{i,q}| = \delta^q}} |\mathcal{P}_{J_{i,q}} g|^2 \right)^{\frac{1}{6}} \right\|_{L_\sharp^p(w_{B^s})}.
$$

As was the case with the analogous expressions for the paraboloid, the $A_p$ terms are auxiliary terms that encode trilinear Kakeya-type information. Their use will allow us to take advantage of the two trilinear Kakeya inequalities in Section 11.6. At the end of the argument, the $A_p$ terms will be replaced with $D_p$ terms.

The following multiscale inequality will play the same role in the argument for the twisted cubic as the role played by the two-scale inequality (10.18) in the case of the paraboloid. In particular, note that (11.4) is the analog of the requirement $l(1 - \kappa) \geq 1$, cf. Exercise 10.24.

**Theorem 11.4** *For each $p$ smaller than and sufficiently close to the critical exponent 12, there is $r = r(p)$ and there are constants $u_0, \gamma_0, \dots, \gamma_r \in (0, 1)$ and $b_0, \dots, b_r > 1$ satisfying*

$$
\sum_{j=0}^{r} \gamma_j < 1,
$$

$$
\sum_{j=0}^{r} \gamma_j b_j > 1, \tag{11.4}
$$

*and such that the following holds: for each $\delta < 1, 0 < u \leq u_0$, each ball $B^3$ of radius $\delta^{-3}$, and each $g$ with supp $(\widehat{g}) \subset \Gamma(\delta)$, we have*

$$
A_p(u, B^3, g) \lesssim_{\epsilon, K} \delta^{-\epsilon} V_p(\delta)^{1 - \sum_0^r \gamma_j} D_p(1, B^3, g)^{1 - \sum_0^r \gamma_j} \prod_{j=0}^{r} A_p(b_j u, B^3, g)^{\gamma_j}.
$$

Note that intervals of length $\delta^u$ that enter the $A_p$ term on the left are decoupled into intervals of smaller scales $\delta^{b_j u}$. The number $r + 1$ of such scales that is needed in order to guarantee the critical inequality (11.4) depends on $p$. The value of $r$ will be determined at the end of Section 11.9.

The proof of this theorem will be presented in Section 11.9, building on the material from Sections 11.6 through 11.8. Before getting there, we will show how this multiscale inequality leads to the proof of Theorem 11.1.

## 11.5 Iteration and the Proof of Theorem 11.1

In this section, we will prove the inequality $V_{12}(\delta) \lesssim_\epsilon \delta^{-\epsilon}$. Invoking interpolation (Exercise 9.21), it will suffice to prove that $V_p(\delta) \lesssim_\epsilon \delta^{-\epsilon}$, for all $p$ smaller than and sufficiently close to 12. This restriction places us in the range of Theorem 11.4.

The first step is to iterate Theorem 11.4.

**Theorem 11.5** *Let $p, r, u_0, \gamma_j, b_j, g$ be as in Theorem 11.4. Given an integer $M \geq 1$, let $u > 0$ be such that*

$$u \leq u_0 (\max b_j)^{1-M}.$$

*Then for each ball $B^3$ of radius $\delta^{-3}$, we have*

$$A_p(u, B^3, g) \lesssim_{\epsilon, K, M} \delta^{-\epsilon} V_p(\delta)^{1 - (\sum_0^r \gamma_j)^M} D_p(1, B^3, g)^{1 - (\sum_0^r \gamma_j)^M} \tag{11.5}$$

$$\times \prod_{j_1 = 0}^r \cdots \prod_{j_M = 0}^r A_p \left( u \prod_{l=1}^M b_{j_l}, B^3, g \right)^{\prod_{l=1}^M \gamma_{j_l}}.$$

*Proof* The proof follows by induction on $M$. The case $M = 1$ is provided by Theorem 11.4. Assume (11.5) holds for some $M$, and let us prove it for $M + 1$. Let $u$ satisfy $u \leq u_0 (\max b_j)^{-M}$. Since each term $u \prod_{l=1}^M b_{j_l}$ is smaller than $u_0$, we may apply Theorem 11.4 to write

$$A_p \left( \prod_{l=1}^M b_{j_l} u, B^3, g \right) \lesssim_{\epsilon, K} \delta^{-\epsilon} V_p(\delta)^{1 - \sum_0^r \gamma_j} D_p(1, B^3, g)^{1 - \sum_0^r \gamma_j}$$

$$\times \prod_{j_{M+1} = 0}^r A_p \left( u b_{j_{M+1}} \prod_{l=1}^M b_{j_l}, B^3, g \right)^{\gamma_{j_{M+1}}}.$$

Combining this with (11.5), we arrive at the desired inequality

$$A_p(u, B^3, g) \lesssim_{\epsilon, K, M} \delta^{-\epsilon} [V_p(\delta) D_p(1, B^3, g)]^{1 - (\sum_0^r \gamma_j)^M + (1 - \sum_0^r \gamma_j)(\sum_0^r \gamma_j)^M}$$

$$\times \prod_{j_1 = 0}^r \cdots \prod_{j_{M+1} = 0}^r A_p \left( u \prod_{l=1}^{M+1} b_{j_l}, B^3, g \right)^{\prod_{l=1}^{M+1} \gamma_{j_l}}. \qquad \square$$

Let $\eta_p \geq 0$ be the unique number such that

$$\lim_{\delta \to 0} V_p(\delta) \delta^{\eta_p + \epsilon} = 0, \text{ for each } \epsilon > 0 \tag{11.6}$$

and

$$\limsup_{\delta \to 0} V_p(\delta)\delta^{\eta_p - \sigma} = \infty, \text{ for each } \sigma > 0. \tag{11.7}$$

Our goal is to argue that $\eta_p = 0$ for each $p$ slightly less than 12. The remaining part of the argument is similar to the one we used for the parabola (Sections 10.2.3 and 10.4.4).

We start by rewriting Theorem 11.5 as follows.

**Theorem 11.6** *Assume $p$ is smaller than and sufficiently close to the critical exponent 12. Then for each $W > 0$, each sufficiently small $u > 0$ and each $g$ with supp $(\widehat{g}) \subset \Gamma(\delta)$, we have*

$$A_p(u, B^3, g) \lesssim_{\epsilon, K, W} \delta^{-(1-uW)(\eta_p + \epsilon)} D_p(1, B^3, g).$$

*Proof* Theorem 11.5 with $M$ large enough provides weights $b_I, \gamma_I$ satisfying

$$\sum_I \gamma_I < 1,$$

$$\sum_I b_I \gamma_I > W,$$

so that for each sufficiently small $u > 0$ (how small will only depend on $W$, not on $\delta$), the following inequality holds:

$$A_p(u, B^3, g) \lesssim_{\epsilon, K, W} \delta^{-\epsilon} V_p(\delta)^{1 - \sum_I \gamma_I} D_p(1, B^3, g)^{1 - \sum_I \gamma_I} \tag{11.8}$$

$$\times \prod_I A_p(ub_I, B^3, g)^{\gamma_I}.$$

Assume, in addition, that $u$ is so small that $uW < 1$ and $ub_I < 1$ for each $I$. First, an argument as in Lemma 10.15 shows that

$$A_p(b_I u, B^3, g) \lesssim D_p(b_I u, B^3, g). \tag{11.9}$$

Next, using Proposition 11.2, we get

$$D_p(b_I u, B^3, g) \lesssim V_p(\delta^{1-ub_I}) D_p(1, B^3, g). \tag{11.10}$$

It suffices now to combine (11.6), (11.8), (11.9), and (11.10). $\qquad\square$

**Corollary 11.7** *Let $g, W, u, p$ be as in Theorem 11.6. Then*

$$\left\| \left( \prod_{i=1}^3 \mathcal{P}_{I_i} g \right)^{1/3} \right\|_{L^p_\sharp(B^3)} \lesssim_{\epsilon, K, W} \delta^{-(\frac{u}{2} + (\eta_p + \epsilon)(1-uW))} D_p(1, B^3, g). \tag{11.11}$$

*Proof* Using the Cauchy–Schwarz inequality and the fact that

$$1_{B^3} \lesssim \sum_{B^u \in \mathcal{B}_u(B^3)} 1_{B^u},$$

we may write

$$\left\|\left(\prod_{i=1}^{3}\mathcal{P}_{I_i}g\right)^{1/3}\right\|_{L^p_{\sharp}(B^3)} \leq \delta^{-\frac{u}{2}}\left\|\left(\prod_{i=1}^{3}\sum_{\substack{J_{i,u}\subset I_i \\ |J_{i,u}|=\delta^u}}|\mathcal{P}_{J_{i,u}}g|^2\right)^{\frac{1}{6}}\right\|_{L^p_{\sharp}(B^3)}$$

$$\lesssim \delta^{-\frac{u}{2}}\left(\frac{1}{|\mathcal{B}_u(B^3)|}\sum_{B^u\in\mathcal{B}_u(B^3)}\left\|\left(\prod_{i=1}^{3}\sum_{\substack{J_{i,u}\subset I_i \\ |J_{i,u}|=\delta^u}}|\mathcal{P}_{J_{i,u}}g|^2\right)^{\frac{1}{6}}\right\|_{L^p_{\sharp}(B^u)}^{p}\right)^{\frac{1}{p}}.$$

$$(11.12)$$

Using Minkowski's inequality, the term in (11.12) is dominated by

$$\delta^{-\frac{u}{2}}\left(\frac{1}{|\mathcal{B}_u(B^3)|}\sum_{B^u\in\mathcal{B}_u(B^3)}D_{p,\mathrm{loc}}(u,B^u,g)^p\right)^{1/p},$$

where

$$D_{p,\mathrm{loc}}(u,B^u,g) = \left[\prod_{i=1}^{3}\left(\sum_{\substack{J_{i,u}\subset I_i \\ |J_{i,u}|=\delta^u}}\|\mathcal{P}_{J_{i,u}}g\|_{L^t_{\sharp}(B^u)}^2\right)^{1/2}\right]^{\frac{1}{3}}.$$

On the other hand, Lemma 9.19 shows that $D_{p,\mathrm{loc}}(u,B^u,g) \lesssim D_2(u,B^u,g)$, and consequently

$$\left(\frac{1}{|\mathcal{B}_u(B^3)|}\sum_{B^u\in\mathcal{B}_u(B^3)}D_{p,\mathrm{loc}}(u,B^u,g)^p\right)^{1/p} \lesssim A_p(u,B^3,g).$$

We conclude that

$$\left\|\left(\prod_{i=1}^{3}\mathcal{P}_{I_i}g\right)^{1/3}\right\|_{L^p_{\sharp}(B^3)} \lesssim \delta^{-\frac{u}{2}}A_p(u,B^3,g),$$

which when combined with Theorem 11.6 finishes the proof of the corollary. $\qquad\square$

Fix $p$ smaller than and sufficiently close to 12. We will show that $\eta_p = 0$ using a bootstrapping argument that is very similar to (albeit written a bit differently) the one we have used for the parabola.

Note that (11.11) holds uniformly over all $\delta$, $B^3$, $g$ with supp $(\widehat{g}) \subset \Gamma(\delta)$ and $(I_1, I_2, I_3) \in \mathcal{I}_K$. As a consequence, we have

$$V_p(\delta, K) \lesssim_{\epsilon, K, W} \delta^{-(\frac{u}{2} + (\eta_p + \epsilon)(1 - uW))}.$$

Combining this with (11.3) and (11.7), we may write

$$\delta_l^{-\eta_p + \epsilon_p(K) + \sigma} \lesssim_{\sigma, \epsilon, K, W} \delta_l^{-(\frac{u}{2} + (\eta_p + \epsilon)(1 - uW))},$$

for $\sigma > 0$ and some sequence $\delta_l$ converging to zero. This further leads to

$$\eta_p - \epsilon_p(K) - \sigma \le \frac{u}{2} + (\eta_p + \epsilon)(1 - uW),$$

which in turn can be rearranged as follows

$$\eta_p \le \frac{1}{2W} + \frac{\sigma + \epsilon_p(K) + \epsilon(1 - uW)}{uW}.$$

This inequality holds true for each $\epsilon, \sigma > 0$ and each large enough $K$. Recalling that

$$\lim_{K \to \infty} \epsilon_p(K) = 0,$$

we can further write

$$\eta_p \le \frac{1}{2W}.$$

Letting now $W \to \infty$ leads to $\eta_p = 0$, as desired.

## 11.6 Two Trilinear Kakeya Inequalities

As a first step toward the proof of the multiscale inequality in Theorem 11.4, we present two applications of Theorem 6.16 in the case $m = n = 3$. The first one is for $d = 1$ and is rather simple, the second one is for $d = 2$ and is more complicated.

Let us fix $K$ and $(I_1, I_2, I_3) \in \mathcal{I}_K$. Recall that $\gamma(t) = (t, t^2, t^3)$. We will consider the following family of one-dimensional linear subspaces of $\mathbb{R}^3$:

$$V_1(t) = \mathrm{span}(\gamma'(t)), \quad t \in [0, 1].$$

Note that

$$|\gamma'(t_1) \wedge \gamma'(t_2) \wedge \gamma'(t_3)| \gtrsim_K 1$$

uniformly over $t_j \in I_j$. Let $\mathcal{V}_1$ consist of all triples $(V_1(t_1), V_1(t_2), V_1(t_3))$ with $t_j \in I_j$. Exercise 6.10 implies that

$$\mathrm{BL}(\mathcal{V}_1) \lesssim_K 1.$$

The following result is then an immediate consequence of Theorem 6.16. We remind the reader that this inequality has an easier proof, explored in Exercise 6.21.

**Theorem 11.8** (Trilinear Kakeya for thin boxes/plates) *For $W > 0$, $S \ge 1$ and $1 \le j \le 3$, let $\mathcal{T}_{j, S, W}$ be the collection of all rectangular boxes $T$ in $\mathbb{R}^3$ with* *two sides parallel to some $V_1(t_j)^\perp$ ($t_j \in I_j$) and having length $SW$, and with the remaining side parallel to $V_1(t_j)$ and of length $W$.*

*Then for each finite collection $\mathcal{T}_j \subset \mathcal{T}_{j,S,W}$ and each $c_T \geq 0$, we have*

$$\left\| \prod_{j=1}^{3} \left( \sum_{T \in \mathcal{T}_j} c_T 1_T \right)^{\frac{1}{3}} \right\|_{L^3(\mathbb{R}^3)} \lesssim_K W \prod_{j=1}^{3} \left( \sum_{T \in \mathcal{T}_j} c_T \right)^{\frac{1}{3}}.$$

Let us also define the osculating planes

$$V_2(t) = \mathrm{span}(\gamma'(t), \gamma''(t)), \quad t \in [0,1].$$

A normal vector to $V_2(t)$ is

$$B(t) = (3t^2, -3t, 1),$$

and an easy computation shows that

$$|B(t_1) \wedge B(t_2) \wedge B(t_3)| \gtrsim_K 1,$$

uniformly over $t_j \in I_j$.

Let $\mathcal{V}_2$ consist of all triples $(V_2(t_1), V_2(t_2), V_2(t_3))$ with $t_j \in I_j$. Exercise 6.9 implies that

$$\mathrm{BL}(\mathcal{V}_2) \lesssim_K 1.$$

This yields another consequence of Theorem 6.16.

**Theorem 11.9** (Trilinear Kakeya for tall boxes/tubes) *For $W > 0$, $S \geq 1$ and $1 \leq j \leq 3$, let $\mathcal{T}_{j,S,W}$ be the collection of all rectangular boxes $T$ in $\mathbb{R}^3$ with the long side parallel to some $V_2(t_j)^\perp$ ($t_j \in I_j$) and having length $SW$, and with the shorter sides of length $W$.*

*Then for all finite collections $\mathcal{T}_j \subset \mathcal{T}_{j,S,W}$ and each $c_T \geq 0$, we have*

$$\left\| \prod_{j=1}^{3} \left( \sum_{T \in \mathcal{T}_j} c_T 1_T \right)^{\frac{1}{3}} \right\|_{L^{\frac{3}{2}}(\mathbb{R}^3)} \lesssim_{\epsilon,K} S^\epsilon W^2 \prod_{j=1}^{3} \left( \sum_{T \in \mathcal{T}_j} c_T \right)^{\frac{1}{3}}.$$

## 11.7 Ball Inflations

We will recast the trilinear Kakeya inequalities from the previous section in the language of decoupling norms. The process we are about to describe will be called *ball inflation*, as it facilitates the transition from smaller to larger spatial balls. While Theorem 11.10 is strictly speaking not a decoupling, it leads to one when combined with lower-dimensional decoupling and $L^2$ orthogonality. This is analogous to Theorem 10.22 for the parabola.

**Theorem 11.10** *Let* $(I_1, I_2, I_3) \in \mathcal{I}_K$, $\rho \in (0, 1)$, $1 \leq k \leq 2$ *and* $p \geq 6$. *Let $B$ be an arbitrary ball in $\mathbb{R}^3$ with radius $\rho^{-(k+1)}$, and let $\mathcal{B}$ be a finitely overlapping cover of $B$ with balls $\Delta$ of radius $\rho^{-k}$. Then for each $g \colon \mathbb{R}^3 \to \mathbb{C}$ with its Fourier transform supported on $\Gamma(\rho)$, we have*

$$\frac{1}{|\mathcal{B}|} \sum_{\Delta \in \mathcal{B}} \prod_{i=1}^{3} \left( \sum_{\substack{J_i \subset I_i \\ |J_i| = \rho}} \| \mathcal{P}_{J_i} g \|_{L_\#^{\frac{pk}{3}}(w_\Delta)}^2 \right)^{\frac{p}{6}} \lesssim_{\epsilon, K} \rho^{-\epsilon} \prod_{i=1}^{3} \left( \sum_{\substack{J_i \subset I_i \\ |J_i| = \rho}} \| \mathcal{P}_{J_i} g \|_{L_\#^{\frac{pk}{3}}(w_B)}^2 \right)^{\frac{p}{6}}.$$

(11.13)

*Moreover, the implicit constant is independent of $g, \rho, B$.*

Let us first briefly sketch the proof of this theorem, in anticipation of the more detailed one that follows. At the heart of the argument lies the fact that $\Gamma_J(\rho)$ is an almost rectangular box with dimensions $\rho, \rho^2, \rho^3$, whenever $|J| = \rho$. Thus, if supp $(\widehat{g}) \subset \Gamma(\rho)$, then $|\mathcal{P}_J g|$ is essentially constant on planks $P$ with dimensions $\rho^{-1}, \rho^{-2}, \rho^{-3}$, dual to $\Gamma_J(\rho)$. See Figure 11.2.

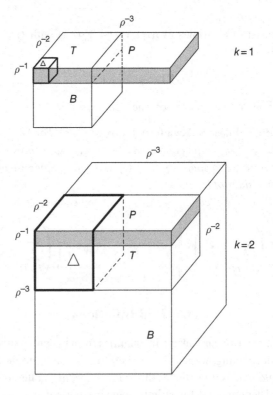

Figure 11.2 Planks, plates, and tubes.

By doing standard manipulations, we may replace $l^2$ norms with $l^{pk/3}$ norms in (11.13). Roughly speaking, only the part of the plank $P$ that intersects the ball $B$ has relevance to (11.13).

When $k = 1$, $T = P \cap B$ is essentially a plate with dimensions $\rho^{-1}, \rho^{-2}, \rho^{-2}$. We write for each $J_i \subset I_i$

$$|\mathcal{P}_{J_i} g|^{\frac{p}{3}} \approx \sum_{T \in \mathcal{T}_{J_i}} c_T 1_T, \quad \text{on } B,$$

with $\mathcal{T}_{J_i}$ consisting of pairwise disjoint plates as before. Thus, for each ball $\Delta \in \mathcal{B}$ of radius $\rho^{-1}$, we have

$$\prod_{i=1}^{3} \left( \sum_{\substack{J_i \subset I_i \\ |J_i| = \rho}} \|\mathcal{P}_{J_i} g\|^{\frac{p}{3}}_{L^{\frac{p}{3}}_{\sharp}(w_\Delta)} \right) \approx \prod_{i=1}^{3} \left( \frac{1}{|\Delta|} \int_\Delta \left( \sum_{\substack{J_i \subset I_i \\ |J_i| = \rho}} \sum_{T \in \mathcal{T}_{J_i}} c_T 1_{T_{J_i}} \right) \right)$$

$$\approx \frac{1}{|\Delta|} \int_\Delta \prod_{i=1}^{3} \left( \sum_{\substack{J_i \subset I_i \\ |J_i| = \rho}} \sum_{T \in \mathcal{T}_{J_i}} c_T 1_T \right).$$

To commute the product with the integral, we have used the fact that each of the functions $\sum_{J_i \subset I_i} \sum_{T \in \mathcal{T}_{J_i}} c_T 1_T$ is essentially constant on $\Delta$. Summation over $\Delta \in \mathcal{B}$ leads to

$$\frac{1}{|\mathcal{B}|} \sum_{\Delta \in \mathcal{B}} \prod_{i=1}^{3} \left( \sum_{\substack{J_i \subset I_i \\ |J_i| = \rho}} \|\mathcal{P}_{J_i} g\|^{\frac{p}{3}}_{L^{\frac{p}{3}}_{\sharp}(w_\Delta)} \right) \approx \frac{1}{|\mathcal{B}|} \int_B \prod_{i=1}^{3} \left( \sum_{\substack{J_i \subset I_i \\ |J_i| = \rho}} \sum_{T \in \mathcal{T}_{J_i}} c_T 1_T \right).$$

This quantity will be evaluated using Theorem 11.8 with $S = W = \rho^{-1}$.

When $k = 2$, we focus attention on the planks $P$ that are inside the ball $B$ of radius $\rho^{-3}$. Write for $1 \le i \le 3$

$$\sum_{\substack{J_i \subset I_i \\ |J_i| = \rho}} |\mathcal{P}_{J_i} g|^{\frac{2p}{3}} \approx \sum_{P \in \mathcal{P}_i} c_P 1_P, \quad \text{on } B.$$

Each $P$ sits inside a tube $T = T_P$ with dimensions $\rho^{-2}, \rho^{-2}, \rho^{-3}$, and for each ball $\Delta \in \mathcal{B}$ of radius $\rho^{-2}$, we have

$$|P \cap \Delta| \approx \rho |T_P \cap \Delta|.$$

Call $\mathcal{T}_i$ the collection of all these tubes corresponding to $P \in \mathcal{P}_i$. For each such $T$, define

$$c_T = \sum_{P \subset T} c_P.$$

Then for each ball $\Delta \in \mathcal{B}$, we have

$$\prod_{i=1}^{3} \left( \sum_{\substack{J_i \subset I_i \\ |J_i| = \rho}} \|\mathcal{P}_{J_i} g\|_{L_{\sharp}^{\frac{2p}{3}}(w_\Delta)}^{\frac{2p}{3}} \right)^{\frac{1}{2}} \approx \prod_{i=1}^{3} \left( \frac{1}{|\Delta|} \int_\Delta \sum_{P \in \mathcal{P}_i} c_P 1_P \right)^{\frac{1}{2}}$$

$$\approx \prod_{i=1}^{3} \left( \frac{\rho}{|\Delta|} \int_\Delta \sum_{T \in \mathcal{T}_i} c_T 1_T \right)^{\frac{1}{2}}$$

$$\approx \frac{\rho^{\frac{3}{2}}}{|\Delta|} \int_\Delta \prod_{i=1}^{3} \left( \sum_{T \in \mathcal{T}_i} c_T 1_T \right)^{\frac{1}{2}}.$$

Summation over $\Delta \in \mathcal{B}$ leads to

$$\frac{1}{|\mathcal{B}|} \sum_{\Delta \in \mathcal{B}} \prod_{i=1}^{3} \left( \sum_{\substack{J_i \subset I_i \\ |J_i| = \rho}} \|\mathcal{P}_{J_i} g\|_{L_{\sharp}^{\frac{2p}{3}}(w_\Delta)}^{\frac{2p}{3}} \right)^{\frac{1}{2}} \approx \frac{\rho^{\frac{3}{2}}}{|\mathcal{B}|} \int_\mathcal{B} \prod_{i=1}^{3} \left( \sum_{T \in \mathcal{T}_i} c_T 1_T \right)^{\frac{1}{2}}.$$

The right-hand side is evaluated using Theorem 11.9.

The more detailed argument follows.

*Proof of Theorem 11.10*   To be able to apply the trilinear Kakeya inequalities, we need to first replace the $l^2$ norm with the $l^{pk/3}$ norm. Since we can afford logarithmic losses in $\rho$, it suffices to prove the inequality with the summation on both sides restricted to families of intervals $J_i$ for which $\|\mathcal{P}_{J_i} g\|_{L_{\sharp}^{pk/3}(w_B)}$ have comparable size (up to a factor of 2), for each $i$. Indeed, the intervals $J_i' \subset I_i$ satisfying (for some large enough $C = O(1)$)

$$\|\mathcal{P}_{J_i'} g\|_{L_{\sharp}^{\frac{pk}{3}}(w_B)} \le \rho^C \max_{J_i \subset I_i} \|\mathcal{P}_{J_i} g\|_{L_{\sharp}^{\frac{pk}{3}}(w_B)}$$

can be easily dealt with by using the triangle inequality, since this guarantees the favorable bound (for some large enough $C'$)

$$\max_{\Delta \in \mathcal{B}} \|\mathcal{P}_{J_i'} g\|_{L_{\sharp}^{\frac{pk}{3}}(w_\Delta)} \le \rho^{C'} \max_{J_i \subset I_i} \|\mathcal{P}_{J_i} g\|_{L_{\sharp}^{\frac{pk}{3}}(w_B)}.$$

This leaves only $\log_2(\rho^{-O(1)})$ sizes to consider.

Let us now assume that for each $i$ we have $N_i$ intervals $J_i \subset I_i$, with $\|\mathcal{P}_{J_i} g\|_{L_{\sharp}^{pk/3}(w_B)}$ of comparable size. Since $p \ge 6$, by Hölder's inequality the left-hand side in (11.13) is at most

$$\left(\prod_{i=1}^{3} N_i^{\frac{1}{2}-\frac{3}{pk}}\right)^{\frac{p}{3}} \frac{1}{|\mathcal{B}|} \sum_{\Delta \in \mathcal{B}} \left[\prod_{i=1}^{3} \left(\sum_{J_i} \|\mathcal{P}_{J_i} g\|_{L_\sharp^{\frac{pk}{3}}(w_\Delta)}^{\frac{pk}{3}}\right)\right]^{\frac{1}{k}}. \tag{11.14}$$

Let $B$ and $\mathcal{B}$ be as in our hypothesis. Recall the definition of the spaces $V_k(t)$ from the previous section. For each interval $J = J_i$ of the form $[t_J - \rho/2, t_J + \rho/2]$, we cover $\cup_{\Delta \in \mathcal{B}} \Delta$ with a family $\mathcal{T}_J$ of pairwise disjoint tiles (rectangular boxes) $T_J$ with $k$ short sides parallel to $V_k(t_J)$ and of length $\rho^{-k}$, and $3 - k$ sides of length $\rho^{-k-1}$. Moreover, we can assume these tiles to be inside $4B$. We let $T_J(x)$ be the tile containing $x$, and we let $2T_J$ be the dilation of $T_J$ by a factor of 2 around its center.

Let us use $q$ to abbreviate $pk/3$. Our goal is to control the expression

$$\frac{1}{|\mathcal{B}|} \sum_{\Delta \in \mathcal{B}} \prod_{i=1}^{3} \left(\sum_{J_i} \|\mathcal{P}_{J_i} g\|_{L_\sharp^q(w_\Delta)}^q\right)^{\frac{1}{k}}.$$

We now define $F_J$ for $x \in \cup_{T_J \in \mathcal{T}_J} T_J$ by

$$F_J(x) := \sup_{y \in 2T_J(x)} \|\mathcal{P}_J g\|_{L_\sharp^q\left(w_{B(y,\rho^{-k})}\right)}.$$

For any point $x \in \Delta$, we have $\Delta \subset 2T_J(x)$, and so we also have

$$\|\mathcal{P}_J g\|_{L_\sharp^q(w_\Delta)} \le F_J(x).$$

Therefore,

$$\frac{1}{|\mathcal{B}|} \sum_{\Delta \in \mathcal{B}} \prod_{i=1}^{3} \left(\sum_{J_i} \|\mathcal{P}_{J_i} g\|_{L_\sharp^q(w_\Delta)}^q\right)^{\frac{1}{k}} \lesssim \fint_{4B} \prod_{i=1}^{3} \left(\sum_{J_i} F_{J_i}^q\right)^{\frac{1}{k}},$$

where the last expression denotes the average value of the integral over $4B$.

The function $F_J$ is constant on each tile $T_J \in \mathcal{T}_J$. Applying Theorem 11.8 when $k = 1$ and Theorem 11.9 when $k = 2$, both with $S = \rho^{-1}$, we get the bound

$$\fint_{4B} \prod_{i=1}^{3} \left(\sum_{J_i} F_{J_i}^q\right)^{\frac{1}{k}} \lesssim_{\epsilon,K} \rho^{-\epsilon} \prod_{i=1}^{3} \left(\sum_{J_i} \fint_{4B} F_{J_i}^q\right)^{\frac{1}{k}}.$$

It remains to check that for each $J = J_i$, we have

$$\|F_J\|_{L_\sharp^q(4B)} \lesssim \|\mathcal{P}_J g\|_{L_\sharp^q(w_B)}. \tag{11.15}$$

Once this is established, it follows that (11.14) is dominated by

$$\rho^{-\epsilon} \left(\prod_{i=1}^{3} N_i^{\frac{1}{2}-\frac{3}{pk}}\right)^{p/3} \prod_{i=1}^{3} \left(\sum_{J_i} \|\mathcal{P}_{J_i} g\|_{L_\sharp^{\frac{pk}{3}}(w_B)}^{\frac{pk}{3}}\right)^{\frac{1}{k}}. \tag{11.16}$$

Recalling the restriction we have made on $J_i$, (11.16) is comparable to

$$\rho^{-\epsilon} \prod_{i=1}^{3} \left( \sum_{J_i} \| \mathcal{P}_{J_i} g \|^2_{L_\sharp^{\frac{pk}{3}}(w_B)} \right)^{\frac{p}{6}},$$

as desired.

To prove (11.15), we may translate and assume that $J = [-\rho/2, \rho/2]$ and that $\widehat{\mathcal{P}_J g}$ is supported on $\prod_{j=1}^{3}[-\rho^j, \rho^j]$. Fix $x = (x_1, x_2, x_3)$ with $T_J(x) \in \mathcal{T}_J$ and $y \in 2T_J(x)$. Note that $T_J(x)$ has sides parallel to the coordinate axes. In particular, $y = x + y'$ with $|y'_j| < 4\rho^{-k}$ for $1 \le j \le k$ and $|y'_j| < 4\rho^{-k-1}$, for $k+1 \le j \le 3$. Then

$$\| \mathcal{P}_J g \|^q_{L^q(w_{B(y,\rho^{-k})})} \lesssim \int |\mathcal{P}_J g(x_1 + u_1, x_k + u_k, x_{k+1} + u_{k+1} \qquad (11.17)$$

$$+ y'_{k+1}, x_3 + u_3 + y'_3)|^q w_{B(0,\rho^{-k})}(u)\, du.$$

We are abusing notation here and in the following lines of computation. The expressions have an obvious interpretation depending on whether $k = 1$ or $k = 2$.

Using Taylor series, we may write

$$|\mathcal{P}_J g(x_1 + u_1, x_k + u_k, x_{k+1} + u_{k+1} + y'_{k+1}, x_3 + u_3 + y'_3)|$$

$$= \left| \int \widehat{\mathcal{P}_J g}(\lambda) e(\lambda \cdot (x + u)) e(\lambda_{k+1} y'_{k+1}) e(\lambda_3 y'_3)\, d\lambda \right|$$

$$\le \sum_{s_{k+1}=0}^{\infty} \frac{100^{s_{k+1}}}{s_{k+1}!}$$

$$\times \sum_{s_3=0}^{\infty} \frac{100^{s_3}}{s_3!} \left| \int \widehat{\mathcal{P}_J g}(\lambda) e(\lambda \cdot (x + u)) \left( \frac{\lambda_{k+1}}{2\rho^{k+1}} \right)^{s_{k+1}} \left( \frac{\lambda_3}{2\rho^{k+1}} \right)^{s_3} d\lambda \right|$$

$$= \sum_{s_{k+1}=0}^{\infty} \frac{100^{s_{k+1}}}{s_{k+1}!} \sum_{s_3=0}^{\infty} \frac{100^{s_3}}{s_3!} |M_{s_{k+1},s_3}(\mathcal{P}_J g)(x + u)|.$$

Here $M_{s_{k+1},s_3}$ is the operator with the Fourier multiplier $m_{s_{k+1},s_3}(\lambda/(2\rho^{k+1}))$, where

$$m_{s_{k+1},s_3}(\lambda) = (\lambda_{k+1})^{s_{k+1}}(\lambda_3)^{s_3} 1_{[-1/2,1/2]}(\lambda_{k+1}) 1_{[-1/2,1/2]}(\lambda_3).$$

We are able to insert the cut-off because of our initial restriction on the support of $\widehat{\mathcal{P}_J g}$.

Plugging this estimate into (11.17), we obtain

$$\|\mathcal{P}_J g\|^q_{L^q_\sharp(w_{B(y,\rho^{-k})})} \lesssim \sum_{s_{k+1}=0}^{\infty} \frac{100^{s_{k+1}}}{s_{k+1}!} \sum_{s_3=0}^{\infty} \frac{100^{s_3}}{s_3!} \|M_{s_{k+1},s_3}(\mathcal{P}_J g)\|^q_{L^q_\sharp(w_{B(x,\rho^{-k})})}.$$

Recalling the definition of $F_J$ and the fact that

$$\int_{4B} w_{B(x,\rho^{-k})}(z)\, dx \lesssim w_B(z), \quad z \in \mathbb{R}^3,$$

we conclude that

$$\|F_J\|^q_{L^q_\sharp(4B)} \lesssim \sum_{s_{k+1}=0}^{\infty} \frac{100^{s_{k+1}}}{s_{k+1}!} \sum_{s_3=0}^{\infty} \frac{100^{s_3}}{s_3!} \|M_{s_{k+1},s_3}(\mathcal{P}_J g)\|^q_{L^q_\sharp(w_B)}. \qquad (11.18)$$

Note that $m_{s_{k+1},s_3}$ admits a smooth extension $m^*_{s_{k+1},s_3}$ to the whole $\mathbb{R}^3$, supported on $\mathbb{R}^k \times [-1,1]^{3-k}$ and with derivatives of any given order uniformly bounded over $s_{k+1}, s_3$. It follows that

$$|\widehat{m^*_{s_{k+1},s_3}}(x)| \lesssim \xi(x_{k+1}, x_3),$$

with implicit constants independent of $s_{k+1}, s_3$, where

$$\xi(x_{k+1}, x_3) \lesssim_M (1 + (x_{k+1}^2 + x_3^2)^{1/2})^{-M}$$

for all $M > 0$. We will let $M^*_{s_{k+1},s_3}$ denote the operator with multiplier $m^*_{s_{k+1},s_3}(\lambda/(2\rho^{k+1}))$. We can now write

$$|M_{s_{k+1},s_3}(\mathcal{P}_J g)(x)| = |M^*_{s_{k+1},s_3}(\mathcal{P}_J g)(x)| \lesssim |\mathcal{P}_J g| \odot \xi_{\rho^{k+1}}(x),$$

where $\odot$ denotes the convolution with respect to the last $3 - k$ variables $\bar{x}$, and

$$\xi_{\rho^{k+1}}(\bar{x}) = \rho^{(3-k)(k+1)} \xi(\rho^{k+1}\bar{x}).$$

Using this, one can easily check that

$$\|M_{s_{k+1},s_3}(\mathcal{P}_J g)\|^q_{L^q(w_B)} \lesssim \langle |\mathcal{P}_J g|^q \odot \xi_{\rho^{k+1}}, w_B \rangle$$
$$= \langle |\mathcal{P}_J g|^q, \xi_{\rho^{k+1}} \odot w_B \rangle \lesssim \langle |\mathcal{P}_J g|^q, w_B \rangle.$$

Combining this with (11.18) leads to the proof of (11.15):

$$\|F_J\|^q_{L^q_\sharp(4B)} \lesssim \sum_{s_{k+1}=0}^{\infty} \frac{100^{s_{k+1}}}{s_{k+1}!} \sum_{s_3=0}^{\infty} \frac{100^{s_3}}{s_3!} \|\mathcal{P}_J g\|^q_{L^q_\sharp(w_B)}$$
$$\lesssim \|\mathcal{P}_J g\|^q_{L^q_\sharp(w_B)}.$$

The argument is now complete. $\qquad\qquad\qquad\qquad\qquad\qquad\qquad\square$

## 11.8 Lower-Dimensional Decoupling

The following inequality is a very efficient tool when decoupling small arcs of the twisted cubic with length $\sim \rho^{2/3}$, on spatial balls of radius $\rho^{-2}$. When $|I| = \rho^{2/3}$, the $\rho^2$-neighborhood of $\Gamma_I(\rho)$ is a curved planar tube, essentially the $\rho^2$-neighborhood (in $\mathbb{R}^3$) of an arc on $\mathbb{P}^1$ (of length $\sim \rho^{2/3}$). The strongest decoupling available in this setup is in $L^6$.

**Proposition 11.11** *Assume* $\text{supp}(\widehat{g}) \subset \Gamma_I(\rho)$, *for some interval $I$ of length greater than $\rho$. Then for each ball $B$ of radius at least $\rho^{-2}$, we have*

$$\|g\|_{L^6(w_B)} \lesssim_\epsilon \rho^{-\epsilon} \left( \sum_{\substack{J \subset I \\ |J| = \rho}} \|\mathcal{P}_J g\|_{L^6(w_B)}^2 \right)^{1/2}.$$

*Proof* Let

$$S_J = \{(\xi, \xi^2 + s) : \xi \in J, |s| \le \rho^2\}.$$

Note that $\widehat{\mathcal{P}_J g}$ is supported on $S_J \times \mathbb{R}$. Combining this observation with cylindrical decoupling (Exercise 9.22) and the $l^2(L^6)$ decoupling for the parabola (Theorem 10.1), we find that

$$\|g\|_{L^6(\mathbb{R}^3)} \lesssim_\epsilon \rho^{-\epsilon} \left( \sum_{\substack{J \subset I \\ |J| = \rho}} \|\mathcal{P}_J g\|_{L^6(\mathbb{R}^3)}^2 \right)^{1/2}.$$

The desired weighted inequality follows by adapting the argument from Proposition 9.15. $\qquad\qquad\qquad\qquad\qquad\qquad\qquad\qquad\qquad\qquad\square$

## 11.9 Proof of the Multiscale Inequality

Let $\delta \ll_K 1$. Recall that $B^s$ denotes an arbitrary ball of radius $\delta^{-s}$ in $\mathbb{R}^3$.

We are finally in a position to prove Theorem 11.4. We will achieve this in several stages. First, we perform a ball inflation, shifting from balls $B^1$ to balls $B^2$.

For $p \ge 6$, let $\alpha_1 = (p - 6)/(p - 3)$ satisfy

$$\frac{1}{\frac{p}{3}} = \frac{\alpha_1}{\frac{2p}{3}} + \frac{1 - \alpha_1}{2}.$$

**Proposition 11.12** *For each g with* supp $(\widehat{g}) \subset \Gamma(\delta)$, *we have*

$$A_p(1, B^2, g) \lesssim_{\epsilon, K} \delta^{-\epsilon}$$

$$\times A_p(2, B^2, g)^{1-\alpha_1} \tag{11.19}$$

$$\times D_{\frac{2p}{3}}(1, B^2, g)^{\alpha_1}. \tag{11.20}$$

*Proof* Since $p \geq 6$, we have the following consequence of Hölder's inequality

$$\|\mathcal{P}_{J_{i,1}} g\|_{L^2_\sharp(w_{B^1})} \lesssim \|\mathcal{P}_{J_{i,1}} g\|_{L^{\frac{p}{3}}_\sharp(w_{B^1})},$$

for each $B^1 \in \mathcal{B}_1(B^2)$. Using this and Theorem 11.10 with $k = 1$, we find a first upper bound

$$A_p(1, B^2, g) \lesssim_{\epsilon, K} \delta^{-\epsilon} \left( \prod_{i=1}^{3} \sum_{\substack{J_{i,1} \subset I_i \\ |J_{i,1}| = \delta}} \|\mathcal{P}_{J_{i,1}} g\|^2_{L^{\frac{p}{3}}_\sharp(w_{B^2})} \right)^{\frac{1}{6}}.$$

Using Hölder's inequality on $B^2$, we may further dominate this by

$$\left( \prod_{i=1}^{3} \sum_{\substack{J_{i,1} \subset I_i \\ |J_{i,1}| = \delta}} \|\mathcal{P}_{J_{i,1}} g\|^2_{L^2_\sharp(w_{B^2})} \right)^{\frac{1-\alpha_1}{6}} \left( \prod_{i=1}^{3} \sum_{\substack{J_{i,1} \subset I_i \\ |J_{i,1}| = \delta}} \|\mathcal{P}_{J_{i,1}} g\|^2_{L^{\frac{2p}{3}}_\sharp(w_{B^2})} \right)^{\frac{\alpha_1}{6}}.$$

It suffices now to apply $L^2$ decoupling (Proposition 9.18) to the first term in the preceding:

$$\left( \prod_{i=1}^{3} \sum_{\substack{J_{i,1} \subset I_i \\ |J_{i,1}| = \delta}} \|\mathcal{P}_{J_{i,1}} g\|^2_{L^2_\sharp(w_{B^2})} \right)^{\frac{1-\alpha_1}{6}} \lesssim \left( \prod_{i=1}^{3} \sum_{\substack{J_{i,2} \subset I_i \\ |J_{i,2}| = \delta^2}} \|\mathcal{P}_{J_{i,2}} g\|^2_{L^2_\sharp(w_{B^2})} \right)^{\frac{1-\alpha_1}{6}}.$$

This is possible since the $\delta^2$-neighborhoods of the supports $\Gamma_{J_{i,2}}(\delta)$ of $\widehat{\mathcal{P}_{J_{i,2}} g}$ have bounded overlap. □

Note that we have interpolated between $L^2$ and $L^{2p/3}$. We used $L^2$ because on this space we have a very efficient decoupling tool. The choice of $2p/3$ on the other hand requires a more careful explanation. If we had interpolated instead between $L^2$ and $L^p$, we would have obtained the inequality

$$A_p(1, B^2, g) \lesssim_{\epsilon, K} \delta^{-\epsilon} A_p(2, B^2, g)^{1-\beta} D_p(1, B^2, g)^\beta,$$

with

$$\frac{1}{\frac{p}{3}} = \frac{\beta}{p} + \frac{1-\beta}{2}.$$

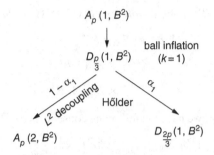

Figure 11.3 First ball inflation: from $B^1$ to $B^2$.

This is a two-scale inequality analogous to inequality (10.18) for $\mathbb{P}^{n-1}$. Iterating it as before leads to $V_p(\delta) \lesssim_\epsilon \delta^{-\epsilon}$ for $p \le 10$. This is the range for which $2(1 - \beta) = 8/(p - 2) \ge 1$ (cf. Exercise 10.24). See Figure 11.3.

To get closer to the critical exponent $p = 12$, we will interpolate instead with $L^{2p/3}$. This is the space corresponding to $k = 2$ in Theorem 11.10. In doing so, we prepare the ground for the second type of ball inflation.

In the next stage, we process the term (11.20). We proceed with a second ball inflation, shifting from balls $B^2$ to balls $B^3$ (see Figure 11.4). We continue with a lower-dimensional decoupling followed by $L^2$ decoupling. The term (11.19) will not undergo any serious modification; it will simply become $A_p(2, B^3, g)^{1-\alpha_1}$.

For $p \ge 9$, let $\alpha_2 = (p - 9)/(p - 6)$, $\beta_2 = 2p/[3(p - 3)] \in [0, 1]$ satisfy

$$\frac{1}{\frac{2p}{3}} = \frac{1 - \alpha_2}{6} + \frac{\alpha_2}{p},$$

$$\frac{1}{6} = \frac{1 - \beta_2}{2} + \frac{\beta_2}{\frac{2p}{3}}.$$

**Proposition 11.13** *For each $g$ with supp $(\widehat{g}) \subset \Gamma(\delta^{3/2})$, we have*

$$A_p(1, B^3, g) \lesssim_{\epsilon, K} \delta^{-\epsilon}$$

$$\times A_p(2, B^3, g)^{1-\alpha_1} \tag{11.21}$$

$$\times A_p(3, B^3, g)^{\alpha_1(1-\alpha_2)(1-\beta_2)} \tag{11.22}$$

$$\times D_{\frac{2p}{3}}\left(\frac{3}{2}, B^3, g\right)^{\alpha_1(1-\alpha_2)\beta_2} \tag{11.23}$$

$$\times D_p(1, B^3, g)^{\alpha_1\alpha_2}. \tag{11.24}$$

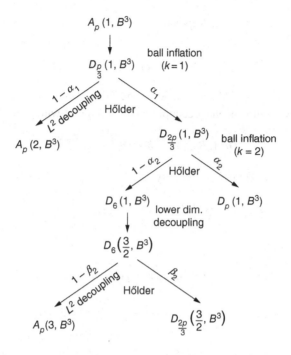

Figure 11.4 Second ball inflation: from $B^2$ to $B^3$.

*Proof* The argument consists of four steps.

Note that $\Gamma(\delta^{3/2}) \subset \Gamma(\delta)$. We first raise inequality (11.19) and (11.20) to the power $p$ and average it over all balls $B^2 \in \mathcal{B}_2(B^3)$. We use Hölder's inequality and Theorem 11.10 with $k = 2$ to get

$$A_p(1, B^3, g)^p \lesssim_{\epsilon, K} \delta^{-\epsilon} A_p(2, B^3, g)^{p(1-\alpha_1)}$$

$$\times \left[ \frac{1}{|\mathcal{B}_2(B^3)|} \sum_{B^2 \in \mathcal{B}_2(B^3)} D_{\frac{2p}{3}}(1, B^2, g)^p \right]^{\alpha_1}$$

$$\lesssim_{\epsilon, K} \delta^{-\epsilon} A_p(2, B^3, g)^{p(1-\alpha_1)} D_{\frac{2p}{3}}(1, B^3, g)^{p\alpha_1}.$$

Thus

$$A_p(1, B^3, g) \lesssim_{\epsilon, K} \delta^{-\epsilon} A_p(2, B^3, g)^{1-\alpha_1} D_{\frac{2p}{3}}(1, B^3, g)^{\alpha_1}. \tag{11.25}$$

The second step is an interpolation between $L^p$ and $L^6$ using Hölder's inequality:

$$D_{\frac{2p}{3}}(1, B^3, g) \leq D_6(1, B^3, g)^{1-\alpha_2} D_p(1, B^3, g)^{\alpha_2}. \tag{11.26}$$

To justify the use of $L^6$, we move to the third stage of the argument. By applying lower-dimensional decoupling (Lemma 11.11 with $\rho = \delta^{3/2}$), we get

$$D_6(1, B^3, g) \lesssim_\epsilon \delta^{-\epsilon} D_6\left(\frac{3}{2}, B^3, g\right). \tag{11.27}$$

Finally, we apply Hölder's inequality

$$D_6\left(\frac{3}{2}, B^3, g\right) \leq D_2\left(\frac{3}{2}, B^3, g\right)^{1-\beta_2} D_{\frac{2p}{3}}\left(\frac{3}{2}, B^3, g\right)^{\beta_2} \tag{11.28}$$

and $L^2$ decoupling

$$D_2\left(\frac{3}{2}, B^3, g\right) \lesssim D_2(3, B^3, g). \tag{11.29}$$

To close the argument, we combine (11.25) through (11.29). $\qquad\square$

Compared to Proposition 11.12, we now have three terms $A_p(m, B^3, g)$, one for each of the values $m = 1, 2, 3$. Let us assess the improvement arising from this second ball inflation. If in the last step we had interpolated between $L^2$ and $L^p$ (rather than between $L^2$ and $L^{2p/3}$)

$$\frac{1}{6} = \frac{1-\beta_2'}{2} + \frac{\beta_2'}{p},$$

and if no $l^2(L^6)$ decoupling had been invoked, we would have obtained the inequality

$$A_p(1, B^3, g) \lesssim_{\epsilon, K} \delta^{-\epsilon} A_p(2, B^3, g)^{1-\alpha_1}$$
$$\times A_p(3, B^3, g)^{\alpha_1(1-\alpha_2)(1-\beta_2')} D_p(1, B^3, g)^{\alpha_1(1-\alpha_2)\beta_2' + \alpha_1\alpha_2}.$$

A slight variation of Exercise 10.24 can be applied to this three-scale inequality and leads to $V_p(\delta) \lesssim_\epsilon \delta^{-\epsilon}$ as long as

$$2(1-\alpha_1) + 3\alpha_1(1-\alpha_2)(1-\beta_2') = \frac{27p - 90}{3p^2 - 15p + 18} \geq 1.$$

A direct computation shows that this is equivalent to $p \leq 7 + \sqrt{13} \approx 10.6$. While this is better than what we have achieved after the first ball inflation ($p \leq 10$), we will see that this range improves further if we perform more ball inflations, each time increasing the radius of the ball with a logarithmic factor of $3/2$. Note that the actual role of using $2p/3$ (rather than $p$) as an interpolation endpoint and of using lower-dimensional decoupling was to introduce the term $D_{2p/3}$ in (11.23). It is precisely this term that will make possible the next ball inflation.

**Proposition 11.14** *For $p \geq 9$, $r \geq 1$, each $g$ with supp $(\widehat{g}) \subset \Gamma(\delta^{(3/2)^r})$ and each ball $B \subset \mathbb{R}^3$ with radius $\delta^{-2(3/2)^r}$, we have*

$$A_p(1, B, g) \lesssim_{\epsilon, K, r} \delta^{-\epsilon} A_p(2, B, g)^{1-\alpha_1}$$

$$\times \prod_{i=1}^{r} A_p \left( 2 \left( \frac{3}{2} \right)^i, B, g \right)^{\alpha_1(1-\alpha_2)(1-\beta_2)[(1-\alpha_2)\beta_2]^{i-1}}$$

$$\times D_{\frac{2p}{3}} \left( \left( \frac{3}{2} \right)^r, B, g \right)^{\alpha_1[(1-\alpha_2)\beta_2]^r}$$

$$\times \prod_{i=0}^{r-1} D_p \left( \left( \frac{3}{2} \right)^i, B, g \right)^{\alpha_1 \alpha_2[(1-\alpha_2)\beta_2]^i} .$$

*Proof* The proof follows by induction of $r$. The base case $r = 1$ is covered by Proposition 11.13. The transition from $r$ to $r + 1$ follows by repeating the four steps in the proof of Proposition 11.13. The details are left to the reader. $\square$

We summarize the ball inflations as follows:

$$B^1 \mapsto B^2 \mapsto B^3 \mapsto B^{\frac{9}{2}} \mapsto \cdots \mapsto B^{2\left(\frac{3}{2}\right)^r} .$$

To simplify notation, we will write

$$\gamma_0 = 1 - \alpha_1,$$

$$\gamma_i = \alpha_1(1 - \alpha_2)(1 - \beta_2)[(1 - \alpha_2)\beta_2]^{i-1}, \quad 1 \leq i \leq r,$$

$$b_i = 2 \left( \frac{3}{2} \right)^i, \quad 0 \leq i \leq r.$$

The inequality in Proposition 11.14 can now be rewritten as follows.

**Proposition 11.15** *Let $p \geq 9$. Let $u > 0$ be such that*

$$\left( \frac{3}{2} \right)^r u \leq 1.$$

*Then for each ball $B^3$ of radius $\delta^{-3}$ and each $g$ with supp $(\widehat{g}) \subset \Gamma(\delta)$, we have*

$$A_p(u, B^3, g) \lesssim_{\epsilon, K, r} \delta^{-\epsilon} V_p(\delta)^{1 - \sum_0^r \gamma_j} D_p(1, B^3, g)^{1 - \sum_0^r \gamma_j}$$

$$\times \prod_{j=0}^{r} A_p(b_j u, B^3, g)^{\gamma_j} .$$

*Proof*  The proof relies on a few observations. The first one is that Proposition 11.14 holds true for all balls $B = B_R$ with radius $R \geq \delta^{-3(3/2)^r}$. This follows by covering $B_R$ with balls $B$ of radius $\delta^{-2(3/2)^r}$, applying Proposition 11.14 to each of these balls and summing up the terms with the help of Hölder's inequality.

Using this observation and replacing $\delta$ with $\delta^u$, the inequality in Proposition 11.14 can be rewritten as follows:

$$A_p(u, B^3, g) \lesssim_{\epsilon, K, r} \delta^{-\epsilon} \prod_{j=0}^{r} A_p(b_j u, B^3, g)^{\gamma_j}$$

$$\times D_{\frac{2p}{3}} \left( \left( \frac{3}{2} \right)^r u, B^3, g \right)^{\alpha_1 [(1-\alpha_2)\beta_2]^r}$$

$$\times \prod_{i=0}^{r-1} D_p \left( \left( \frac{3}{2} \right)^i u, B^3, g \right)^{\alpha_1 \alpha_2 [(1-\alpha_2)\beta_2]^i}.$$

To close the argument, we combine this inequality with

$$D_{\frac{2p}{3}} \left( \left( \frac{3}{2} \right)^r u, B^3, g \right) \lesssim D_p \left( \left( \frac{3}{2} \right)^r u, B^3, g \right)$$

and

$$D_p \left( \left( \frac{3}{2} \right)^i u, B^3, g \right) \leq V_p(\delta) D_p(1, B^3, g), \ 0 \leq i \leq r.$$

The reader may check that the final exponent of $D_p(1, B^3, g)$ is precisely $1 - \sum_0^r \gamma_j$, as desired.                                                   □

An easy computation shows that

$$\lim_{r \to \infty} \sum_{j=0}^{r} b_j \gamma_j = \frac{9}{p-3} \left( 1 + \frac{12-p}{p^2 - 12p + 18} \right).$$

A quick verification reveals that

$$\sum_{j=0}^{\infty} b_j \gamma_j > 1$$

for $p$ sufficiently close to but less than the critical exponent 12. Thus, for each such $p$, we may choose an $r$ satisfying

$$\sum_{j=0}^{r} b_j \gamma_j > 1.$$

It is now obvious that the multiscale inequality in Theorem 11.4 follows from Proposition 11.15, by choosing $u_0 = (2/3)^r$.

This finishes the proof of Theorem 11.1. In the next section, we will briefly discuss the extension of this argument to higher-dimensional moment curves.

## 11.10 Decoupling for the Higher-Dimensional Moment Curve

Let us consider the moment curve $\Gamma_n$ in $\mathbb{R}^n$, given by the graph of

$$\gamma_n(t) = (t, t^2, \ldots, t^n), \quad t \in [0, 1].$$

To state the main theorem, we start by defining the analog of (11.1) for $n \geq 4$. This will be a hypersurface $\mathcal{H}_n$ that approximates $\gamma_n$ locally.

More precisely, each point $\xi = (\xi_1, \ldots, \xi_n)$ on $\mathcal{H}_n$ satisfies the equation

$$\det[\xi - \gamma_n(\xi_1), \gamma_n'(\xi_1), \gamma_n''(\xi_1), \ldots, \gamma_n^{(n-1)}(\xi_1)] = 0.$$

Let us denote by $\psi_n$ the function whose graph is $\mathcal{H}_n$. For an interval $J \subset [0, 1]$ and $n \geq 4$, we define inductively the anisotropic neighborhoods

$$\Gamma_{n, J}(\delta) = \{(\bar{\xi}, \xi_n) : \bar{\xi} \in \Gamma_{n-1, J}(\delta) \text{ and } |\xi_n - \psi_n(\bar{\xi})| \leq \delta^n\}.$$

Here $\Gamma_{3, J}(\delta)$ is the previously introduced neighborhood of the twisted cubic, which was denoted simply by $\Gamma_J(\delta)$ in the earlier sections.

Let $V_{p,n}(\delta)$ be the smallest constant such that the inequality

$$\|g\|_{L^p(\mathbb{R}^n)} \leq V_{p,n}(\delta) \left( \sum_{\substack{J \subset [0,1] \\ |J| = \delta}} \|\mathcal{P}_J g\|_{L^p(\mathbb{R}^n)}^2 \right)^{1/2}$$

holds for each $g \colon \mathbb{R}^n \to \mathbb{C}$ with its Fourier transform supported on $\Gamma_n(\delta)$. Here, $\mathcal{P}_J$ denotes the Fourier projection operator onto the infinite strip $J \times \mathbb{R}^{n-1}$.

The following is the extension of Theorem 11.1 to all dimensions.

**Theorem 11.16** *For $n \geq 2$, we have*

$$V_{n(n+1), n}(\delta) \lesssim_\epsilon \delta^{-\epsilon}.$$

The proof of the theorem for $n \geq 4$ is similar to the one for $n = 3$ that we have described in the previous sections. Here is a brief overview.

We use induction on $n$. The induction hypothesis enters the proof when using lower-dimensional decoupling, the same way we have used $l^2(L^6)$ decoupling in the proof for the twisted cubic. There will be $n - 1$ kinds of ball inflation, with logarithmic increments $(d+1)/d$, for $1 \le d \le n - 1$. These correspond to separate applications of Theorem 6.16, one for each of these values of $d$.

As in the case $n = 3$, one can easily show that there is an essential equivalence between the $n$-linear and the linear decoupling constants. At the heart of this equivalence lie the good scaling properties of the anisotropic neighborhoods. This is the only noteworthy addition to the argument from [33], so we present the details in the following.

Let $\sigma < 1$ and $a \in [0, 1]$. Consider the affine transformation in $\mathbb{R}^n$

$$T_{n,\sigma,a}(\xi) = \xi',$$

where

$$
\begin{cases}
\xi'_1 = \frac{\xi_1 - a}{\sigma} \\
\xi'_2 = \frac{\xi_2 - 2a\xi_1 + a^2}{\sigma^2} \\
\quad \cdots \cdots \cdots \\
\xi'_n = \frac{\xi_n - na\xi_{n-1} + \ldots + (-1)^n a^n}{\sigma^n}.
\end{cases}
$$

**Proposition 11.17** *For each $n \ge 3$, we have*

$$T_{n,\sigma_2,a}\left(\Gamma_{n,[b,b+\sigma_1]}(\delta)\right) = \Gamma_{n,\left[\frac{b-a}{\sigma_2}, \frac{b-a}{\sigma_2} + \frac{\sigma_1}{\sigma_2}\right]}\left(\frac{\delta}{\sigma_2}\right).$$

*Proof* The proof follows by induction on $n$. The case $n = 3$ has been proved earlier. We first observe that the first $n - 1$ entries of $T_{n,\sigma,a}(\bar{\xi}, \xi_n)$ coincide with $T_{n-1,\sigma,a}(\bar{\xi})$. It thus suffices to prove that each $T = T_{n,\sigma,a}$ leaves $\mathcal{H}_n$ invariant. Let

$$L(\xi_1) = \frac{\xi_1 - a}{\sigma}.$$

We will prove that each point $\xi = (\xi_1, \ldots, \xi_n)$ satisfying

$$\det[\xi - \gamma_n(\xi_1), \gamma'_n(\xi_1), \ldots, \gamma_n^{(n-1)}(\xi_1)] = 0$$

is mapped by $T$ to a point that satisfies

$$\det[T(\xi) - \gamma_n(L(\xi_1)), \gamma'_n(L(\xi_1)), \ldots, \gamma_n^{(n-1)}(L(\xi_1))] = 0.$$

To see this, note first that

$$T \circ \gamma_n = \gamma_n \circ L.$$

Differentiating $j$ times leads to the relation

$$\tilde{T} \circ \gamma_n^{(j)} = \sigma^{-j} \gamma_n^{(j)} \circ L,$$

where $\tilde{T}(\xi) = T(\xi) - T(0)$ is the linear part of $T$. It follows that

$$\det[T(\xi) - \gamma_n(L(\xi_1)), \gamma_n'(L(\xi_1)), \ldots, \gamma_n^{(n-1)}(L(\xi_1))]$$
$$= \sigma^{\frac{n(n-1)}{2}} \det[\tilde{T}(\xi - \gamma_n(\xi_1)), \tilde{T}(\gamma_n'(\xi_1)), \ldots, \tilde{T}(\gamma_n^{(n-1)}(\xi_1))]$$
$$= \sigma^{\frac{n(n-1)}{2}} \det[\tilde{T}] \det[\xi - \gamma_n(\xi_1), \gamma_n'(\xi_1), \ldots, \gamma_n^{(n-1)}(\xi_1)]$$
$$= 0. \qquad \qquad \square$$

**Exercise 11.18** Use Theorem 11.16 to prove that

$$V_{p,n}(\delta) \lesssim_\epsilon \begin{cases} \delta^{-\epsilon}, & 2 \le p \le n(n+1) \\ \delta^{\frac{n(n+1)}{2p} - \frac{1}{2} - \epsilon}, & n(n+1) < p. \end{cases}$$

Prove that these estimates are sharp, apart from $\epsilon$ losses.

Hint: For the upper bound, interpolate with $L^2$ and $L^\infty$, using Exercise 9.21. To prove the lower bound, use $g$ with $\hat{g} \approx 1_{\Gamma(\delta)}$.

**Exercise 11.19** Give a direct proof of the inequality $V_{p,n}(\delta) \lesssim_\epsilon \delta^{-\epsilon}$ in the range $2 \le p \le 2n$, by combining the equivalence between linear and multilinear decoupling with Theorem 9.6.

# 12

# Decouplings for Other Manifolds

In this chapter, we demonstrate how to extend the sharp decouplings from the previous chapters to more general manifolds. The unifying theme for hypersurfaces will be the nonzero Gaussian curvature, while nonzero torsion will play the same role for curves. We also touch base with the case of zero curvature (the cone) and the case of real analytic curves containing points of zero torsion. The decoupling theory for manifolds of intermediate codimension is largely unexplored, and will not be discussed in this book. In Section 12.4 we present a refinement of the $l^p$ decoupling for the parabola.

## 12.1 Hypersurfaces with Positive Principal Curvatures

Let $\mathcal{M}$ be a hypersurface in $\mathbb{R}^n$

$$\mathcal{M} = \{(\xi, \psi(\xi)) : \xi \in U\},$$

with $\psi \in C^3(U)$. We assume $\mathcal{M}$ has all $n-1$ principal curvatures positive at each point.

For each $\mu$, let $\mathcal{U}_\mu$ be a cover of $U$ with pairwise disjoint cubes of diameter $\mu$. Let $\Theta_{\mathcal{M}}(\delta)$ be the partition of $\mathcal{N}_{\mathcal{M}}(\delta)$ into almost rectangular boxes of the form $\theta_\tau = \mathcal{N}_\tau(\delta)$, with $\tau \in \mathcal{U}_{\delta^{1/2}}$. Let $\text{Dec}_{\mathcal{M}}(\delta, p)$ be the $l^2(L^p)$ decoupling constant associated with this partition.

Theorem 10.1 admits the following extension to this class of hypersurfaces.

**Theorem 12.1 ([29])** *We have*

$$\text{Dec}_{\mathcal{M}}(\delta, p) \lesssim_{\epsilon, \mathcal{M}} \begin{cases} \delta^{-\epsilon}, & \text{if } 2 \leq p \leq \frac{2(n+1)}{n-1} \\ \delta^{-\epsilon + \frac{n+1}{2p} - \frac{n-1}{4}}, & \text{if } \frac{2(n+1)}{n-1} \leq p \leq \infty. \end{cases}$$

*Proof* We may, as before, restrict attention to dyadic values of the parameter $\mu$ and assume that each $\tau' \subset \mathcal{U}_{\mu'}$ is inside some $\tau \in \mathcal{U}_\mu$, whenever $\mu' \leq \mu$.

The argument uses an elegant induction on scales originating in the work of Pramanik–Seeger [91]. The key idea is to approximate $\mathcal{M}$ locally with elliptic paraboloids, and to use Theorem 10.1. More precisely, for each $\alpha \in \mathcal{U}_{\delta^{1/3}}$, Taylor's formula with cubic error term gives

$$\psi(\xi) = \psi_\alpha(\xi) + O(\delta), \ \xi \in \alpha,$$

where $\psi_\alpha$ is a quadratic function with positive definite Hessian matrix. It follows that $\mathcal{N}_{\mathcal{M}|_\alpha}(\delta) \subset \mathcal{N}_{\mathcal{M}^{\psi_\alpha}}(C\delta)$. Here $\mathcal{M}|_\alpha$ is the part of $\mathcal{M}$ above $\alpha$ and $\mathcal{M}^{\psi_\alpha}$ is a nonsingular affine image of a subset $S_\alpha$ of $\mathbb{P}^{n-1}$.

We may thus write for each $\alpha$ and each $F$ with $\mathrm{supp}\,(\widehat{F}) \subset \mathcal{N}_{\mathcal{M}|_\alpha}(\delta)$,

$$\|F\|_{L^p(\mathbb{R}^n)} \lesssim \mathrm{Dec}_{\mathcal{M}^{\psi_\alpha}}(\delta, p) \left( \sum_{\substack{\tau \in \mathcal{U}_{\delta^{1/2}} \\ \tau \subset \alpha}} \|\mathcal{P}_{\theta_\tau} F\|_{L^p(\mathbb{R}^n)}^2 \right)^{\frac{1}{2}}.$$

Exercise 9.9 shows that

$$\mathrm{Dec}_{\mathcal{M}^{\psi_\alpha}}(\delta, p) \lesssim_{\mathcal{M}} \mathrm{Dec}_{S_\alpha}(\delta, p).$$

Since $S_\alpha$ is a subset of $\mathbb{P}^{n-1}$ with diameter $\sim \delta^{1/3}$, Proposition 10.2 implies that

$$\mathrm{Dec}_{S_\alpha}(\delta, p) \lesssim \mathrm{Dec}_{\mathbb{P}^{n-1}}(\delta^{\frac{1}{3}}, p).$$

Combining the last three inequalities, we conclude that

$$\mathrm{Dec}_{\mathcal{M}}(\delta, p) \leq C\mathrm{Dec}_{\mathbb{P}^{n-1}}(\delta^{\frac{1}{3}}, p)\mathrm{Dec}_{\mathcal{M}}(\delta^{\frac{2}{3}}, p), \tag{12.1}$$

with $C$ independent of $\delta$.

It remains to show that an upper bound of the form $\mathrm{Dec}_{\mathbb{P}^{n-1}}(\delta, p) \lesssim \delta^{-A}$ forces a similar upper bound $\mathrm{Dec}_{\mathcal{M}}(\delta, p) \lesssim_\epsilon \delta^{-A-\epsilon}$ for each $\epsilon > 0$. We choose $m \sim \log\log(\delta^{-1})$, so that $\delta^{(2/3)^m} \sim 1$. Iterating (12.1) $m$ times leads to the inequality

$$\mathrm{Dec}_{\mathcal{M}}(\delta, p) \leq C^m \mathrm{Dec}_{\mathcal{M}}\left(\delta^{(\frac{2}{3})^m}, p\right) \prod_{j=0}^{m-1} \mathrm{Dec}_{\mathbb{P}^{n-1}}\left(\delta^{\frac{1}{3}(\frac{2}{3})^j}, p\right)$$

$$\lesssim (\log \delta^{-1})^{O(1)} \delta^{-A(1-(\frac{2}{3})^m)}$$

$$\lesssim_\epsilon \delta^{-A-\epsilon}. \qquad \square$$

Other than the elliptic paraboloid, one canonical manifold covered by Theorem 12.1 is the sphere $\mathbb{S}^{n-1}$. Another interesting class consists of the $C^3$ curves $\phi: I \to \mathbb{R}$ satisfying $\inf_{t \in I} |\phi''(t)| > 0$. One such example is

$$\{(t, t^\beta): t \in [1, 2]\}, \quad \beta > 1. \tag{12.2}$$

Section 12.6 will further clarify the decoupling theory of such curves.

## 12.2  The Cone

We will next examine how to decouple the truncated cone

$$\mathbb{C}\text{one}^{n-1} = \{(\xi, |\xi|): 1 < |\xi| < 2\}.$$

It is worth pointing out that in [119] Wolff investigated the problem for $\mathbb{C}\text{one}^2$ using wave packets spatially concentrated on thin planks of width $\sim 1$. His proof exploits the aperture of the cone by means of estimating the number of special incidences between planks. Our argument, on the other hand, makes no serious distinction between the cone and the cylinder, and it reduces the problem to repeated applications of the decoupling inequality for $\mathbb{P}^{n-2}$. A dissection of the whole argument reveals that its only component with incidence geometry flavor is the $n - 1$ linear Kakeya estimate (which goes into the proof of the decoupling for $\mathbb{P}^{n-2}$). Interestingly, when $n = 3$ this component collapses to the (trivial) bilinear Kakeya in $\mathbb{R}^2$. Thus, in sharp contrast to [119], our approach provides an (essentially) incidence geometry-free argument for the sharp decoupling for $\mathbb{C}\text{one}^2$.

Let us partition the sphere $\mathbb{S}^{n-2}$ into coin-shaped caps $\alpha$ with diameter $\sim \delta^{1/2}$. For each such $\alpha$, let us define the corresponding subset of the cone

$$\tau = \left\{ (\xi, |\xi|): 1 < |\xi| < 2, \ \frac{\xi}{|\xi|} \in \alpha \right\}.$$

We will use $\mathcal{N}$ to denote neighborhoods of the cone. Note that each $\theta = \mathcal{N}_\tau(\delta)$ is an almost rectangular box with $n - 2$ sides of length $\sim \delta^{1/2}$ and the remaining two sides of lengths $\sim \delta$ and $\sim 1$.

Let $\text{Dec}_{\mathbb{C}\text{one}^{n-1}}(\delta, p)$ be the $l^2(L^p)$ decoupling constant associated with the partition $\Theta_{\mathbb{C}\text{one}^{n-1}}(\delta)$ into sets $\theta$. The choice of these sets is enforced by Proposition 9.5. In particular, note that each line segment on $\mathbb{C}\text{one}^{n-1}$ is fully contained in some $\theta$.

**Theorem 12.2** ([29]) *For each $p \geq 2$, we have*

$$\mathrm{Dec}_{\mathbb{Cone}^{n-1}}(\delta, p) \lesssim_{\epsilon} \begin{cases} \delta^{-\epsilon}, & \text{if } 2 \leq p \leq \frac{2n}{n-2} \\ \delta^{-\epsilon + \frac{n}{2p} - \frac{n-2}{4}}, & \text{if } \frac{2n}{n-2} \leq p \leq \infty. \end{cases}$$

The reader will notice that the numerology for $\mathbb{Cone}^{n-1}$ matches that of $\mathbb{P}^{n-2}$. This is consistent with the fact that we do not decouple in the direction of zero principal curvature. Our proof will accommodate this feature by locally approximating the cone with parabolic cylinders and using cylindrical decoupling.

To simplify the notation, we present the argument for $n = 3$. The extension to higher dimensions can be obtained with no serious modifications.

*Proof* Invoking interpolation (Exercise 9.21), it will suffice to prove that

$$\mathrm{Dec}_{\mathbb{Cone}^2}(\delta, 6) \lesssim_{\epsilon} \delta^{-\epsilon}.$$

Let $\arg(\xi) \in [-\pi, \pi)$ be the angle between $\xi$ and the $\xi_2$ axis. We fix $\nu = 1/(2 + m)$, $m \in \mathbb{N}$, and restrict attention to the subset of the cone given by

$$\tau_0 = \{(\xi, |\xi|) : 1 \leq |\xi| \leq 1 + \delta^{\nu}, |\arg(\xi)| \leq \delta^{\nu}\}.$$

Note that the cone can be covered with $O(\delta^{-2\nu})$ similar regions. Let us denote by $\theta_m$ the intersection of a generic $\theta \in \Theta_{\mathbb{Cone}^2}(\delta)$ with $\mathcal{N}_{\tau_0}(\delta)$. The collection of these $\theta_m$ partition $\mathcal{N}_{\tau_0}(\delta)$.

Assume that $F$ has its Fourier transform supported on $\mathcal{N}_{\tau_0}(\delta)$. We will prove that

$$\|F\|_{L^6(\mathbb{R}^3)} \lesssim_{\epsilon, m} \delta^{-\epsilon m} \left( \sum_{\theta_m} \|\mathcal{P}_{\theta_m} F\|_{L^6(\mathbb{R}^3)}^2 \right)^{1/2}. \tag{12.3}$$

Let us assume for a moment that this inequality holds. Using the triangle inequality over the sets (similar to) $\tau_0$ covering the cone, we find that whenever $\mathrm{supp}(\widehat{F}) \subset \mathcal{N}_{\mathbb{Cone}^2}(\delta)$, we have

$$\|F\|_{L^6(\mathbb{R}^3)} \lesssim_{\epsilon, m} \delta^{-\epsilon m - O(\nu)} \left( \sum_{\theta_m} \|\mathcal{P}_{\theta_m} F\|_{L^6(\mathbb{R}^3)}^2 \right)^{1/2}.$$

The summation here is over segments $\theta_m$ of $\theta \in \Theta_{\mathbb{Cone}^2}(\delta)$ of length $\sim \delta^{\nu}$. Each $\theta$ contributes with $O(\delta^{-\nu})$ such $\theta_m$, and

$$\|\mathcal{P}_{\theta_m} F\|_{L^6(\mathbb{R}^3)} \lesssim \|\mathcal{P}_{\theta} F\|_{L^6(\mathbb{R}^3)}.$$

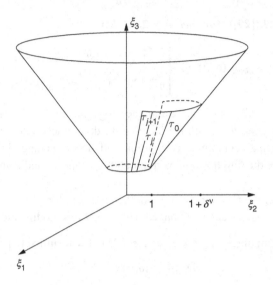

Figure 12.1 Caps $\tau$ of three different scales.

We conclude that

$$\|F\|_{L^6(\mathbb{R}^3)} \lesssim_{\epsilon,m} \delta^{-\epsilon m - O(m^{-1})} \left( \sum_{\theta \in \Theta_{\text{Cone}^2}(\delta)} \|\mathcal{P}_\theta F\|_{L^6(\mathbb{R}^3)}^2 \right)^{1/2}.$$

Choosing $m \sim \epsilon^{-1/2}$ leads to the desired estimate.

It remains to prove (12.3). We will decouple in a few stages. For each $0 \leq j \leq m$, we partition $(-\delta^\nu, \delta^\nu)$ into a collection $\{I_j\}$ of intervals of length $2\delta^{\nu+j\nu/2}$. For each such interval, we write

$$\tau_{I_j} = \{(\xi, |\xi|): 1 < |\xi| < 1 + \delta^\nu, \arg(\xi) \in I_j\}$$

and $\theta_{I_j} = \mathcal{N}_{\tau_{I_j}}(\delta)$. See Figure 12.1. Since decouplings can be iterated, inequality (12.3) will follow if we prove that for each $0 \leq j \leq m - 1$ and each $\theta_j := \theta_{I_j}$, we have

$$\|\mathcal{P}_{\theta_j} F\|_{L^6(\mathbb{R}^3)} \lesssim_\epsilon \delta^{-\epsilon} \left( \sum_{\theta_{j+1} \subset \theta_j} \|\mathcal{P}_{\theta_{j+1}} F\|_{L^6(\mathbb{R}^3)}^2 \right)^{1/2}. \qquad (12.4)$$

Due to symmetry, it suffices to prove this when $\theta_j = \mathcal{N}_{\tau_{I_j}}(\delta)$ with $I_j = [-\delta^{\nu+j\nu/2}, \delta^{\nu+j\nu/2}]$.

Using Taylor's formula with third-order error terms near $(\xi_1, \xi_2) = (0, 1)$, we find that each $(\xi_1, \xi_2, \xi_3) \in \tau_{I_j}$ satisfies

$$\xi_3 = \xi_2 + \frac{1}{2}\xi_1^2 + O(\xi_1^2 |\xi_2 - 1|) = \xi_2 + \frac{1}{2}\xi_1^2 + O(\delta^{(j+3)\nu}).$$

Since $(j + 3)\nu \leq 1$, we conclude that $\theta_j$ lies inside the $O(\delta^{(j+3)\nu})$-neighborhood of the cylinder:

$$\xi_3 = \xi_2 + \frac{1}{2}\xi_1^2.$$

The cross section of this cylinder is a parabola with curvature $\sim 1$. Inequality (12.4) follows by combining the $l^2(L^6)$ decoupling for this parabola (with the thickness parameter $\delta$ replaced with $O(\delta^{(j+3)\nu})$) with cylindrical decoupling. Indeed, note that $(\delta^{(j+3)\nu})^{1/2}$ matches the angular size $\delta^{\nu+(j+1)\nu/2}$ of the sets $\theta_{j+1}$. □

**Exercise 12.3** Prove that Theorem 12.2 implies Theorem 10.1.

**Exercise 12.4** Prove that Theorem 12.2 is sharp, apart from $\epsilon$ losses.

**Exercise 12.5** (General conical surfaces) Let $\phi \in C^3([-1, 1])$ satisfy $\inf_\xi |\phi''(\xi)| > 0$. Consider the conical surface

$$S_\phi = \left\{ t(\xi_1, 1, \phi(\xi_1)), \ t \in \left[\frac{1}{2}, 1\right], \ |\xi_1| \leq 1 \right\}.$$

Find a partition of $\mathcal{N}_{S_\phi}(\delta)$ similar to that of $\mathbb{C}one^2$ and prove the $l^2(L^6)$ decoupling for it.

**Exercise 12.6** Adapt the argument from Section 10.4 to prove the $l^2(L^p)$ decoupling for $\mathbb{C}one^2$ when $p = 4$. Reflect on why this argument is not strong enough to cover the range $4 < p \leq 6$.

Hint: Use a trilinear approach. Apply a Bourgain–Guth-type argument relying on Lorentz transformations to argue that linear decoupling is equivalent with trilinear decoupling in this context.

At each scale, use wave packets $W_{P_k}$ spatially concentrated on rectangular boxes (planks) of dimensions $\sim 1, R^{2^{-k+1}}, R^{2^{-k}}$. Each plank sits inside an $(R^{2^{-k}}, R^{2^{-k+1}})$-tube.

For each category of parameters, there are now three that need to be selected, e.g., $M_k, N_k, O_k$ will represent the number of directions for the tubes used in each of the three families. There is a new category of parameters (and an additional pigeonholing), counting the number of planks inside each tube.

For example, the first family will have $\sim M_k U_k A_k$ planks, with $\sim A_k$ planks inside a tube, and there are $\sim U_k$ tubes corresponding to each of the $M_k$ distinct directions.

Use trilinear Kakeya for tubes (Theorem 6.16) to estimate the number of cubes with side length $R^{2^{-k}}$ that intersect a given fraction of the tubes from each family.

## 12.3 $l^p$ Decouplings: The Case of Nonzero Gaussian Curvature

Given $v = (v_1, \ldots, v_{n-1}) \in (\mathbb{R} \setminus \{0\})^{n-1}$, we introduce the paraboloid

$$\mathbb{H}_v^{n-1} := \{(\xi_1, \ldots, \xi_{n-1}, v_1 \xi_1^2 + \cdots + v_{n-1} \xi_{n-1}^2) : |\xi_i| < 1\}.$$

When all entries of $v$ have the same sign, $\mathbb{H}_v^{n-1}$ is a (nonsingular) affine image of the *elliptic* paraboloid $\mathbb{P}^{n-1}$. In all other cases, $\mathbb{H}_v^{n-1}$ will be referred to as *hyperbolic* paraboloid. The multiruled nature of hyperbolic paraboloids prevents them from hosting any interesting genuine $l^2(L^p)$ decoupling (cf. Proposition 9.5 and the discussion following it). However, we will see that there is a satisfactory substitute for $l^2$ decoupling. Let us start by generalizing Definition 9.1.

**Definition 12.7** ($l^q(L^p)$ decoupling) Let $p, q \geq 2$. For a family $\mathcal{S}$ consisting of pairwise disjoint sets $S_i \subset \mathbb{R}^n$, let $\mathrm{Dec}(\mathcal{S}, p, q)$ be the smallest constant such that the inequality (with an obvious interpretation if $q = \infty$)

$$\|F\|_{L^p(\mathbb{R}^n)} \leq \mathrm{Dec}(\mathcal{S}, p, q) |\mathcal{S}|^{\frac{1}{2} - \frac{1}{q}} \left( \sum_i \|\mathcal{P}_{S_i} F\|_{L^p(\mathbb{R}^n)}^q \right)^{\frac{1}{q}}$$

holds for each $F$ with the Fourier transform supported on $\cup S_i$.

We will call such an inequality $l^q(L^p)$ decoupling (or simply $l^q$ decoupling, if the value of $p$ is clear from the context ) and we will refer to $\mathrm{Dec}(\mathcal{S}, p, q)$ as the $l^q(L^p)$ decoupling constant associated with $\mathcal{S}$.

Recall that we have previously denoted $\mathrm{Dec}(\mathcal{S}, p, 2)$ by $\mathrm{Dec}(\mathcal{S}, p)$. Note that $\mathrm{Dec}(\mathcal{S}, p, q)$ is a nonincreasing function of $q$. Exercise 9.7 shows that $\mathrm{Dec}(\mathcal{S}, p, q) \gtrsim 1$. The ideal scenario – we will refer to it as *genuine* $l^q(L^p)$ decoupling – is when

$$\mathrm{Dec}(\mathcal{S}, p, q) \lesssim_\epsilon |\mathcal{S}|^\epsilon,$$

for the collections of sets $\mathcal{S}$ under consideration.

The question formulated in Problem 9.4 about $l^2(L^p)$ decoupling for manifolds can also be asked about $l^q(L^p)$ decoupling. To keep the discussion more concise, we will restrict attention to the case $q = p$. It turns out that in many applications, a genuine $l^p(L^p)$ decoupling is as good as a genuine $l^2(L^p)$ decoupling. The hyperbolic paraboloids are examples of manifolds for which the latter is false, while the former holds.

Let us partition $[-1,1]^{n-1}$ into almost cubes $\tau$ of diameter $\sim \delta^{1/2}$. Let $\Theta_{\mathbb{H}_v^{n-1}}(\delta)$ be the partition of $\mathcal{N}_{\mathbb{H}_v^{n-1}}(\delta)$ into almost rectangular boxes $\theta_\tau = \mathcal{N}_\tau(\delta)$.

**Theorem 12.8** ([27]) *For each $p \geq 2$ and each $v \in (\mathbb{R} \setminus \{0\})^{n-1}$, we have*

$$\text{Dec}(\Theta_{\mathbb{H}_v^{n-1}}(\delta), p, p) \lesssim_{\epsilon, v} \begin{cases} \delta^{-\epsilon}, & \text{if } 2 \leq p \leq \frac{2(n+1)}{n-1} \\ \delta^{-\epsilon + \frac{n+1}{2p} - \frac{n-1}{4}}, & \text{if } \frac{2(n+1)}{n-1} \leq p \leq \infty. \end{cases}$$

Apart from the $\epsilon$ losses, these upper bounds are sharp for all $v$. They match the values of $\text{Dec}_{\mathbb{P}^{n-1}}(\delta, p)$ from Theorem 10.1.

Theorem 12.8 combined with induction on scales as in the proof of Theorem 12.1 leads to a unified $l^p(L^p)$ theory for all manifolds with nonzero Gaussian curvature.

**Corollary 12.9** *Let $\mathcal{M}$ be a $C^3$ hypersurface in $\mathbb{R}^n$ with nonzero Gaussian curvature, and let $\Theta_{\mathcal{M}}(\delta)$ be a partition of $\mathcal{N}_{\mathcal{M}}(\delta)$ into almost rectangular boxes with dimensions $\delta^{1/2}, \ldots, \delta^{1/2}, \delta$.*

*The following essentially sharp estimates hold:*

$$\text{Dec}(\Theta_{\mathcal{M}}(\delta), p, p) \lesssim_{\epsilon, \mathcal{M}} \begin{cases} \delta^{-\epsilon}, & \text{if } 2 \leq p \leq \frac{2(n+1)}{n-1} \\ \delta^{-\epsilon + \frac{n+1}{2p} - \frac{n-1}{4}}, & \text{if } \frac{2(n+1)}{n-1} \leq p \leq \infty. \end{cases}$$

The proof of Theorem 12.8 is similar to the one for $\mathbb{P}^{n-1}$ from Section 10.3, in particular, it relies on induction on the dimension. There is one notable difference, though, that is ultimately responsible for the lack of a genuine $l^2$ decoupling in this generalized setting. This has to do with certain losses coming from lower-dimensional contributions. We explain this difference when $n = 3$ in the case of

$$\mathbb{H}_{(1,-1)}^2 = \{(\xi_1, \xi_2, \xi_1^2 - \xi_2^2) : |\xi_1|, |\xi_2| < 1\}.$$

In what follows, all neighborhoods and decoupling constants will be with respect to $\mathbb{H}_{(1,-1)}^2$. Following the notation from Section 10.3, we introduce the trilinear decoupling constant $\text{TriDec}(\delta, p, p, K)$ to be the smallest number for which the following inequality holds for each $(\alpha_1, \alpha_2, \alpha_3) \in \mathcal{I}_K$ and each $F$ with its Fourier transform supported on $\mathcal{N}_{\alpha_1}(\delta) \cup \mathcal{N}_{\alpha_2}(\delta) \cup \mathcal{N}_{\alpha_3}(\delta)$

$$\left\| \left( \prod_{i=1}^{3} \mathcal{P}_{\alpha_i} F \right)^{1/3} \right\|_{L^p(\mathbb{R}^3)} \leq \mathrm{TriDec}(\delta, p, p, K) \delta^{\frac{1}{p} - \frac{1}{2}}$$

$$\times \left( \prod_{i=1}^{3} \sum_{\tau \in \mathcal{C}_{\delta^{-1/2}}(\alpha_i)} \|\mathcal{P}_\tau F\|_{L^p(\mathbb{R}^3)}^p \right)^{\frac{1}{3p}}.$$

It is immediate that $\mathrm{TriDec}(\delta, p, p, K) \lesssim \mathrm{Dec}(\delta, p, p)$. The reverse inequality is also essentially true, with some negligible losses.

**Proposition 12.10** *There is $\epsilon_p(K) > 0$ with $\lim_{K \to \infty} \epsilon_p(K) = 0$ such that*

$$\mathrm{Dec}(\delta, p, p) \lesssim_K \delta^{-\epsilon_p(K)} \left( 1 + \sup_{\delta \leq \delta' \leq 1} \mathrm{TriDec}(\delta', p, p, K) \right).$$

The proof of this proposition is very similar to that of Proposition 10.25; it relies on the Bourgain–Guth decomposition and induction on scales. The only difference comes when estimating the lower-dimensional contribution. While every vertical slice of $\mathbb{P}^2$ is a parabola, some of the vertical slices of $\mathbb{H}^2_{(1,-1)}$ are lines. The subset of $\mathbb{H}^2_{(1,-1)}$ near such a line is essentially flat, and the best that can be said about it is that it satisfies a certain "trivial decoupling" (Exercise 12.12). This issue is addressed in the following analog of Lemma 10.26.

**Lemma 12.11** (Trivial lower-dimensional decoupling) *Let $L$ be a line in $\mathbb{R}^2$. Let $\mathcal{F}$ be a collection of squares $\beta \subset \mathcal{C}_{K^{1/2}}$ with $\mathrm{dist}(\beta, L) \lesssim K^{-1/2}$. For each $F$ with the Fourier transform supported on $\cup_{\beta \in \mathcal{F}} \mathcal{N}_\beta(1/K)$ and each ball $B_K \subset \mathbb{R}^3$ with radius $K$, we have*

$$\|F\|_{L^p(w_{B_K})} \lesssim K^{\frac{1}{2}\left(1 - \frac{2}{p}\right)} \left( \sum_{\beta \in \mathcal{F}} \|\mathcal{P}_\beta F\|_{L^p(w_{B_K})}^p \right)^{1/p}.$$

The use of trivial decoupling may at first seem wasteful, since the exponent $1 - 2/p$ of $K^{1/2}$ is twice as large as the exponent $1/2 - 1/p$ associated with a genuine $l^p$ decoupling. However, this loss ends up not being costly since we only use trivial decoupling for a small ("one-dimensional") subset of $\mathbb{H}^2_{(1,-1)}$. This observation also applies to the hyperbolic paraboloids in higher dimensions. More precisely, it is easy to see that $\mathbb{H}^{n-1}_\nu$ may contain affine subspaces of dimension as high as $m = [(n+1)/2] - 1$, but not higher than that. For example,

$$\xi_5 = \xi_1^2 + \xi_2^2 - \xi_3^2 - \xi_4^2$$

contains the planes

$$\text{span}((1,0,1,0,0),(0,1,0,1,0))$$

and

$$\text{span}((1,0,0,1,0),(0,1,1,0,0))$$

through the origin. Whenever this happens, there are $\delta^{-m/2}$ sets $\theta \in \Theta_{\mathbb{H}^{n-1}_v}(\delta)$ intersecting an $m$-dimensional affine space. Calling this collection $\mathcal{S}_\delta$, the combination of Exercises 12.12 and 12.13 shows that estimating this lower-dimensional contribution using trivial decoupling is sharp:

$$\text{Dec}(\mathcal{S}_\delta, p, p) \sim \delta^{-\frac{1}{2}([\frac{n+1}{2}]-1)(\frac{1}{2}-\frac{1}{p})}.$$

It turns out, however, that this local contribution is not large enough to surpass the global contribution, since $[(n+1)/2] - 1 \le (n-1)/2$.

**Exercise 12.12** (Trivial $l^p$ decoupling) Let $\mathcal{S}$ be a family of rectangular boxes $R$ such that the boxes $2R$ are pairwise disjoint. Prove that for each $p \ge 2$,

$$\text{Dec}(\mathcal{S}, p, p) \lesssim |\mathcal{S}|^{\frac{1}{2}-\frac{1}{p}}.$$

Hint: Interpolate as in the proof of Lemma 4.5.

The next exercise shows that trivial decoupling is sharp in the flat setting.

**Exercise 12.13** Let $T$ be a tube in $\mathbb{R}^n$. Consider a partition $\mathcal{T}_N$ of $T$ into $N$ shorter congruent tubes. Prove that $\text{Dec}(\mathcal{T}_N, p, p) \gtrsim N^{1/2-1/p}$.

**Exercise 12.14** Assume $\widehat{F}$ is supported on $\mathcal{N}_{\mathbb{P}^1}(\delta)$. Prove that for each $4 \le p \le 6$ and $q = 2p/(8-p)$

$$\|F\|_{L^p(\mathbb{R}^2)} \lesssim_\epsilon \delta^{\frac{2}{p}-\frac{1}{2}-\epsilon} \left\| \left( \sum_{\theta \in \Theta_{\mathbb{P}^1}(\delta)} |\mathcal{P}_\theta F|^q \right)^{\frac{1}{q}} \right\|_{L^p(\mathbb{R}^2)}.$$

Hint: Interpolate a reverse square function estimate with the $l^6(L^6)$ decoupling as in the proof of Proposition 5.20.

**Exercise 12.15** Let $\mathcal{M}$ be a $C^3$ hypersurface in $\mathbb{R}^n$ with positive principal curvatures. Use Theorem 12.1 to prove the Stein–Tomas endpoint restriction estimate (with $\epsilon$ losses) $R^*_{\mathcal{M}}(2 \mapsto 2(n+1)/(n-1); \epsilon)$.

**Exercise 12.16** ($l^p$ decoupling and the Restriction Conjecture) Let $\mathcal{M}$ be a hypersurface in $\mathbb{R}^n$. For each $\delta$, we partition $\mathcal{N}_{\mathcal{M}}(\delta)$ into almost cubes of the form $\mathcal{N}_Q(\delta)$, where $Q \subset \mathbb{R}^{n-1}$ is an almost cube of diameter $\sim \delta$. Assume that for each $F: \mathbb{R}^n \to \mathbb{C}$ with supp $(\widehat{F}) \subset \mathcal{N}_\mathcal{M}(\delta)$, the following genuine $l^p(L^p)$ decoupling holds for $p = 2n/(n-1)$:

$$\|F\|_{L^p(\mathbb{R}^n)} \lesssim_\epsilon \delta^{-\epsilon + (n-1)(\frac{1}{p} - \frac{1}{2})} \left( \sum_Q \|\mathcal{P}_{\mathcal{N}_Q(\delta)} F\|^p_{L^p(\mathbb{R}^n)} \right)^{\frac{1}{p}}.$$

Prove that the endpoint restriction estimate $R^*_\mathcal{M}(\infty \mapsto 2n/(n-1); \epsilon)$ holds for each $\epsilon > 0$.

## 12.4 A Refined $l^p$ Decoupling for the Parabola

Throughout this section, a tube $T$ will be a parallelogram in the plane with horizontal short sides. The vertical distance $H$ between the short sides of $T$ is its height, while the length $W$ of the short sides is its width. We will refer to $T$ as an $(W, H)$-*tube*. The direction of the long sides gives the direction of the tube. For each tube $T$ and each $r > 1$, $rT$ will denote the tube with the same center as $T$ and dimensions $r$ times as large as those of $T$.

Assume $F: \mathbb{R}^2 \to \mathbb{C}$ has the Fourier transform supported on $\mathcal{N}_{\mathbb{P}^1}(R^{-1})$. Consider the wave packet decomposition (Exercise 2.7):

$$F = \sum_{T \in \mathcal{T}_R(F)} F_T = \sum_{T \in \mathcal{T}_R(F)} a_T W_T.$$

Each $T$ is an $(R^{1/2}, R)$-tube whose direction is normal to the parabola. We assume $L^\infty$ normalization for $W_T$

$$|W_T(x)| \lesssim_l (\chi_T(x))^l$$

for each $l \geq 1$, and

$$\text{supp}(\widehat{W_T}) \subset 2\theta$$

for some $\theta \in \Theta_{\mathbb{P}^1}(R^{-1})$.

The following theorem refines the $l^p(L^p)$ decoupling for the parabola, by replacing $R^{1/4 - 1/2p}$ with the smaller quantity $M^{1/2 - 1/p}$. It is a particular case of theorem 4.2 from [64].

Fix $\delta > 0$ and consider the weight on $\mathbb{R}^2$

$$w(x) = \frac{1}{(1 + |x|)^{\frac{1000}{\delta}}}.$$

If $P = A([-1, 1]^2)$ is the image of $[-1, 1]^2$ under the nonsingular affine map $A$, we define $w_P = w \circ A^{-1}$.

**Theorem 12.17** *Let $\mathcal{Q}$ be a collection of pairwise disjoint squares $q$ in $\mathbb{R}^2$ with side length $R^{1/2}$. Assume that each $q$ intersects at most $M$ fat tubes $R^\delta T$ with $T \in \mathcal{T}_R(F)$, for some $M \geq 1$.*

*Then for each $2 \leq p \leq 6$ and $\epsilon > 0$, we have*

$$\left( \sum_{q \in \mathcal{Q}} \|F\|_{L^p(w_q)}^p \right)^{\frac{1}{p}} \lesssim_{\delta, \epsilon} R^\epsilon M^{\frac{1}{2} - \frac{1}{p}} \left( \sum_{T \in \mathcal{T}_R(F)} \|F_T\|_{L^p(\mathbb{R}^2)}^p \right)^{\frac{1}{p}}.$$

*Proof* Let $D_R$ be the smallest constant such that

$$\left( \sum_{q \in \mathcal{Q}} \|F\|_{L^p(w_q)}^p \right)^{\frac{1}{p}} \leq D_R M^{\frac{1}{2} - \frac{1}{p}} \left( \sum_{T \in \mathcal{T}_R(F)} \|F_T\|_{L^p(\mathbb{R}^2)}^p \right)^{\frac{1}{p}}$$

holds for each $F$ and $\mathcal{Q}$ as before. It suffices to prove that

$$D_R \leq O_\delta(1) + C_R D_{R^{1/2}}, \tag{12.5}$$

for some $C_R = O(R^\epsilon)$. We may assume $R$ is large enough.

Let us partition $\mathbb{P}^1$ into arcs $\omega \in \Omega$ of length $\sim R^{-1/4}$. We write $T \perp \omega$ if the direction of $T$ is normal to the arc $\omega$. For each $T$, there is a unique such $\omega$.

For each $q \in \mathcal{Q}$ and $j \geq 0$, we let $\Omega(q, j)$ consist of those $\omega \in \Omega$ satisfying

$$2^j - 1 \leq |\{T : T \perp \omega \text{ and } R^\delta T \cap q \neq \emptyset\}| < 2^{j+1} - 1.$$

Note that if $j \geq 1$,

$$|\Omega(q, j)| \lesssim M 2^{-j}. \tag{12.6}$$

We define

$$\mathcal{T}(q, j) = \{T : T \perp \omega \text{ for some } \omega \in \Omega(q, j)\}.$$

Letting

$$F_{q,j} = \sum_{T \in \mathcal{T}(q,j)} F_T,$$

it is clear that for each $q \in \mathcal{Q}$, we have

$$F = \sum_{0 \leq j \lesssim \log M} F_{q,j}.$$

We use the triangle inequality on both $L^p(w_q)$ and $l^p$ to write

$$\left( \sum_{q \in \mathcal{Q}} \|F\|_{L^p(w_q)}^p \right)^{\frac{1}{p}} \leq \sum_{0 \leq j \lesssim \log M} \left( \sum_{q \in \mathcal{Q}} \|F_{q,j}\|_{L^p(w_q)}^p \right)^{\frac{1}{p}}. \tag{12.7}$$

We first deal with the term corresponding to $j = 0$. Let $\mathcal{T}'(q, 0)$ be a collection of tubes in $\mathcal{T}(q, 0)$ with fixed direction. Recall that $R^\delta T \cap q = \emptyset$ for each

$T \in \mathcal{T}'(q,0)$. Elementary inequalities show that for large enough $l$ depending on $\delta$, we have

$$\left\| \sum_{T \in \mathcal{T}'(q,0)} F_T \right\|_{L^p(w_q)}^p$$

$$\lesssim_l \left\| \left( \sum_{T \in \mathcal{T}'(q,0)} \chi_T^l \right)^{\frac{p}{p'}} \sqrt{w_q} \right\|_{L^\infty(\mathbb{R}^2)} \int_{\mathbb{R}^2} \sum_{T \in \mathcal{T}'(q,0)} |a_T|^p \chi_T \sqrt{w_q}$$

$$\lesssim_\delta R^{-10} \int_{\mathbb{R}^2} \sum_{T \in \mathcal{T}'(q,0)} |a_T|^p \chi_T \sqrt{w_q}.$$

Since there are $O(R^{1/2})$ directions, an application of the triangle inequality shows that

$$\left\| \sum_{T \in \mathcal{T}(q,0)} F_T \right\|_{L^p(w_q)}^p \lesssim_\delta \int_{\mathbb{R}^2} \sum_{T \in \mathcal{T}(q,0)} |a_T|^p \chi_T \sqrt{w_q}.$$

Summing this over all $q \in \mathcal{Q}$ leads to the inequality

$$\sum_{q \in \mathcal{Q}} \| F_{q,0} \|_{L^p(w_q)}^p \lesssim_\delta \int_{\mathbb{R}^2} \sum_{T \in \mathcal{T}_R(F)} |a_T|^p \chi_T$$

$$\sim_\delta \int_{\mathbb{R}^2} \sum_{T \in \mathcal{T}_R(F)} |F_T|^p.$$

This contribution accounts for the term $O_\delta(1)$ in (12.5).

Next, fix $j \geq 1$. First, combining the local version of $l^2(L^p)$ decoupling (Theorem 10.1 and Proposition 9.15) at scale $R^{1/2}$ with Hölder's inequality and (12.6), we may write for each $q$

$$\| F_{q,j} \|_{L^p(w_q)} \lesssim_\epsilon R^\epsilon \left( \sum_{\omega \in \Omega(q,j)} \left\| \sum_{T \perp \omega} F_T \right\|_{L^p(w_q)}^2 \right)^{\frac{1}{2}}$$

$$\lesssim_{\epsilon,\delta} R^\epsilon (M2^{-j})^{\frac{1}{2}-\frac{1}{p}} \left( \sum_{\omega \in \Omega(q,j)} \left\| \sum_{T \perp \omega} F_T \right\|_{L^p(w_q)}^p \right)^{\frac{1}{p}}.$$

Summing over $q \in \mathcal{Q}$, we get

$$\left( \sum_{q \in \mathcal{Q}} \| F_{q,j} \|_{L^p(w_q)}^p \right)^{\frac{1}{p}} \lesssim_{\epsilon,\delta} R^\epsilon (M2^{-j})^{\frac{1}{2}-\frac{1}{p}} \left( \sum_{\omega \in \Omega} \left\| \sum_{T \perp \omega} F_T \right\|_{L^p(\sum_{q \in \mathcal{Q}(\omega,j)} w_q)}^p \right)^{\frac{1}{p}},$$

where

$$\mathcal{Q}(\omega,j) = \{q \in \mathcal{Q} : \omega \in \Omega(q,j)\}.$$

Let us now fix the arc $\omega \in \Omega$. We will argue that

$$\left\| \sum_{T \perp \omega} F_T \right\|_{L^p(\sum_{q \in \mathcal{Q}(\omega, j)} w_q)} \lesssim D_{R^{1/2}} 2^{j\left(\frac{1}{2}-\frac{1}{p}\right)} \left( \sum_{T \perp \omega} \|F_T\|_{L^p(\mathbb{R}^2)}^r \right)^{\frac{1}{p}}. \quad (12.8)$$

Combining this with the previous inequality shows that

$$\left( \sum_{q \in \mathcal{Q}} \|F_{q,j}\|_{L^p(w_q)}^p \right)^{\frac{1}{p}} \lesssim_\epsilon R^\epsilon D_{R^{1/2}} M^{\frac{1}{2}-\frac{1}{p}} \left( \sum_{T \in \mathcal{T}_R(F)} \|F_T\|_{L^p(\mathbb{R}^2)}^p \right)^{\frac{1}{p}}.$$

Note that this together with (12.7) completes the proof of (12.5).

The proof of (12.8) will rely on parabolic rescaling. Namely, if

$$\omega = \{(\xi, \xi^2) : |\xi - c| \leq R^{-\frac{1}{4}}\},$$

then the affine transformation

$$(\xi, \eta) \mapsto \left( \frac{\xi - c}{R^{-\frac{1}{4}}}, \frac{\eta - 2c\xi + c^2}{R^{-\frac{1}{2}}} \right)$$

maps $\mathcal{N}_\omega(R^{-1})$ to $\mathcal{N}_{\mathbb{P}^1}(R^{-1/2})$.

Let us partition $\mathbb{R}^2$ into $(R^{1/2}, R^{3/4})$-tubes $\Delta$ with direction $(-2c, 1)$. Note that

$$\sum_{\substack{q \in \mathcal{Q} \\ q \cap \Delta \neq \emptyset}} w_q \lesssim_\delta w_\Delta. \quad (12.9)$$

The affine transformation

$$A(x_1, x_2) = \left( \frac{x_1 + 2cx_2}{R^{1/4}}, \frac{x_2}{R^{1/2}} \right)$$

maps $(R^{1/2}, R)$-tubes $T$ to $(R^{1/4}, R^{1/2})$-tubes $\tilde{T}$, and each $\Delta$ to a square $\tilde{q}$ with side length $R^{1/4}$. Let $\tilde{\mathcal{T}}$ be the collection of all the tubes $\tilde{T} = A(T)$, with $T \perp \omega$. Let $\tilde{\mathcal{Q}}$ be the family of all the squares $\tilde{q} = A(\Delta)$, where $\Delta$ intersects at least one $q \in \mathcal{Q}(\omega, j)$. See Figure 12.2. The proof of (12.8) follows by applying the induction hypothesis to the family $\tilde{\mathcal{Q}}$. The details follow.

Let $\tilde{q} = A(\Delta)$ and $q \in \mathcal{Q}(\omega, j)$ be such that $\Delta \cap q \neq \emptyset$. The crucial geometric observation is that if $R^{\delta/2}\tilde{T}$ intersects $\tilde{q}$, then $R^{\delta/2}T$ must intersect $\Delta$, where $\tilde{T} = A(T)$. Since $R^\delta \gg 1$, this is easily seen to force $R^\delta T$ to intersect $q$. Thus, since $\omega \in \Omega(q, j)$, we must have that

$$|\{\tilde{T} \in \tilde{\mathcal{T}} : R^{\frac{\delta}{2}}\tilde{T} \cap \tilde{q}\}| < 2^{j+1}$$

for each $\tilde{q} \in \tilde{\mathcal{Q}}$.

Define

$$G_{\tilde{T}} = F_T \circ A^{-1}$$

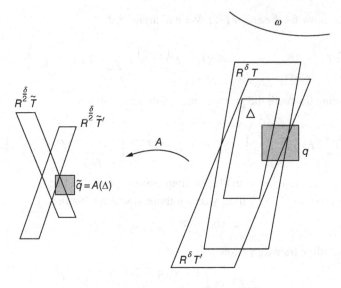

Figure 12.2 From scale $R$ to scale $R^{1/2}$.

and

$$G = \sum_{\tilde{T} \in \tilde{\mathcal{T}}} G_{\tilde{T}}.$$

We may then write, testing the definition of $D_{R^{1/2}}$ with $G$,

$$\sum_{\substack{\Delta : \Delta \cap q \neq \emptyset \\ \text{for some } q \in \mathcal{Q}(\omega, j)}} \left\| \sum_{T \perp \omega} F_T \right\|_{L^p(w_\Delta)}^p = R^{3/4} \sum_{\tilde{q} \in \tilde{\mathcal{Q}}} \left\| \sum_{\tilde{T} \in \tilde{\mathcal{T}}} G_{\tilde{T}} \right\|_{L^p(w_{\tilde{q}})}^p$$

$$\lesssim R^{3/4} (D_{R^{1/2}})^p 2^{j(\frac{p}{2}-1)} \sum_{\tilde{T} \in \tilde{\mathcal{T}}} \| G_{\tilde{T}} \|_{L^p(\mathbb{R}^2)}^p$$

$$= (D_{R^{1/2}})^p 2^{j(\frac{p}{2}-1)} \sum_{T \perp \omega} \| F_T \|_{L^p(\mathbb{R}^2)}^p.$$

The proof of (12.8) now follows from this and (12.9). $\qquad \square$

**Exercise 12.18** Extend the Theorem 12.17 to the higher-dimensional paraboloid $\mathbb{P}^{n-1}$.

## 12.5 Arbitrary Curves with Nonzero Torsion

Let us consider a curve in $\mathbb{R}^n$

$$\Phi(t) = (\phi_1(t), \ldots, \phi_n(t))$$

with $\phi_i \in C^{n+1}(I)$ for some compact interval $I \subset \mathbb{R}$. Assume that its torsion satisfies $\min_{t \in I} \tau(t) > 0$. One example of interest is the generalized moment curve

$$\Phi(t) = (t^{\beta_1}, \ldots, t^{\beta_n}),$$

for pairwise distinct $\beta_i > 0$. A simple computation shows that its torsion is

$$\tau(t) = \prod_{l=1}^{n} |\beta_l| \prod_{i \neq j} |\beta_i - \beta_j| t^{\sum_{i=1}^{n}(\beta_i - i)}. \tag{12.10}$$

If the exponent of $t$ is negative, the interval $I$ must not contain 0.

To keep things simpler in this section, let us see how to decouple the isotropic neighborhoods:

$$\mathcal{N}_{J,\Phi}(\delta) = \{(\phi_1(t), \phi_2(t) + s_2, \ldots, \phi_n(t) + s_n) : t \in J, \ |(s_2, \ldots, s_n)| < \delta\}.$$

Let $\mathrm{Dec}_\Phi(\delta, p)$ be the $l^2(L^p)$ decoupling constant associated with the sets $\theta = \mathcal{N}_{J,\Phi}(\delta)$, where $J$ runs over a partition of $I$ into intervals of size $\sim \delta^{1/n}$. Note that each $\theta$ is a curved tube (rather than an almost rectangular box) when $n \geq 3$.

**Theorem 12.19** *We have*

$$\mathrm{Dec}_\Phi(\delta, p) \lesssim_{\epsilon, \Phi} \begin{cases} \delta^{-\epsilon}, & 2 \leq p \leq n(n+1) \\ \delta^{\frac{n+1}{2p} - \frac{1}{2n} - \epsilon}, & n(n+1) < p \leq \infty. \end{cases} \tag{12.11}$$

*Proof* When $\Phi$ is the moment curve $\Gamma_n$, the upper bound (12.11) is a direct consequence of Theorem 11.16. Indeed, using an affine change of variables, we may assume that $I = [0, 1]$. Moreover,

$$\mathcal{N}_{[0,1],\Gamma_n}(\delta^n) \subset \Gamma_n(\delta).$$

The proof of the general case is very similar to that of Theorem 12.1, using the moment curve as a reference. More precisely, we will prove that for each $p \geq 2$ there is $C$ independent of $\delta$ such that

$$\mathrm{Dec}_\Phi(\delta, p) \leq C \mathrm{Dec}_\Phi(\delta^{\frac{n}{n+1}}, p) \mathrm{Dec}_{\Gamma_n}(\delta^{\frac{1}{n+1}}, p). \tag{12.12}$$

Inequality (12.11) will follow by iterating (12.12) as in the proof of Theorem 12.1.

To prove (12.12), let us observe first that if $\mathrm{supp}(\widehat{F}) \subset \mathcal{N}_\Phi(\delta)$, then

$$\|F\|_{L^p(\mathbb{R}^n)} \lesssim \mathrm{Dec}_\Phi(\delta^{\frac{n}{n+1}}, p) \left( \sum_{\substack{H \subset I \\ |H| = \delta^{\frac{1}{n+1}}}} \|\mathcal{P}_{\mathcal{N}_{H,\Phi}(\delta)} F\|_{L^p(\mathbb{R}^n)}^2 \right)^{1/2}. \tag{12.13}$$

Fix an interval $H = [t_0, t_0 + \delta^{1/(n+1)}] \subset I$. Let $L = L_{t_0}$ be the linear map on $\mathbb{R}^n$ such that $L(\Phi^j(t_0)) = j! \, \mathbf{e}_j$ for each $1 \le j \le n$. Using Taylor's formula, we find that

$$L(\Phi(t)) = (t - t_0, (t - t_0)^2, \ldots, (t - t_0)^n) + L(\Phi(t_0)) + O(\delta), \quad t \in H.$$

This shows that $\mathcal{N}_{H,\Phi}(\delta)$ is a subset of the $O(\delta)$-neighborhood a segment of length $\sim \delta^{1/(n+1)}$ of some nonsingular affine image of $\Gamma_n$. Using rescaling for the moment curve, we find that for each $F$ with $\widehat{F}$ supported on $\mathcal{N}_{H,\Phi}(\delta)$, we have

$$\|F\|_{L^p(\mathbb{R}^n)} \lesssim \mathrm{Dec}_{\Gamma_n}(\delta^{\frac{1}{n+1}}, p) \left( \sum_{\substack{J \subset H \\ |J| = \delta^{1/n}}} \|\mathcal{P}_{\mathcal{N}_{J,\Phi}(\delta)} F\|_{L^p(\mathbb{R}^n)}^2 \right)^{1/2}. \quad (12.14)$$

Inequality (12.12) now follows by combining (12.13) and (12.14). $\qquad \square$

## 12.6 Real Analytic Curves

In this section, we explain how to decouple planar curves in the vicinity of points with zero curvature (same as zero torsion in this context). A motivating example is the polynomial curve $\{(t, t^n), \, t \in [0, 1]\}$. If $n \ge 3$, this has zero curvature at $t = 0$, so Theorem 12.1 is not immediately applicable.

In general, let $\phi \colon [0, 1] \to \mathbb{R}$ be a real analytic function that is not affine. Its second derivative $\phi''$ will have only finitely many zeros $t_1 < t_2 < \cdots < t_N$, each with some finite order. It will suffice to understand how to decouple the graph of $\phi$ on an interval of the form $[t_i - \Delta, t_i + \Delta]$, assuming $t_i$ is the only zero of $\phi''$ in this interval. Indeed, on the rest of the interval $[0, 1]$ we will have $\inf |\phi''| > 0$, and Theorem 12.1 is applicable.

Let us fix $t_0$ with $\phi''(t_0) = \cdots = \phi^{(n-1)}(t_0) = 0$ and $\phi^{(n)}(t_0) \ne 0$, for some $n \ge 3$. By performing a rotation, we may as well assume that $\phi'(t_0) = 0$. Let $I_0 = [t_0 - \Delta, t_0 + \Delta]$ be an interval where $\phi''$ has no zeros other than $t_0$. We will also assume that $\Delta$ is smaller than the radius of convergence of the power series of $\phi$ at $t_0$. In particular, we have

$$\phi(t) = \frac{\phi^{(n)}(t_0)}{n!}(t - t_0)^n + O(|t - t_0|^{n+1}), \quad t \in I_0. \quad (12.15)$$

We will aim to prove a genuine $l^2$ decoupling for $\phi$ on $I_0$. It is slightly more convenient to work with the extension operator:

$$E^\phi f(x_1, x_2) = \int f(t) e(t x_1 + \phi(t) x_2) \, dt.$$

More precisely, given $\delta$ we want to find a partition of $I_0$ into intervals $J$ such that for each $f\colon I_0 \to \mathbb{C}$, $2 \le p \le 6$ and each ball $B \subset \mathbb{R}^2$ with radius at least $\delta^{-2}$, we have

$$\|E^\phi f\|_{L^p(B)} \lesssim_\epsilon \delta^{-\epsilon} \left( \sum_J \|E_J^\phi f\|_{L^p(w_B)}^2 \right)^{1/2}.$$

We are interested in the case when this partition is essentially maximal. This is tantamount to asking that each neighborhood

$$\mathcal{N}_J(\delta^2) = \{(t, \phi(t) + s)\colon t \in J,\ |s| \le \delta^2\}$$

is an almost rectangle. To achieve such a decomposition, we first split $I_0$ dyadically, in order to isolate regions with different curvature. More precisely,

$$I_0 = \bigcup_{k=0}^{O(\log(\delta^{-1}))} J_k$$

with $J_0 = [t_0 - \delta^{2/n}, t_0 + \delta^{2/n}]$ and $J_k = \{t\colon |t - t_0| \sim 2^k \delta^{2/n}\}$ for $k \ge 1$. Note that while $\mathcal{N}_{J_0}(\delta^2)$ is an almost rectangle, the sets $\mathcal{N}_{J_k}(\delta^2)$ are more and more curved as $k$ increases. This can be seen since $|\phi''(t)| \sim (2^k \delta^{2/n})^{n-2}$ for $t \in J_k$. This suggests that $\phi$ can be decoupled further on each $J_k$. Since curvature is still small for small values of $k$, we need to use rescaling.

**Proposition 12.20** *Let $J_{k,l}$ be intervals of size $\delta |J_k|^{(2-n)/2}$ partitioning $J_k$. Then for each $2 \le p \le 6$, each $f$ supported on $J_k$, and each ball $B$ of radius at least $\delta^{-2}$, we have*

$$\|E^\phi f\|_{L^p(B)} \lesssim_\epsilon \delta^{-\epsilon} \left( \sum_l \|E_{J_{k,l}}^\phi f\|_{L^p(w_B)}^2 \right)^{1/2}. \tag{12.16}$$

*Proof* It suffices to consider $f$ supported on the right half of $J_k$, that we denote by $[a_k, a_k + \delta_k]$. Changing variables, we get

$$|E^\phi f(x_1, x_2)| = \delta_k \left| \int_0^1 f(s\delta_k + a_k) e \left( s x_1 \delta_k + \frac{\phi(s\delta_k + a_k)}{\delta_k^n} x_2 \delta_k^n \right) ds \right|.$$

Thus, writing $\phi_k(s) = \phi(s\delta_k + a_k)/\delta_k^n$, $g(s) = f(s\delta_k + a_k)$ and $T_k(x_1, x_2) = (x_1 \delta_k, x_2 \delta_k^n)$, we have

$$\|E^\phi f\|_{L^p(B)} = \delta_k^{1 - \frac{n+1}{p}} \|E^{\phi_k} g\|_{L^p(T_k(B))}.$$

The crucial observation is that the rescaled curve $\phi_k$ satisfies

$$|\phi_k''(s)| \sim 1, \quad s \in [0, 1].$$

Consider now a finitely overlapping cover $\mathcal{B}_k$ of $T_k(B)$ with balls $B'$ of radius $\delta_k^n \delta^{-2}$. Invoking Theorem 12.1 for $\phi_k$ on each $B'$, we get

$$\|E^{\phi_k}g\|_{L^p(B')} \lesssim_\epsilon \delta^{-\epsilon} \left( \sum_{|I|=\delta\delta_k^{-n/2}} \|E_I^{\phi_k}g\|_{L^p(w_{B'})}^2 \right)^{1/2}.$$

Summation leads to

$$\|E^{\phi_k}g\|_{L^p(T_k(B))} \lesssim_\epsilon \delta^{-\epsilon} \left( \sum_{|I|=\delta\delta_k^{-n/2}} \|E_I^{\phi_k}g\|_{L^p(\sum_{B'\in\mathcal{B}_k} w_{B'})}^2 \right)^{1/2}.$$

Finally, (12.16) will follow by rescaling back this inequality.      $\square$

It is easy to check that each $\mathcal{N}_{J_{k,l}}(\delta^2)$ is an almost rectangle. The following result is now a consequence of the triangle inequality and Proposition 12.20. Let $\mathcal{J}_\delta$ be the partition of $I_0$ into the intervals $J_0$ and $J_{k,l}$, for each $1 \le k \le O(\log(\delta^{-1}))$.

**Corollary 12.21** *For each $2 \le p \le 6$, each $f : I_0 \to \mathbb{C}$, and each ball $B \subset \mathbb{R}^2$ with radius at least $\delta^{-2}$, we have*

$$\|E^\phi f\|_{L^p(B)} \lesssim_\epsilon \delta^{-\epsilon} \left( \sum_{J \in \mathcal{J}_\delta} \|E_J^\phi f\|_{L^p(w_B)}^2 \right)^{1/2}.$$

A more systematic study of decoupling for planar curves appears in [15]. The analysis in this section may be easily extended to real analytic curves in $\mathbb{R}^n$, having zero torsion at finitely many points. Decouplings for real analytic surfaces of revolution in $\mathbb{R}^3$ are proved in [25].

Let us close with a discussion on the optimality of $L^6$ in the decoupling theory of planar curves. Theorem 12.1 (see also (9.7)) shows that for each $\phi \in C^3([0,1])$ with $\inf_{t\in[0,1]} |\phi''(t)| > 0$, we have the following genuine $l^2$ decoupling on balls of radius at least $\delta^{-2}$:

$$\|E^\phi f\|_{L^6(B)} \lesssim_\epsilon \delta^{-\epsilon} \left( \sum_{|J|=\delta} \|E_J^\phi f\|_{L^6(w_B)}^2 \right)^{1/2}.$$

A natural question is whether there is a curve with genuine $l^q(L^p)$ decoupling for some $p > 6$ and some $q \ge 2$. Could we enlarge the range of $p$ by decoupling on balls with radius much larger than $\delta^{-2}$? This is partly motivated by questions from number theory regarding exponential sums; see (13.14). The following result provides a negative answer.

**Proposition 12.22** *Let $q \geq 2$, $p > 6$ and $\phi$ be as before. Let $D_\phi(\delta, p, q)$ be such that the following inequality holds for each $f : [0, 1] \to \mathbb{C}$*

$$\|E^\phi f\|_{L^p(B(0,1))} \leq D_\phi(\delta, p, q)\delta^{\frac{1}{q}-\frac{1}{2}} \left( \sum_{|J|=\delta} \|E_J^\phi f\|_{L^p(\mathbb{R}^2)}^q \right)^{1/q}.$$

*Then there is $\epsilon_p > 0$ such that*

$$D_\phi(\delta, p, q) \gtrsim \delta^{-\epsilon_p}.$$

*Proof* We test the inequality with $f = 1_{[0,1]}$. First, note that

$$|E^\phi f(x_1, x_2)| \gtrsim 1$$

if $|(x_1, x_2)| \ll 1$, and thus $\|E^\phi f\|_{L^p(B(0,1))} \gtrsim 1$. Also, a change of variables reveals that

$$|E_{[a,a+\delta]}^\phi f(x_1, x_2)| = \delta|E^\psi g(x_1\delta, x_2\delta^2)|,$$

where $\psi(s) = \delta^{-2}\phi(a + s\delta)$ and $g(s) = f(a + s\delta)$. Applying the Restriction Theorem 1.14 to $E^\psi$ and $g$, we get

$$\|E_{[a,a+\delta]}^\phi f\|_{L^p(\mathbb{R}^2)} = \delta^{1-\frac{3}{p}}\|E^\psi g\|_{L^p(\mathbb{R}^2)} \lesssim \delta^{1-\frac{3}{p}}.$$

Combining these estimates, it follows that we may take $\epsilon_p = 1/2 - 3/p$. $\quad\square$

**Exercise 12.23** Let $\phi$, $I_0$ and $n$ be as in (12.15) and assume that $2 \leq n \leq 4$. Prove that the following genuine $l^6(L^6)$ decoupling holds on each ball $B$ of radius at least $\delta^{-2}$:

$$\|E^\phi f\|_{L^p(B)} \lesssim_\epsilon \delta^{-\epsilon-\frac{1}{3}} \left( \sum_{|J|=\delta} \|E_J^\phi f\|_{L^6(w_B)}^6 \right)^{1/6}.$$

# 13

## Applications of Decoupling

The main value of decoupling lies in its extraordinary potential for applications to a wide range of topics. On the discrete side, it often implies sharp $L^p$ estimates for exponential sums. Examples from this category presented in this chapter include Strichartz estimates for the Schrödinger equation on tori and the proof of the main conjecture in Vinogradov's Mean Value Theorem. On the continuous side, we will explore connections to the Restriction Conjecture and will close the book with discussing local smoothing for the wave equation, a topic of historical significance for decoupling.

### 13.1 Decoupling and Exponential Sums

Given a $d$-dimensional manifold $\mathcal{M}$ in $\mathbb{R}^n$, we consider its neighborhood $\tilde{\mathcal{N}}_{\mathcal{M}}(\delta) \subset \mathbb{R}^n$ and the partition $\Theta_{\mathcal{M}}(\delta)$ into sets $\theta$ as in Problem 9.4. We denote as before by $\mathrm{Dec}(\Theta_{\mathcal{M}}(\delta), p, q)$ the $l^q(L^p)$ decoupling constant associated with this partition.

Recall that

$$\theta = \left(\tau \times \mathbb{R}^{n-d}\right) \cap \tilde{\mathcal{N}}_{\mathcal{M}}(\delta),$$

for some almost rectangular box $\tau \subset \mathbb{R}^d$.

For each $\theta$, let us pick a point $\xi_\theta \in \mathcal{M}$ such that $B(\xi_\theta, \delta) \subset \theta$. In particular, the points $\xi_\theta$ will be $\delta$-separated. If $\bar{\xi}_\theta \in \mathbb{R}^d$ represents the first $d$ coordinates of $\xi_\theta$, this requirement is equivalent with $B_d(\bar{\xi}_\theta, \delta) \subset \tau$. See Figure 13.1.

The following result was proved in [20] for $q = 2$. It opened the door to a large host of applications to number theory, as we shall see in the next few sections.

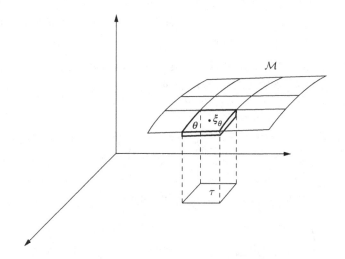

Figure 13.1 The partition $\Theta_{\mathcal{M}}(\delta)$.

**Theorem 13.1** *For each $p, q \geq 2$, each $a_\theta \in \mathbb{C}$, and each ball $B_R \subset \mathbb{R}^n$ with radius $R \geq \delta^{-1}$, we have*

$$\left( \frac{1}{|B_R|} \int_{B_R} \left| \sum_\theta a_\theta e(\xi_\theta \cdot x) \right|^p dx \right)^{\frac{1}{p}} \lesssim \mathrm{Dec}(\Theta_{\mathcal{M}}(\delta), p, q) |\Theta_{\mathcal{M}}(\delta)|^{\frac{1}{2} - \frac{1}{q}} \|a_\theta\|_{l^q}.$$

*Proof* It suffices to consider the case $R = \delta^{-1}$ and $B_R = B(0, R)$. Let $\phi \colon \mathbb{R}^n \to \mathbb{C}$ satisfy $\mathrm{supp}\,(\widehat{\phi}) \subset B(0, 1)$ and $|\phi(x)| \geq 1$ for $x \in B(0, 1)$. Denote by $\phi_R$ the rescaled function $\phi(\cdot / R)$ and write

$$F(x) = \phi_R(x) \sum_\theta a_\theta e(\xi_\theta \cdot x).$$

Since the Fourier transform of $\phi_R(x) e(\xi_\theta \cdot x)$ is supported on $\theta$, we have

$$\left( \int_{B(0,R)} \left| \sum_\theta a_\theta e(\xi_\theta \cdot x) \right|^p dx \right)^{\frac{1}{p}}$$

$$\leq \|F\|_{L^p(\mathbb{R}^n)}$$

$$\leq \mathrm{Dec}(\Theta_{\mathcal{M}}(\delta), p, q) |\Theta_{\mathcal{M}}(\delta)|^{\frac{1}{2} - \frac{1}{q}} \left( \sum_\theta \|a_\theta \phi_R\|_{L^p(\mathbb{R}^n)}^q \right)^{\frac{1}{q}}$$

$$\lesssim R^{\frac{n}{p}} \mathrm{Dec}(\Theta_{\mathcal{M}}(\delta), p, q) |\Theta_{\mathcal{M}}(\delta)|^{\frac{1}{2} - \frac{1}{q}} \|a_\theta\|_{l^q}. \qquad \square$$

As an immediate consequence, we get the following example of $L^p$-square root cancellation for exponential sums.

**Corollary 13.2** (Reverse Hölder) *Assume either that* $\mathrm{Dec}(\Theta_{\mathcal{M}}(\delta), p, 2)$ $\lesssim_\epsilon \delta^{-\epsilon}$ *or that* $\mathrm{Dec}(\Theta_{\mathcal{M}}(\delta), p, q) \lesssim_\epsilon \delta^{-\epsilon}$ *(for some* $q > 2$*) and that* $|a_\theta|$ *is essentially constant. Let* $w_{B_R}$ *be the weight from (9.6). Then for* $R \geq \delta^{-1}$ *we have*

$$\left( \frac{1}{|B_R|} \int_{\mathbb{R}^n} \left| \sum_\theta a_\theta e(\xi_\theta \cdot x) \right|^p w_{B_R}(x)\, dx \right)^{\frac{1}{p}} \lesssim_\epsilon \delta^{-\epsilon} \|a_\theta\|_{l^2}. \tag{13.1}$$

A slight variant of Exercise 1.31 implies that

$$\left( \frac{1}{|B_R|} \int_{\mathbb{R}^n} \left| \sum_\theta a_\theta e(\xi_\theta \cdot x) \right|^2 w_{B_R}(x)\, dx \right)^{\frac{1}{2}} \sim \|a_\theta\|_{l^2},$$

with implicit constant independent of $R$ and $B_R$. Thus, apart from the $\delta^{-\epsilon}$ term, (13.1) is a reverse Hölder's inequality. Moreover, it follows that the left-hand side of (13.1) is essentially independent of $B_R$, in the sense that it is comparable to $\|a_\theta\|_{l^2}$, up to a $O(\delta^{-\epsilon})$ multiplicative factor. This is a strong manifestation of the almost periodicity of the exponential sums.

In the following sections, we will present a few concrete applications of Corollary 13.2.

## 13.2 A Number Theoretic Approach to Lower Bounds

Let $\Theta_{\mathbb{P}^{n-1}}(\delta)$ be the standard partition of the $\delta$-neighborhood $\mathcal{N}_{\mathbb{P}^{n-1}}(\delta)$ of the paraboloid introduced in Chapter 10. Theorem 10.1 proves that

$$\mathrm{Dec}(\Theta_{\mathbb{P}^{n-1}}(\delta), p) \lesssim_\epsilon \delta^{-\epsilon}$$

for $2 \leq p \leq 2(n+1)/(n-1)$. The reader may have wondered whether $\mathrm{Dec}(\Theta_{\mathbb{P}^{n-1}}(\delta), p)$ does indeed depend on $\delta$. We will now prove that the answer is "yes" when $p = 2(n+1)/(n-1)$. We expect the decoupling constants to be independent of $\delta$ in the range $2 < p < 2(n+1)/(n-1)$, but proving this remains a difficult open problem.

To achieve our goal, we will reduce matters to exponential sum estimates, with the aid of Theorem 13.1. Throughout this section, we assume $q$ is an odd integer. For $1 \leq a < q$ with $(a, q) = 1$, $0 \leq b < q$ and $N \geq 1$, we introduce the incomplete Gauss sum

$$G_N(a,b,q) = \sum_{k=1}^{N} e\left(\frac{bk}{q} + \frac{ak^2}{q}\right).$$

Recall one of the classical examples of square root cancellation:

$$|G_q(a,0,q)| = q^{1/2}. \tag{13.2}$$

While this simple equation would suffice for our purposes, we choose to present things with slightly greater generality. In particular, we will also use the following result of Hardy–Littlewood [65]

$$\max_{N \leq q} |G_N(a,0,q)| \leq Cq^{1/2}. \tag{13.3}$$

Since $q$ is odd, $2a$ is invertible mod $q$ with inverse $c$. Then, using (13.2), (13.3), and periodicity, we have

$$\left|\sum_{k=1}^{N} e\left(\frac{bk}{q} + \frac{ak^2}{q}\right)\right| = \left|\sum_{k=1}^{N} e\left(\frac{a}{q}(k+bc)^2\right)\right| \leq \left(C + \frac{N}{q}\right)q^{\frac{1}{2}} \tag{13.4}$$

for each $N \geq 1$, and

$$\left|\sum_{k=1}^{N} e\left(\frac{bk}{q} + \frac{ak^2}{q}\right)\right| = \left|\sum_{k=1}^{N} e\left(\frac{a}{q}(k+bc)^2\right)\right| \in \left[\frac{N}{4q^{\frac{1}{2}}}, \frac{4N}{q^{\frac{1}{2}}}\right] \tag{13.5}$$

for each $N \geq (4C+1)q$.

We use these estimates to prove a few auxiliary results.

**Lemma 13.3** *For each $l \geq 0$ and $N \geq Cq$, we have*

$$\left|\sum_{k=1}^{N} k^l e\left(\frac{bk}{q} + \frac{ak^2}{q}\right)\right| < \frac{10N^{l+1}}{q^{\frac{1}{2}}}.$$

*Proof* Using partial summation and (13.4), we get

$$\left|\sum_{k=1}^{N} k^l e\left(\frac{bk}{q} + \frac{ak^2}{q}\right)\right| \leq N^l |G_N(a,b,q)| + \sum_{n=1}^{N-1} |G_n(a,b,q)|((n+1)^l - n^l)$$

$$\leq \frac{10N^{l+1}}{q^{\frac{1}{2}}}. \qquad \square$$

**Proposition 13.4** *Let $N \geq (4C+1)q$ and assume that $|x_1 - b/q| \leq 1/10^{10}N$ and $|x_2 - a/q| < 1/10^{10}N^2$. Then*

$$\left| \sum_{k=1}^{N} e(x_1 k + x_2 k^2) \right| \sim \frac{N}{q^{1/2}}$$

*uniformly over $a, b, q, x_1, x_2$, and $N$.*

*Proof* We use Taylor's formula for

$$f(x_1, x_2) = \sum_{k=1}^{N} e(x_1 k + x_2 k^2)$$

near $(b/q, a/q)$. Lemma 13.3 implies that

$$\left| \partial_{x_1}^{l_1} \partial_{x_2}^{l_2} f\left(\frac{b}{q}, \frac{a}{q}\right) \right| \leq 10(2\pi)^{l_1 + l_2} \frac{N^{l_1 + 2l_2 + 1}}{q^{\frac{1}{2}}}.$$

It follows that if $(x_1, x_2)$ are like in our hypothesis, we have

$$\left| f(x_1, x_2) - f\left(\frac{b}{q}, \frac{a}{q}\right) \right| \leq \frac{N}{q^{\frac{1}{2}}} \sum_{(l_1, l_2) \neq (0,0)} \frac{1}{l_1! l_2!} \left(\frac{2\pi}{10^{10}}\right)^{l_1 + l_2} \leq \frac{N}{10 q^{\frac{1}{2}}}.$$

It suffices to combine this with (13.5).                              $\square$

**Lemma 13.5** *Let $\varphi$ be Euler's totient function. Then*

$$\sum_{\substack{q \leq N \\ q \text{ odd}}} \frac{\varphi(q)}{q^2} \sim \log N.$$

*Proof* For $Q$ odd, let

$$S_Q = \sum_{\substack{q=1 \\ q \text{ odd}}}^{Q} \varphi(q).$$

Using crude estimates, we write

$$S_Q = \sum_{\substack{q=1 \\ q \text{ odd}}}^{Q} \sum_{a=1}^{q-1} 1 - \sum_{\substack{q=1 \\ q \text{ odd}}}^{Q} \sum_{\substack{1 \le a \le q-1 \\ (a,q)>1}} 1$$

$$\ge \frac{(Q+1)(Q-1)}{4} - \sum_{\substack{p \text{ prime} \\ p \ge 3}} \sum_{\substack{1 \le q \le Q \\ p|q \\ q \text{ odd}}} \sum_{\substack{1 \le a \le q-1 \\ p|a}} 1$$

$$\ge \frac{(Q+1)(Q-1)}{4} - \sum_{\substack{p \text{ prime} \\ p \ge 3}} \sum_{\substack{1 \le q \le Q \\ p|q \\ q \text{ odd}}} \frac{Q-1}{p}$$

$$\ge \frac{(Q+1)(Q-1)}{4} - \frac{(Q+1)(Q-1)}{2} \sum_{\substack{p \text{ prime} \\ p \ge 3}} \frac{1}{p^2}$$

$$> \frac{(Q+1)(Q-1)}{2} \left( \frac{1}{2} - \frac{1}{9} - \int_4^\infty \frac{1}{x^2} \, dx \right)$$

$$\gtrsim Q^2.$$

Since $\varphi(q) < q$, we conclude that in fact $S_Q \sim Q^2$. Using this and summation by parts, we get for each odd $N$

$$\sum_{\substack{q \le N \\ q \text{ odd}}} \frac{\varphi(q)}{q^2} = \frac{S_N}{N^2} + \sum_{\substack{Q=1 \\ Q \text{ odd}}}^{N-2} S_Q \left( \frac{1}{Q^2} - \frac{1}{(Q+2)^2} \right) \sim \log N. \qquad \square$$

We now combine all these ingredients to prove the main result of this section.

**Theorem 13.6** *For each $n \ge 2$, we have*

$$\left\| \sum_{k_1=1}^{N} \cdots \sum_{k_{n-1}=1}^{N} e(k_1 x_1 + \cdots + k_{n-1} x_{n-1} + (k_1^2 + \cdots + k_{n-1}^2) x_n) \right\|_{L^{\frac{2(n+1)}{n-1}}(\mathbb{T}^n)}$$

$$\gtrsim N^{\frac{n-1}{2}} (\log N)^{\frac{n-1}{2(n+1)}}.$$

*Proof* Let $1 \le a < q$, $(a,q) = 1$, $0 \le b_1, \ldots, b_{n-1} < q$, with $q$ an odd number satisfying $N \ge (4C + 1)q$. Define the major arc:

$$\mathcal{M}(a, b_1, \ldots, b_{n-1}, q)$$

$$= \left\{ (x_1, \ldots, x_n) : \max_{1 \le i \le n-1} \left| x_i - \frac{b_i}{q} \right| \le \frac{1}{10^{10} N}, \left| x_n - \frac{a}{q} \right| \le \frac{1}{10^{10} N^2} \right\}.$$

Proposition 13.4 shows that

$$\left| \sum_{k_1=1}^{N} \cdots \sum_{k_{n-1}=1}^{N} e(k_1 x_1 + \cdots + k_{n-1} x_{n-1} + (k_1^2 + \cdots + k_{n-1}^2) x_n) \right|$$

$$= \prod_{i=1}^{n-1} \left| \sum_{k=1}^{N} e(x_i k + x_n k^2) \right|$$

$$\gtrsim \frac{N^{n-1}}{q^{\frac{n-1}{2}}}.$$

Moreover, the major arcs are easily seen to be pairwise disjoint. Combining these with Lemma 13.5, we conclude as follows:

$$\left\| \sum_{k_1=1}^{N} \cdots \sum_{k_{n-1}=1}^{N} e(k_1 x_1 + \cdots + k_{n-1} x_{n-1} + (k_1^2 + \cdots + k_{n-1}^2) x_n) \right\|_{L^{\frac{2(n+1)}{n-1}}(\mathbb{T}^n)}^{\frac{2(n+1)}{n-1}}$$

$$\gtrsim \sum_{\substack{q \lesssim N \\ q \text{ odd}}} \varphi(q) q^{n-1} \left( \frac{N^{n-1}}{q^{\frac{n-1}{2}}} \right)^{\frac{2(n+1)}{n-1}} \frac{1}{N^{n+1}}$$

$$= N^{n+1} \sum_{\substack{q \lesssim N \\ q \text{ odd}}} \frac{\varphi(q)}{q^2}$$

$$\sim N^{n+1} \log N. \qquad \square$$

The lower bound in Theorem 13.6 is known to be sharp when $n = 2, 3$. When $n = 2$, very precise asymptopics are known for $\| \sum_{k=1}^{N} e(kx_1 + k^2 x_2) \|_{L^6(\mathbb{T}^2)}$: see [16]. The case $n = 3$ will be discussed in Corollary 13.26.

**Corollary 13.7** *For each $n \ge 2$, we have the lower bound*

$$\mathrm{Dec}\left( \Theta_{\mathbb{P}^{n-1}}(\delta), \frac{2(n+1)}{n-1} \right) \gtrsim \left( \log \left( \frac{1}{\delta} \right) \right)^{\frac{n-1}{2(n+1)}}.$$

*Proof* We let $\delta = 1/N^2$. Each point of the form

$$\left(\frac{k_1}{N}, \ldots, \frac{k_{n-1}}{N}, \frac{k_1^2 + \cdots + k_{n-1}^2}{N^2}\right)$$

with $1 \leq k_1, \ldots, k_{n-1} \leq N$ is contained in some $\theta \in \Theta_{\mathbb{P}^{n-1}}(\delta)$. Let us denote this point by $\xi_\theta$. Using the inequality in Theorem 13.6 and invoking periodicity shows that

$$\left\|\sum_\theta e(\xi_\theta \cdot x)\right\|_{L_\sharp^{\frac{2(n+1)}{n-1}}([-N^2, N^2]^n)} \gtrsim \left(\log\left(\frac{1}{\delta}\right)\right)^{\frac{n-1}{2(n+1)}} |\Theta_{\mathbb{P}^{n-1}}(\delta)|^{\frac{1}{2}}.$$

Combining this with Theorem 13.1 finishes the proof.    $\square$

## 13.3 Strichartz Estimates for the Schrödinger Equation on Tori

An application of Theorem 10.1 and Corollary 13.2 to $\mathcal{M} = \mathbb{P}^{n-1}$ leads to the following result.

**Theorem 13.8** *Let* $2 \leq p \leq 2(n+1)/(n-1)$. *For each collection* $\Lambda$ *consisting of* $1/R$-*separated points on* $\mathbb{P}^{n-1}$, *each ball* $B_{R'}$ *of radius* $R' \geq R^2$ *in* $\mathbb{R}^n$, *and each* $a_\lambda \in \mathbb{C}$, *we have*

$$\left\|\sum_{\lambda \in \Lambda} a_\lambda e(\lambda \cdot x)\right\|_{L_\sharp^p(B_{R'})} \lesssim_\epsilon R^\epsilon \|a_\lambda\|_{l^2}. \tag{13.6}$$

Let us compare this result with the discrete restriction estimate from Exercise 1.31. That estimate was concerned with averages over balls of radius $R$. More precisely, for each $2 \leq p \leq 2(n+1)/(n-1)$, we have

$$\left\|\sum_{\lambda \in \Lambda} a_\lambda e(\lambda \cdot x)\right\|_{L_\sharp^p(B_R)} \lesssim R^{\frac{n-1}{2}(\frac{1}{2} - \frac{1}{p})} \|a_\lambda\|_{l^2}, \tag{13.7}$$

and the exponent of $R$ is sharp.

We next show that Theorem 13.8 implies (essentially) sharp Strichartz estimates for the Schrödinger equation in the periodic and quasiperiodic case. It may help to start by describing the general framework in which this type of estimate is investigated.

Let $(M, g)$ be a compact Riemannian manifold of dimension $n - 1 \geq 1$, and let $\Delta_g$ be the associated Laplace–Beltrami operator. The free Schrödinger equation in this context reads as follows:

$$\begin{cases} Cu_t(x,t) = \Delta_g u(x,t), & (x,t) \in M \times \mathbb{R} \\ u(x,0) = \phi(x), & x \in M. \end{cases}$$

Here $C$ is an arbitrary nonzero complex number.

Let $I \subset \mathbb{R}$ be an interval. A Strichartz estimate is one of the form

$$\|u\|_{L_x^p L_t^r(M \times I)} \lesssim \|\phi\|_{L^q(M)}.$$

Of particular interest is the case $p = r$, $q = 2$. When $M = \mathbb{R}^{n-1}$ (with the Euclidean metric and standard Laplace operator $\Delta$), $I = \mathbb{R}$, and $\hat{\phi}$ is compactly supported in frequency, Corollary 1.17 shows that for $p \geq 2(n+1)/(n-1)$

$$\|u\|_{L^p(\mathbb{R}^n)} \lesssim \|\phi\|_{L^2(\mathbb{R}^{n-1})}.$$

Moreover, Exercise 1.24 shows that when $p = 2(n+1)/(n-1)$ this inequality holds true for arbitrary $\phi \in L^2(\mathbb{R}^{n-1})$.

Let us now investigate the family of flat tori

$$\mathbb{T}_\alpha = \prod_{i=1}^{n-1} \mathbb{R}/(\alpha_i \mathbb{Z}),$$

for $\alpha = (\alpha_1, \dots, \alpha_{n-1})$ with $1/2 < \alpha_1, \dots, \alpha_{n-1} < 2$. When $\alpha = (1, \dots, 1)$, we recover the standard torus $\mathbb{T}_\alpha = \mathbb{T}^{n-1}$. If we identify the set $\mathbb{T}_\alpha$ with $\mathbb{T}^{n-1}$, the Laplacian of $\phi \colon \mathbb{T}^{n-1} \to \mathbb{C}$ in the new coordinates is given by

$$\Delta_\alpha \phi(x_1, \dots, x_{n-1}) = -4\pi^2 \sum_{\xi_1, \dots, \xi_{n-1} \in \mathbb{Z}} (\xi_1^2 \alpha_1 + \cdots + \xi_{n-1}^2 \alpha_{n-1}) \hat{\phi}$$

$$\times (\xi_1, \dots, \xi_{n-1}) e(\xi_1 x_1 + \cdots + \xi_{n-1} x_{n-1}),$$

and the free Schrödinger equation on $\mathbb{T}_\alpha$ reads as follows:

$$\begin{cases} 2\pi i u_t(x,t) = \Delta_\alpha u(x,t), & (x,t) \in \mathbb{T}^{n-1} \times \mathbb{R} \\ u(x,0) = \phi(x), & x \in \mathbb{T}^{n-1}. \end{cases}$$

The solution is easily seen to be

$$u(x_1, \dots, x_{n-1}, t) = \sum_{\xi_1, \dots, \xi_{n-1} \in \mathbb{Z}} \hat{\phi}(\xi_1, \dots, \xi_{n-1})$$

$$\times e(x_1 \xi_1 + \cdots + x_{n-1} \xi_{n-1} + t(\xi_1^2 \alpha_1 + \cdots + \xi_{n-1}^2 \alpha_{n-1})).$$

We have the following consequence of Theorem 13.8.

**Theorem 13.9** (Strichartz estimates for flat tori, [29]) *Let* $\phi \in L^2(\mathbb{T}^{n-1})$ *with* supp $(\hat{\phi}) \subset [-N, N]^{n-1}$. *Then for each* $\epsilon > 0$, *we have*

$$\|u\|_{L^p(\mathbb{T}^{n-1}\times[0,1])} \lesssim_\epsilon \begin{cases} N^\epsilon \|\phi\|_2, & 2 \le p \le \frac{2(n+1)}{n-1} \\ N^{\frac{n-1}{2} - \frac{n+1}{p} + \epsilon} \|\phi\|_2, & p \ge \frac{2(n+1)}{n-1}. \end{cases} \tag{13.8}$$

*Proof* The cases $p = 2$ and $p = \infty$ are trivial. Due to Hölder's inequality, it suffices to prove the estimate for the critical exponent $p = 2(n + 1)/(n - 1)$

$$\int_{\mathbb{T}^{n-1}\times[0,1]} |u(x,t)|^{\frac{2(n+1)}{n-1}} \, dxdt \lesssim_\epsilon N^\epsilon \|\phi\|_2^{\frac{2(n+1)}{n-1}}.$$

For $-N \le \xi_1, \ldots, \xi_{n-1} \le N$, define $\eta_i = \alpha_i^{1/2}\xi_i/2N$. Writing $\xi = (\xi_1, \ldots, \xi_{n-1})$ and $\eta = (\eta_1, \ldots, \eta_{n-1})$, we let $a_\eta = \hat{\phi}(\xi)$. Note that the points $\eta$ lie inside $[-1, 1]^{n-1}$ and are $1/10N$-separated.

A simple change of variables $(x, t) \mapsto y$ shows that

$$\int_{\mathbb{T}^{n-1}\times[0,1]} |u(x,t)|^{\frac{2(n+1)}{n-1}} \, dxdt$$

$$\sim \frac{1}{N^{n+1}} \int_{T_N} \left| \sum_{\eta_1,\ldots,\eta_{n-1}} a_\eta e(y_1\eta_1 + \cdots + y_{n-1}\eta_{n-1} + y_n(\eta_1^2 + \cdots \eta_{n-1}^2)) \right|^{\frac{2(n+1)}{n-1}} dy,$$

$$\tag{13.9}$$

where $T_N$ is a rectangular box with dimensions $\sim N \times \cdots \times N \times N^2$.

There are essentially two different ways to estimate this expression. Let us start by describing the one that leads to a suboptimal result. The average integral over $T_N$ is dominated by averages over balls of radius $N$. Using (13.7) with $R \sim N$ leads to

$$(13.9) \lesssim N^{\frac{n-1}{2}(\frac{1}{2} - \frac{n-1}{2(n+1)})\frac{2(n+1)}{n-1}} \|\phi\|_2^{\frac{2(n+1)}{n-1}}.$$

Note that the exponent of $N$ is positive, which makes this estimate unfavorable.

Let us now describe the correct approach. Due to periodicity in the $y_1, \ldots, y_{n-1}$ variables, we can rewrite (13.9) as

$$\frac{1}{N^{2n}} \int_B \left| \sum_{\eta_1,\ldots,\eta_{n-1}} a_\eta e(y_1\eta_1 + \cdots + y_{n-1}\eta_{n-1} + y_n(\eta_1^2 + \cdots \eta_{n-1}^2)) \right|^{\frac{2(n+1)}{n-1}} dy,$$

for some box $B$ with side lengths $\sim N^2$. The desired Strichartz estimate now follows from Theorem 13.8 with $R' \sim N^2$. $\qquad\square$

The systematic treatment of Strichartz estimates for $\mathbb{T}^{n-1}$ was initiated by Bourgain in his groundbreaking paper [**21**]. He proved (13.8) for $n = 2$ and

$n = 3$ using purely arithmetic methods that are applicable due to the fact that the critical exponent $p = 2(n + 1)/(n - 1)$ is an even integer in these cases. Moreover, he also made progress on the much more difficult case $n \geq 4$, by combining the Stein–Tomas method with Gauss sum estimates (see Exercise 13.40 for a related computation). While both of these ingredients are sharp in nature, they have not proved to be effective in solving the problem completely, in spite of many subsequent refinements. A possible explanation of this relative inefficiency has to do with the fact the Stein–Tomas method is tailored for proving restriction estimates. As we have seen in Section 1.6, these estimates amount to averaging exponential sums over spatial balls of radius $R \sim \delta^{-1}$, where $\delta$ is the separation between frequencies. Instead, the final resolution of (13.8) came through the use of decoupling, which controls averages over balls of larger radii $R \sim \delta^{-2}$. The proof of Theorem 13.9 shows the necessity for considering such large radii.

When $p > 2(n + 1)/(n - 1)$, the $N^\epsilon$ loss in (13.8) can be removed with the aid of a bit of number theory; see [29] and [75]. On the other hand, Theorem 13.6 shows that there is a logarithmic loss when $p = 2(n + 1)/(n - 1)$. Whether or not there is some loss in $N$ in the range $2 < p < 2(n + 1)/(n - 1)$ remains an open and probably very difficult problem.

## 13.4 Eigenfunction Estimates for the Laplacian on Tori

Let $(M, g)$ be a compact Riemannian manifold of dimension $n \geq 2$, and let $\Delta_g$ be the associated Laplace–Beltrami operator. We denote by $\lambda^2$ a typical eigenvalue of $-\Delta_g$ and by $e_\lambda$ a corresponding eigenfunction

$$-\Delta_g e_\lambda = \lambda^2 e_\lambda.$$

For $p \geq 2$, we define

$$\mu(p) = \max\left(\frac{n-1}{2}\left(\frac{1}{2} - \frac{1}{p}\right), n\left(\frac{1}{2} - \frac{1}{p}\right) - \frac{1}{2}\right).$$

The following inequality was proved by Sogge in [101].

**Theorem 13.10** *We have*

$$\|e_\lambda\|_{L^p(M)} \lesssim \lambda^{\mu(p)} \|e_\lambda\|_{L^2(M)},$$

*with an implicit constant independent of $\lambda$. Moreover, the inequality is saturated when $M$ is the round sphere.*

The upper bound may be improved by logarithmic factors when $M$ has non-positive curvature. See [99] for a recent result and some historical background.

Note that $\mu(p) > 0$ for all $p \geq 2$. We will show how to improve this exponent when $M$ is the flat torus $\mathbb{T}^n$. Given integers $n \geq 2$ and $\lambda^2 \geq 1$, consider the discrete sphere

$$\mathcal{F}_{n-1,\lambda} = \{\xi = (\xi_1, \ldots, \xi_n) \in \mathbb{Z}^n : |\xi_1|^2 + \cdots + |\xi_n|^2 = \lambda^2\}.$$

We recall the following well-known estimates; see [54]:

$$|\mathcal{F}_{n-1,\lambda}| \lesssim_\epsilon \lambda^{n-2+\epsilon}, \quad n = 2, 3, 4,$$

$$|\mathcal{F}_{n-1,\lambda}| \sim \lambda^{n-2}, \quad n \geq 5.$$

When $n = 2, 3$, $\mathcal{F}_{n-1,\lambda}$ is empty for a sequence of integers $\lambda^2 \to \infty$. When $n \geq 4$, an old theorem of Lagrange guarantees that all discrete spheres $\mathcal{F}_{n-1,\lambda}$ are nonempty.

The (nonzero) eigenvalues of the (normalized) minus Laplacian on $\mathbb{T}^n$

$$-\Delta\phi(x_1, \ldots, x_n)$$
$$= \sum_{\xi_1, \ldots, \xi_n \in \mathbb{Z}} (\xi_1^2 + \cdots + \xi_n^2)\hat{\phi}(\xi_1, \ldots, \xi_n)e(\xi_1 x_1 + \cdots + \xi_n x_n)$$

are those positive integers $\lambda^2$ for which $\mathcal{F}_{n-1,\lambda}$ is nonempty. The corresponding eigenspace consists of functions of the form

$$\sum_{\xi \in \mathcal{F}_{n-1,\lambda}} a_\xi e(\xi \cdot x).$$

In [23], Bourgain made the following conjecture and proved a partial result toward it, combining a Stein–Tomas-type argument with estimates for Kloosteman sums. Further improvements were obtained in [30] and [31] using incidence geometry and Siegel's mass formula.

**Conjecture 13.11** *Let $n \geq 3$ and $a_\xi \in \mathbb{C}$. For each $\epsilon > 0$, we have*

$$\left\| \sum_{\xi \in \mathcal{F}_{n-1,\lambda}} a_\xi e(\xi \cdot x) \right\|_{L^p([0,1]^n)} \lesssim_\epsilon \begin{cases} \lambda^\epsilon \|a_\xi\|_{l^2}, & 2 \leq p \leq \frac{2n}{n-2} \\ \lambda^{\frac{n-2}{2}-\frac{n}{p}+\epsilon} \|a_\xi\|_{l^2}, & p \geq \frac{2n}{n-2} \end{cases}.$$

$$\tag{13.10}$$

Note that the exponent of $\lambda$ is smaller than $\mu(p)$ for all $p \geq 2$. We only consider $n \geq 3$, as $\mathcal{F}_{1,\lambda}$ contains $O(\lambda^\epsilon)$ points. Let us now show how decoupling enabled further progress on this problem.

**Theorem 13.12** ([29]) *The inequality (13.10) holds for $2 \leq p \leq 2(n + 1)/(n - 1)$ when $n \geq 3$, and for $p \geq 2(n - 1)/(n - 3)$ when $n \geq 4$.*

*Proof* The first statement is an immediate application of Theorem 12.1 and Corollary 13.2 to $\mathcal{M} = \mathbb{S}^{n-1}$, using rescaling and periodicity.

To see the second inequality, assume that $\|a_\xi\|_2 = 1$ and let

$$F(x) = \sum_{\xi \in \mathcal{F}_{n-1,\lambda}} a_\xi e(\xi \cdot x).$$

We start by recalling the following estimate from [31], valid for $n \geq 4$ and $\alpha \gtrsim_\epsilon N^{(n-1)/4+\epsilon}$

$$|\{x \in \mathbb{T}^n : |F(x)| > \alpha\}| \lesssim_\epsilon \alpha^{-\frac{2(n-1)}{n-3}} N^{\frac{2}{n-3}+\epsilon}. \tag{13.11}$$

The proof of this relies on a Stein–Tomas-type argument and on standard estimates for Kloosterman sums. By invoking interpolation with the trivial $L^\infty$ bound, it suffices to consider the endpoint $p = p_n = 2(n-1)/(n-3)$. Note that $\|F\|_\infty \leq N^{C_n}$. It follows that

$$\int |F|^{p_n} \leq \int_{N^{\frac{n-1}{4}+\epsilon} \lesssim_\epsilon \alpha \leq N^{C_n}} \alpha^{p_n - 1} |\{|F| > \alpha\}| \, d\alpha$$

$$+ N^{(\frac{n-1}{4}+\epsilon)(p_n - \frac{2(n+1)}{n-1})} \int |F|^{\frac{2(n+1)}{n-1}}.$$

The result will follow by applying (13.11) to the first term and the result we proved in the first part of the theorem to the second term.    $\square$

Decoupling alone is not enough to completely solve Conjecture 13.11, since the critical exponent $2n/(n-2)$ is larger than $2(n+1)/(n-1)$. This in turn is due to the complex geometry of the lattice points on spheres. In particular, $\mathcal{F}_{n-1,\lambda}$ contains far fewer points than the $n-1$ dimensional paraboloid

$$\{(\xi_1, \ldots, \xi_{n-1}, \xi_1^2 + \cdots + \xi_{n-1}^2) : |\xi_1|, \ldots, |\xi_{n-1}| \leq \lambda\}$$

of comparable diameter. Quite intriguingly, Conjecture 13.11 is open even in the cases $n = 3$ and $n = 4$, when the critical exponents are even integers.

## 13.5 Vinogradov's Mean Value Theorem for Curves with Torsion

Let us consider a curve in $\mathbb{R}^n$

$$\Phi(t) = (\phi_1(t), \ldots, \phi_n(t))$$

with $\phi_j \in C^{n+1}([0,1])$ and nonzero torsion

$$\min_{t \in [0,1]} \tau(t) > 0.$$

Our starting point in this section is the following particular case of Corollary 13.2 for this curve.

**Theorem 13.13** *For each $1 \le k \le N$, let $t_k$ be a point in $((k-1)/N, k/N]$. Then for each ball $B_R$ in $\mathbb{R}^n$ of radius $R \gtrsim N^n$, for each $a_k \in \mathbb{C}$, and for each $p \ge 2$, we have*

$$\left( \frac{1}{|B_R|} \int_{\mathbb{R}^n} \left| \sum_{k=1}^{N} a_k e(x_1 \phi_1(t_k) + \cdots + x_n \phi_n(t_k)) \right|^p w_{B_R}(x)\, dx_1 \ldots dx_n \right)^{\frac{1}{p}}$$
$$\lesssim_\epsilon N^\epsilon \left( 1 + N^{\frac{1}{2}(1 - \frac{n(n+1)}{p})} \right) \|a_k\|_{l^2}.$$

*The implicit constant does not depend on $N$, $R$, and $a_k$.*

*Proof* Use Theorem 12.19 and Corollary 13.2. ☐

Let us now specialize to the case $\phi_j(t) = t^{\beta_j}$, with $\beta_1 < \beta_2 < \cdots < \beta_n$ distinct positive integers. Invoking periodicity leads to the following significant consequence.

**Corollary 13.14** *For $p \ge 2$ and $a_k \in \mathbb{C}$, we have*

$$\left\| \sum_{k=1}^{N} a_k e(x_1 k^{\beta_1} + \cdots + x_n k^{\beta_n}) \right\|_{L^p([0,1]^n)} \lesssim_\epsilon N^\epsilon \left( 1 + N^{\frac{1}{2} - \frac{n(n+1)}{2p}} \right) \|a_k\|_{l^2}.$$

*Proof* Using (12.10) we find that the curve $(t^{\beta_1}, \ldots, t^{\beta_n})$ has nonzero torsion on $[1/2, 1]$. This gives the desired inequality when the summation in $k$ is restricted to $[N/2, N]$, as follows.

Periodicity forces

$$\left\| \sum_{k=[N/2]}^{N} a_k e(x_1 k^{\beta_1} + \cdots + x_n k^{\beta_n}) \right\|_{L^p([0,1]^n)}$$
$$= \left\| \sum_{k=[N/2]}^{N} a_k e(x_1 k^{\beta_1} + \cdots + x_n k^{\beta_n}) \right\|_{L^p_\sharp (\prod_{i=1}^n [0, N^{\beta_n - \beta_i}])}$$
$$= \left\| \sum_{k=[N/2]}^{N} a_k e\left( x_1 \left( \frac{k}{N} \right)^{\beta_1} + \cdots + x_n \left( \frac{k}{N} \right)^{\beta_n} \right) \right\|_{L^p_\sharp ([0, N^{\beta_n}]^n)}.$$

The last expression is dominated using Theorem 13.13.

The case of unrestricted sum follows by summing up $O(\log N)$ terms as before. ☐

The case $\phi_j(t) = t^j$ and $a_k = 1$ was known as (the main conjecture in) Vinogradov's Mean Value Theorem. This problem has a rich history for $n \geq 3$, and has remained open for many decades. For most part, it has been approached with number theoretic tools. Most notably, Trevor Wooley has developed variants of the efficient congruencing method to solve the case $n = 3$ in [**124**]. His approach has also led to significant progress for larger values of $n$. We refer the reader to the survey paper [**125**] for a description of related results, as well as of many of their applications.

A different, purely analytic approach on the problem was initiated by Jean Bourgain and the author in [**27**], using decoupling. The general case of Vinogradov's Mean Value Theorem was eventually proved by Bourgain, Guth, and the author in [**33**]. A second proof was later provided by Wooley in [**123**].

Apart from the $N^\epsilon$ loss, the result of Corollary 13.14 is sharp when $\phi_j(t) = t^j$. It suffices to check this for $p \geq n(n+1)$. Indeed, this follows since

$$\inf_{\substack{|x_1| \ll N^{-1} \\ |x_2| \ll N^{-2} \\ \cdots \\ |x_n| \ll N^{-n}}} \left| \sum_{k=1}^{N} e(x_1 k + x_2 k^2 + \cdots + x_n k^n) \right| \gtrsim N.$$

Let us get a different perspective on the significance of Theorem 13.13, by recasting its content as an upper bound on the number of solutions of a system of polynomial inequalities.

Consider a collection $\mathcal{T} = \{T_1, \ldots, T_N\}$ of real numbers satisfying $k - 1 < T_k \leq k$. To emphasize the generality of our result, we will not insist that $T_k$ are integers. For each integer $s \geq 1$, we denote by $J_{s,n}(\mathcal{T})$ the number of solutions $(T_{k_1}, \ldots, T_{k_{2s}}) \in \mathcal{T}^{2s}$ of the following system of inequalities:

$$\begin{cases} |T_{k_1} + \cdots + T_{k_s} - (T_{k_{s+1}} + \cdots + T_{k_{2s}})| < N^{1-n} \\ |T_{k_1}^2 + \cdots + T_{k_s}^2 - (T_{k_{s+1}}^2 + \cdots + T_{k_{2s}}^2)| < N^{2-n} \\ \quad\quad \cdots\cdots \\ |T_{k_1}^n + \cdots + T_{k_s}^n - (T_{k_{s+1}}^n + \cdots + T_{k_{2s}}^n)| < 1. \end{cases}$$

A trivial solution $(T_{k_1}, \ldots, T_{k_{2s}}) \in \mathcal{T}^{2s}$ for this system is one for which

$$\{T_{k_1}, \ldots, T_{k_s}\} = \{T_{k_{s+1}}, \ldots, T_{k_{2s}}\}.$$

There are $\sim N^s$ trivial solutions. The next proposition shows that when $s \leq n(n+1)/2$, the number of nontrivial solutions is not much larger than $N^s$.

**Proposition 13.15** *For each integer* $1 \le s \le n(n+1)/2$, *we have that*

$$J_{s,n}(\mathcal{T}) \lesssim_\epsilon N^{s+\epsilon},$$

*where the implicit constant does not depend on* $\mathcal{T}$.

*Proof* Let $\eta \colon \mathbb{R}^n \to [0, \infty)$ be a positive Schwartz function with a positive Fourier transform satisfying $\widehat{\eta}(\xi) \ge 1$ for $|\xi| \le \sqrt{n}$. Using Theorem 13.13, we get

$$\frac{1}{N^{n^2}} \int_{\mathbb{R}^n} \left| \sum_{k=1}^{N} e\left( x_1 \frac{T_k}{N} + \cdots + x_n \left( \frac{T_k}{N} \right)^n \right) \right|^{2s} \eta\left( \frac{x}{N^n} \right) dx \lesssim_\epsilon N^{s+\epsilon}. \quad (13.12)$$

After making a change of variables and expanding the product, the term

$$\int_{\mathbb{R}^n} \left| \sum_{k=1}^{N} e\left( x_1 \frac{T_k}{N} + \cdots + x_n \left( \frac{T_k}{N} \right)^n \right) \right|^{2s} \eta\left( \frac{x}{N^n} \right) dx$$

can be written as the sum over all $T_{k_1}, \ldots, T_{k_{2s}} \in \mathcal{T}$ of

$$N^{\frac{n(n+1)}{2}} \int_{\mathbb{R}^n} \eta\left( \frac{x_1}{N^{n-1}}, \frac{x_2}{N^{n-2}}, \ldots, x_n \right) e(x_1 Z_1 + \cdots + x_n Z_n) \, dx,$$

where

$$Z_j = T_{k_1}^j + \cdots + T_{k_s}^j - (T_{k_{s+1}}^j + \cdots + T_{k_{2s}}^j).$$

Each such term is equal to

$$N^{n^2} \widehat{\eta}(N^{n-1} Z_1, N^{n-2} Z_2, \ldots, Z_n).$$

This is always positive, and in fact greater than $N^{n^2}$ at least $J_{s,n}(\mathcal{T})$ many times. It now suffices to use (13.12). $\qquad\square$

When $\mathcal{T} = \{1, 2, \ldots, N\}$, $J_{s,n}(\mathcal{T})$ counts the number of solutions $(k_1, \ldots, k_{2s}) \in \{1, \ldots, N\}^{2s}$ of the system

$$\begin{cases} k_1 + \cdots + k_s = k_{j_{s+1}} + \cdots + k_{2s} \\ (k_1)^2 + \cdots + (k_s)^2 = (k_{s+1})^2 + \cdots + (k_{2s})^2 \\ \qquad \cdots\cdots \\ (k_1)^n + \cdots + (k_s)^n = (k_{s+1})^n + \cdots + (k_{2s})^n. \end{cases}$$

Sharp bounds (modulo $\epsilon$ losses) for higher-dimensional analogs of this system have been derived in [26], [28], and [55], as immediate corollaries of decoupling inequalities for certain manifolds of dimension greater than one.

## 13.6 Exponential Sums for Curves with Torsion

Let $\beta_1 < \beta_2 < \cdots < \beta_n$ be $n$ distinct positive integers. It seems reasonable to make the following conjecture.

**Conjecture 13.16** *For each $p \geq 2$ and $a_k \in \mathbb{C}$,*

$$\left\| \sum_{k=1}^{N} a_k e(x_1 k^{\beta_1} + \cdots + x_n k^{\beta_n}) \right\|_{L^p([0,1]^n)} \lesssim_\epsilon N^\epsilon \left( 1 + N^{\frac{1}{2} - \frac{\beta_1 + \cdots + \beta_n}{p}} \right) \|a_k\|_{l^2}.$$

It is easy to see that the exponents of $N$ are sharp, in particular,

$$\left\| \sum_{k=1}^{N} e(x_1 k^{\beta_1} + \cdots + x_n k^{\beta_n}) \right\|_{L^p([0,1]^n)} \gtrsim N^{\frac{1}{2}} + N^{1 - \frac{\beta_1 + \cdots + \beta_n}{p}}.$$

The first lower bound is due to Hölder's inequality. The second one comes from using the major arc estimate

$$\inf_{\substack{|x_1| \ll N^{-\beta_1} \\ |x_2| \ll N^{-\beta_2} \\ \cdots \\ |x_n| \ll N^{-\beta_n}}} \left| \sum_{k=1}^{N} e(x_1 k^{\beta_1} + \cdots + x_n k^{\beta_n}) \right| \gtrsim N.$$

When $p$ is larger than the critical exponent

$$p_c = 2(\beta_1 + \cdots + \beta_n),$$

the second lower bound is the greater among the two.

There has been a lot of progress on various cases of this conjecture. We will confine ourselves to showing what decoupling has to say about it. The following discussion will reveal both its power and its apparent limitations.

**Theorem 13.17** ([**19**]) *Conjecture 13.16 holds true for $2 \leq p \leq n(n+1)$ and for $p \geq \beta_n(\beta_n + 1)$.*

*Proof* The first part of the theorem was proved in Corollary 13.14.

We will prove the second part of the theorem. It suffices to consider $p = p_n = \beta_n(\beta_n + 1)$. Corollary 13.14 shows that

$$\left\| \sum_{k=1}^{N} a_k e(x_1 k + x_2 k^2 + \cdots + x_{\beta_n} k^{\beta_n}) \right\|_{L^{p_n}([0,1]^{\beta_n})} \lesssim_\epsilon N^\epsilon \|a_k\|_{l^2}. \quad (13.13)$$

Let

$$I = \{1, 2, \ldots, \beta_n\} \setminus \{\beta_1, \ldots, \beta_n\}$$

and

$$F(x) = \sum_{k=1}^{N} a_k e(x_1 k + x_2 k^2 + \cdots + x_{\beta_n} k^{\beta_n}).$$

The following argument provides a rigorous justification of the fact that

$$|F(x)| \approx \left| \sum_{k=1}^{N} a_k e(x_{\beta_1} k^{\beta_1} + x_{\beta_2} k^{\beta_2} + \cdots + x_{\beta_n} k^{\beta_n}) \right|$$

whenever $|x_i| \ll N^{-i}$ for each $i \in I$.

We have the expansion

$$|F(x)|^{p_n} = F(x)^{\frac{p_n}{2}} \bar{F}(x)^{\frac{p_n}{2}}$$

$$= \sum_{Z_1, \ldots, Z_{\beta_n}} A_{Z_1, \ldots, Z_{\beta_n}} e(x_1 Z_1 + x_2 Z_2 + \cdots + x_{\beta_n} Z_{\beta_n}),$$

with

$$Z_i = (k_1)^i + \cdots + (k_{p_n/2})^i - (k_{1+p_n/2})^i - \cdots - (k_{p_n})^i$$

and

$$A_{Z_1, \ldots, Z_{\beta_n}} = a_{k_1} \cdots a_{k_{p_n/2}} \bar{a}_{k_{1+p_n/2}} \cdots \bar{a}_{k_{p_n}},$$

for some $1 \leq k_1, \ldots, k_{p_n} \leq N$. Note that $|Z_i| \leq p_n N^i$.

Let $\eta$ be a Schwartz function on $\mathbb{R}$ with its Fourier transform equal to 1 on $[-1, 1]$. Using the rapid decay of $\eta$, it follows that for each nonnegative 1-periodic function $f$ and each $M \geq 1$, we have

$$\int_0^1 f(x) \, dx \gtrsim \left| \int_{\mathbb{R}} f(x) \eta(Mx) \, dx \right|.$$

Combining this with the previous representation for $|F|^{p_n}$ and the fact that $\hat{\eta}\left(-Z_i/(p_n N^i)\right) = 1$ shows that

$$\|F\|_{L^{p_n}([0,1]^{\beta_n})}^{p_n} \gtrsim \int_{\mathbb{R}^{\beta_n}} |F(x)|^{p_n} \prod_{i \notin I} 1_{[0,1]}(x_i) \prod_{i \in I} \eta(x_i p_n N^i) \, dx$$

$$= \frac{1}{\prod_{i \in I}(p_n N^i)} \left\| \sum_{k=1}^{N} a_k e(x_{\beta_1} k^{\beta_1} + x_{\beta_2} k^{\beta_2} + \cdots + x_{\beta_n} k^{\beta_n}) \right\|_{L^{p_n}([0,1]^n)}^{p_n}.$$

This, together with (13.13), leads to the desired estimate:

$$\left\| \sum_{k=1}^{N} a_k e(x_1 k^{\beta_1} + \cdots + x_n k^{\beta_n}) \right\|_{L^{p_n}([0,1]^n)}^{p_n} \lesssim_\epsilon N^{\epsilon + \sum_{i \in I} i} \|a_k\|_{l^2}^{p_n}$$

$$= N^{\epsilon + \frac{p_n}{2} - (\beta_1 + \cdots + \beta_n)} \|a_k\|_{l^2}^{p_n}.$$

$\square$

The case $n = 1$ is particularly interesting.

**Corollary 13.18** *Let $P$ be a polynomial of degree $l \geq 2$ with integer coefficients. Then for each $p \geq l(l+1)$, we have*

$$\left\| \sum_{k=1}^{N} a_k e(P(k)x) \right\|_{L^p([0,1])} \lesssim_\epsilon N^{\frac{1}{2} - \frac{l}{p} + \epsilon} \|a_k\|_{l^2}.$$

*Proof* Note that the curve

$$\Phi(t) = (t, t^2, \ldots, t^{l-1}, P(t))$$

has everywhere nonzero torsion. Repeat the argument from the previous theorem, using Corollary 13.14 and a change of variables that turns $P$ into $t^l$.   $\square$

The range in this corollary is only sharp when $l = 2$. This case also admits an elementary proof; see Exercise 13.36. The estimate in the range $p \geq 2^l - l$ with constant coefficients is an old influential result of Hua [69].

One of the simplest examples that illustrate the limitations of decoupling for the analysis of incomplete systems is the case $n = 2$, $\beta_1 = 1$, $\beta_2 = 3$ of Conjecture 13.16. The estimate

$$\left\| \sum_{k=1}^{N} e(kx_1 + k^3 x_2) \right\|_{L^p([0,1]^2)} \lesssim_\epsilon N^{\frac{1}{2} + \epsilon} \tag{13.14}$$

is expected to hold in the range $2 \leq p \leq 8$. Proposition 12.22 shows that this cannot be proved in the range $(6, 8]$ by using only decoupling.

One may strengthen Conjecture 13.16 to eliminate $\epsilon$ losses away from the critical exponent. Questions of this type are referred to as $\Lambda(p)$ problems. Even the following simplest case is open, even for $a_k \in \{0, 1\}$.

**Question 13.19** ($\Lambda(p)$ problem for the squares, [97]) *Is it true that for each $a_k \in \mathbb{C}$ and $2 < p < 4$*

$$\left\| \sum_{k=1}^{N} a_k e(k^2 x) \right\|_{L^p([0,1])} \lesssim \|a_k\|_{l^2},$$

*with implicit constant independent of $N$?*

See Exercise 13.37 for a generalization and an application, and [24] and [38] for results related to this problem.

## 13.7 Additive Energies

Given an integer $k \geq 2$ and a finite set $\Lambda \subset \mathbb{R}^n$, we introduce its $k$-additive energy

$$\mathbb{E}_k(\Lambda) = |\{(\lambda_1, \ldots, \lambda_{2k}) \in \Lambda^{2k} : \lambda_1 + \cdots + \lambda_k = \lambda_{k+1} + \cdots + \lambda_{2k}\}|.$$

Note the trivial lower bound $|\mathbb{E}_k(\Lambda)| \geq |\Lambda|^k$. The sets $\Lambda$ for which this lower bound is essentially sharp are of particular interest. A lacunary sequence in $\mathbb{R}$ is such an example, for all $k$. This is a manifestation of multiorthogonality. The results and conjectures in this chapter that have an even integer as the critical exponent provide further examples of such sets. We start with mentioning an open question that is remarkable for its apparent simplicity.

**Question 13.20** *Is it true that for each finite set of squares $\Lambda \subset \{1^2, 2^2, \ldots\}$ and for each $\epsilon > 0$, we have*

$$\mathbb{E}_2(\Lambda) \lesssim_\epsilon |\Lambda|^{2+\epsilon}? \tag{13.15}$$

There is strikingly little known about this, in particular the inequality has not been verified for any $\epsilon < 1$. See [98] for a discussion of recent results. Inequality (13.15) would immediately follow if Question 13.19 were proved to have a positive answer.

The next result explores some rather immediate consequences of decoupling to estimating additive energies for well-separated sets.

**Theorem 13.21** *If $\Lambda$ is a $\delta$-separated subset of either $\mathbb{S}^1$ or $\mathbb{P}^1$, then*

$$\mathbb{E}_3(\Lambda) \lesssim_\epsilon \delta^{-\epsilon} |\Lambda|^3.$$

*If $\Lambda$ is a $\delta$-separated subset of either $\mathbb{S}^2$ or $\mathbb{P}^2$, then*

$$\mathbb{E}_2(\Lambda) \lesssim_\epsilon \delta^{-\epsilon} |\Lambda|^2.$$

*Proof* We will prove the first part; the second part has a virtually identical proof. Theorem 13.8 (and the analogous version for $\mathbb{S}^1$) shows that

$$\frac{1}{R^2} \int_{B(0,R)} \left| \sum_{\lambda \in \Lambda} e(\lambda \cdot x) \right|^6 dx \lesssim_\epsilon \delta^{-\epsilon} |\Lambda|^3,$$

with the implicit constant independent of $R \geq \delta^{-2}$. The left-hand side equals

$$\frac{1}{R^2} \sum_{(\lambda_1,\ldots,\lambda_6) \in T} \int_{B(0,R)} e((\lambda_1 + \lambda_2 + \lambda_3 - \lambda_4 - \lambda_5 - \lambda_6) \cdot x) \, dx$$

$$+ \frac{1}{R^2} \sum_{(\lambda_1,\ldots,\lambda_6) \notin T} \int_{B(0,R)} e((\lambda_1 + \lambda_2 + \lambda_3 - \lambda_4 - \lambda_5 - \lambda_6) \cdot x) \, dx,$$

where

$$T = \{(\lambda_1, \ldots, \lambda_6) \in \Lambda^6 : \lambda_1 + \lambda_2 + \lambda_3 = \lambda_4 + \lambda_5 + \lambda_6\}.$$

The first average equals $\pi \mathbb{E}_3(\Lambda)$. The second average converges to zero as $R \to \infty$. This is because

$$\lim_{R \to \infty} \frac{1}{R^2} \int_{B(0,R)} e(\lambda \cdot x) \, dx = 0$$

whenever $\lambda \neq 0$. The proof is concluded by combining all these observations.
□

Note that the set $\Lambda$ in our theorem satisfies $|\Lambda| \lesssim \delta^{-O(1)}$. We expect the result to be true for arbitrary, not necessarily $\delta$-separated sets $\Lambda$, with $\delta^{-\epsilon}$ replaced with the smaller bound $|\Lambda|^{\epsilon}$. We will next prove that this is indeed the case for $\mathbb{P}^2$, and show some partial progress for $\mathbb{P}^1$ and $\mathbb{S}^1$. At the heart of all these results lies the Szemerédi–Trotter theorem and its slightly more general version due to Székely [106]. The reader may consult [113] for more applications of this result.

**Theorem 13.22** (Generalized Szemerédi–Trotter theorem) *Let $\mathcal{P}$ be a finite collection of points in $\mathbb{R}^2$, and let $\mathcal{C}$ be a finite collection of curves in $\mathbb{R}^2$. Suppose that any two curves in $\mathcal{C}$ intersect in at most $O(1)$ points, and that any two points in $\mathcal{P}$ belong simultaneously to at most $O(1)$ curves in $\mathcal{C}$. Then*

$$|\{(P,C) \in \mathcal{P} \times \mathcal{C} : P \in C\}| \lesssim |\mathcal{P}|^{2/3} |\mathcal{C}|^{2/3} + |\mathcal{P}| + |\mathcal{C}|.$$

The original Szemerédi–Trotter theorem was concerned with the case when $\mathcal{C}$ consists of lines. One of the beautiful applications of this theorem is the following sharp result of Pach and Sharir.

**Theorem 13.23** ([89]) *The number of repetitions of a given angle among $N$ points in the plane is $O(N^2 \log N)$.*

Let us now prove the following strengthening of Theorem 13.21 for $\mathbb{P}^2$.

**Theorem 13.24** *For each finite set $\Lambda \subset \mathbb{P}^2$, we have*

$$\mathbb{E}_2(\Lambda) \lesssim |\Lambda|^2 \log |\Lambda|.$$

*Proof* Let $S$ consist of all points $(\xi, \eta)$ with $\lambda = (\xi, \eta, \xi^2 + \eta^2) \in \Lambda$. Let us assume that for some $\lambda_i = (\xi_i, \eta_i, \xi_i^2 + \eta_i^2)$, $1 \le i \le 4$, we have

$$\lambda_1 + \lambda_2 = \lambda_3 + \lambda_4 = (A, B, C).$$

An easy computation reveals that for each $1 \le i \le 4$,

$$\left(\xi_i - \frac{A}{2}\right)^2 + \left(\eta_i - \frac{B}{2}\right)^2 = \frac{2C - A^2 - B^2}{4}.$$

We also note that

$$(\xi_1, \eta_1) + (\xi_2, \eta_2) = (\xi_3, \eta_3) + (\xi_4, \eta_4).$$

The first equality tells us that the four points $P_i = (\xi_i, \eta_i)$ lie on a circle, while the second one tells us that they determine a parallelogram. We conclude that they in fact determine a rectangle. Thus, each additive quadruple $(\lambda_1, \ldots, \lambda_4)$ determines four distinct right angles in $S$, and it suffices to apply Theorem 13.23. $\qquad\square$

We expect the result of this theorem to also hold for $\mathbb{S}^2$, although there does not appear to be an obvious way to extend the argument to this case. In the case of $\mathbb{P}^1, \mathbb{S}^1$, the best we are able to prove using incidence geometric methods leads to the exponent $7/2$.

**Proposition 13.25** ([**17**], [**29**]) *For each finite subset $\Lambda$ of either $\mathbb{P}^1$ or $\mathbb{S}^1$, we have*

$$\mathbb{E}_3(\Lambda) \lesssim |\Lambda|^{\frac{7}{2}}.$$

*Proof* The proofs for $\mathbb{P}^1$ and $\mathbb{S}^1$ are very similar. We present the details for $\mathbb{P}^1$. Let $N$ be the cardinality of $\Lambda$. We denote by $S$ the collection of those $\xi$ with $(\xi, \xi^2) \in \Lambda$. A simple computation shows that if

$$(\xi_1, \xi_1^2) + (\xi_2, \xi_2^2) + (\xi_3, \xi_3^2) = (i, j), \qquad (13.16)$$

then the point $(3(\xi_1 + \xi_2), \sqrt{3}(\xi_1 - \xi_2))$ belongs to the circle $C_{i,j}$ centered at $(2i, 0)$ and of radius $\sqrt{6j - 2i^2}$. Note that the set

$$\mathcal{P} = \{(3(\xi_1 + \xi_2), \sqrt{3}(\xi_1 - \xi_2)) : \xi_1, \xi_2 \in S\}$$

contains $N^2$ points.

Assume that there are $M_n$ circles $C_{i,j}$ containing roughly $2^n$ points $(3(\xi_1 + \xi_2), \sqrt{3}(\xi_1 - \xi_2)) \in \mathcal{P}$, in such a way that (13.16) is satisfied for some $\xi_3 \in S$. Let us call $\mathcal{C}_n$ the collection of these circles.

Then clearly

$$\mathbb{E}_3(\Lambda) \lesssim \sum_{2^n \leq N} M_n 2^{2n}.$$

To estimate the right-hand side, we derive two upper bounds. First, it is easy to see that

$$M_n 2^n \lesssim N^3, \tag{13.17}$$

as each point in $\mathcal{P}$ can only belong to at most $N$ circles, subject to (13.16).

To find a second upper bound, let us now consider the incidences between the circles in $\mathcal{C}_n$ and the points in $\mathcal{P}$:

$$\mathcal{I}(\mathcal{P}, \mathcal{C}_n) = \{(P, C) \in \mathcal{P} \times \mathcal{C}_n : P \in C\}.$$

Our definition of $\mathcal{C}_n$ implies that

$$M_n 2^n \leq |\mathcal{I}(\mathcal{P}, \mathcal{C}_n)|. \tag{13.18}$$

It is easy to see that any pair of points $(P_1, P_2) \in \mathcal{P}^2$ in the upper (or lower) half-plane determines a unique circle in $\mathcal{C}_n$. Two applications of Theorem 13.22 lead to the estimate

$$|\mathcal{I}(\mathcal{P}, \mathcal{C}_n)| \lesssim (M_n N^2)^{2/3} + N^2 + M_n.$$

Combining this with (13.18) shows that

$$M_n 2^{3n} \lesssim N^4. \tag{13.19}$$

Finally, using (13.17) and (13.19), we may write

$$\sum_{n=1}^{O(\log N)} M_n 2^{2n} \lesssim \sum_{n=1}^{O(\log N)} \min\left(2^n N^3, \frac{N^4}{2^n}\right) \lesssim N^{\frac{7}{2}}.$$

In the case of $\mathbb{S}^1$, similar computations lead to families $\mathcal{C}_n$ consisting of unit circles. It is immediate that two points determine (at most) one unit circle, so Theorem 13.22 is again applicable. $\qquad\square$

Using Theorem 13.24, rescaling and periodicity gives the following corollary. Note that the exponent $1/4$ of $\log |S|$ is sharp when $S = \{1, \ldots, N\}^2$, as proved in Theorem 13.6.

**Corollary 13.26** *For each set $S \subset \mathbb{Z}^2$, we have*

$$\left\| \sum_{(k_1, k_2) \in S} e(k_1 x_1 + k_2 x_2 + (k_1^2 + k_2^2) x_3) \right\|_{L^4(\mathbb{T}^3)} \lesssim |S|^{\frac{1}{2}} (\log |S|)^{\frac{1}{4}}.$$

In light of this result and the lower bound in Theorem 13.6, the following question seems natural.

**Question 13.27** *Let $n = 2$ or $n \geq 4$. Is it true that for each $S \subset \mathbb{Z}^{n-1}$*

$$\left\| \sum_{(k_1,\ldots,k_{n-1}) \in S} e(k_1 x_1 + \cdots + k_{n-1} x_{n-1} + (k_1^2 + \cdots + k_{n-1}^2) x_n) \right\|_{L^{\frac{2(n+1)}{n-1}}(\mathbb{T}^n)}$$

$$\lesssim |S|^{\frac{1}{2}} (\log |S|)^{\frac{n-1}{2(n+1)}} \ ?$$

Even proving slightly weaker upper bounds with $(\log |S|)^{(n-1)/2(n+1)}$ replaced with $|S|^\epsilon$ would be of interest, especially when $n = 2$. The current technology only allows proving upper bounds that depend on the separation of the points in $S$ relative to the diameter of $S$. See Exercise 13.33.

The connections between this and similar questions in the context of the decoupling for the parabola are explored in [**43**].

## 13.8 *l*th Powers in Arithmetic Progressions

For $l \geq 2$, let $\mathcal{Q}_l(N; q, a)$ denote the number of $l$th powers of integers in the arithmetic progression $a + q, a + 2q, \ldots, a + Nq$. Write

$$\mathcal{Q}_l(N) = \sup_{\substack{a, q \in \mathbb{N} \\ q \neq 0}} \mathcal{Q}_l(N; q, a). \tag{13.20}$$

It is an interesting question to determine the sharp order of growth of $\mathcal{Q}_l(N)$ as $N \to \infty$. By considering the arithmetic progression with $a = q = 1$, we see that $\mathcal{Q}_l(N) \gtrsim N^{1/l}$.

Rudin [**97**] has conjectured that $\mathcal{Q}_2(N) \sim N^{1/2}$. The best-known upper bound for $\mathcal{Q}_2(N)$ is due to Bombieri and Zannier [**18**]:

$$\mathcal{Q}_2(N) \lesssim N^{\frac{3}{5}}.$$

We may similarly conjecture that for each $l \geq 2$,

$$\mathcal{Q}_l(N) \lesssim N^{\frac{1}{l}}.$$

Rather than making any significant progress on this problem, we show how decoupling related inequalities can shed light on a related question. The following result shows that while $\mathcal{Q}_2(N; q, a)$ or $\mathcal{Q}_4(N; q, a)$ might (in principle) be occasionally "large", this can only happen for a "small" fraction of the values of $a$.

**Theorem 13.28** *Let $d(q)$ denote the number of divisors of $q$. We have*

$$\sup_{A \in \mathbb{Z}} \sum_{a=A+1}^{A+q} Q_2(N;q,a)(Q_2(N;q,a) - 1) \lesssim d(q)N \log N$$

*and*

$$\sup_{A \in \mathbb{Z}} \sum_{a=A+1}^{A+q} Q_4(N;q,a)(Q_4(N;q,a) - 1) \lesssim d(q)^4 N^{\frac{1}{2}} (\log N)^{\frac{9}{2}}.$$

*Proof* We start with the proof of the first inequality, which is elementary and rather immediate. Let $U_2$ be the collection of all pairs of integers $(t_1, t_2)$ such that $t_1^2 - t_2^2 \in \{q, 2q, \ldots, (N-1)q\}$. Note that for each $A$,

$$2|U_2| \geq \sum_{a=A+1}^{A+q} Q_2(N;q,a)(Q_2(N;q,a) - 1).$$

Each $(t_1, t_2) \in U_2$ is identified by a choice of positive integers $q_-, q_+, n_-, n_+$ such that $q_- q_+ = q, n_- n_+ \leq N - 1$ and

$$\begin{cases} t_1 - t_2 = n_- q_- \\ t_1 + t_2 = n_+ q_+. \end{cases}$$

It suffices to note that there are $O(d(q)N \log N)$ such choices.

Now let $U_4$ be the collection of all pairs of integers $(t_1, t_2)$ such that $t_1^4 - t_2^4 \in \{q, 2q, \ldots, (N-1)q\}$. Note that for each $A$,

$$2|U_4| \geq \sum_{a=A+1}^{A+q} Q_4(N;q,a)(Q_4(N;q,a) - 1).$$

We use the factorization

$$t_1^4 - t_2^4 = (t_1 - t_2)(t_1 + t_2)(t_1^2 + t_2^2). \tag{13.21}$$

Throughout the remainder of the proof, we will denote by $q[-O(A), O(A)]$ all the multiples $m$ of $q$ satisfying $|m| \lesssim qA$.

Write for each $(t_1, t_2) \in U_4$

$$\begin{cases} t_1 - t_2 = q_1 n_1 \\ t_1 + t_2 = q_2 n_2 \\ t_1^2 + t_2^2 = q_3 n_3 \end{cases}$$

with $n_1 n_2 n_3 \leq N - 1$ and $q_1 q_2 q_3 = q$. There are $O(d(q)^2)$ choices for the triple $(q_1, q_2, q_3)$. We will count the pairs $(t_1, t_2)$ satisfying this system for a fixed such triple, and with $n_1, n_2, n_3$ satisfying

$$n_1 \sim 2^{k_1}, \; n_2 \sim 2^{k_2}, \; n_3 \lesssim \frac{N}{2^{k_1+k_2}}$$

for some fixed $k_1, k_2 \lesssim \log N$.

Let $k_1, k_2$ be such that the number of corresponding pairs $(t_1, t_2)$ is $\gtrsim (d(q) \log N)^{-2} |U_4|$. Then

$$((d(q) \log N)^{-2} |U_4|)^4 \lesssim \left| \left\{ ((t_1, t_2), \ldots, (t_7, t_8)) \in U_4 \times U_4 \times U_4 \times U_4 : \right. \right.$$

$$\left. \left. \begin{cases} (t_1 - t_2) - (t_3 - t_4) + (t_5 - t_6) - (t_7 - t_8) \in q_1[-O(2^{k_1}), O(2^{k_1})] \\ (t_1 + t_2) - (t_3 + t_4) + (t_5 + t_6) - (t_7 + t_8) \in q_2[-O(2^{k_2}), O(2^{k_2})] \\ (t_1^2 + t_2^2) - (t_3^2 + t_4^2) + (t_5^2 + t_6^2) - (t_7^2 + t_8^2) \in q_3[-O(\frac{N}{2^{k_1+k_2}}), O(\frac{N}{2^{k_1+k_2}})]. \end{cases} \right\} \right|$$

Let $S$ be the collection consisting of all pairs $(u, v) = (t - t', t + t')$ with $(t, t') \in U_4$. We have $|S| = |U_4|$. There are $\sim N$ triples $(m_1, m_2, m_3)$ in the set

$$q_1[-O(2^{k_1}), O(2^{k_1})] \times q_2[-O(2^{k_2}), O(2^{k_2})] \times q_3 \left[ -O\left(\frac{N}{2^{k_1+k_2}}\right), O\left(\frac{N}{2^{k_1 k_2}}\right) \right].$$

The set

$$\left\{ (u_1, v_1, \ldots, u_4, v_4) \in S^4 : \begin{cases} u_1 + u_3 = u_2 + u_4 + m_1 \\ v_1 + v_3 = v_2 + v_4 + m_2 \\ u_1^2 + v_1^2 + u_3^2 + v_3^2 = u_2^2 + v_2^2 + u_4^2 + v_4^2 + 2m_3 \end{cases} \right\}$$

contains $O(|S|^2 \log |S|)$ elements, as it follows from Corollary 13.26. Putting things together, we find that

$$(d(q) \log N)^{-8} |U_4|^4 \lesssim |U_4|^2 N \log N,$$

which gives what we want. $\qquad \square$

Theorem 13.28 was born out of an idea of Jean Bourgain for squares, and it is stated in a form suggested to us by Igor Shparlinski. Our argument for $l = 4$ relies crucially on the favorable factorization (13.21), and on the fact that Corollary 13.26 does not demand additional separation between points. While there is a favorable factorization for $l = 3$, the extension of our argument to this case would demand a strong, currently unavailable exponential sum estimate for the parabola (see Question 13.27 and the discussion following it). It is likely that a broader spectrum of results can be achieved using purely number theoretic methods.

## 13.9 Decoupling and the Restriction Conjecture

In this section, we revisit the restriction problem for the paraboloid. We denote by $E$ the extension operator for $\mathbb{P}^{n-1}$. Let $C_{\text{lin}}(R,q)$ be the quantity

$$C_{\text{lin}}(R,q) = \sup \left\{ \int_{B(0,R)} |Ef|^q : \|f\|_{L^\infty([-1,1]^{n-1})} \le 1 \right\}.$$

The Restriction Conjecture asserts that for each $q \ge 2n/(n-1)$,

$$C_{\text{lin}}(R,q) \lesssim_\epsilon R^\epsilon. \tag{13.22}$$

For $2 \le k \le n$, let $U_1, \ldots, U_k$ be subsets of $[-1,1]^{n-1}$ such that the corresponding caps on $\mathbb{P}^{n-1}$ are $\nu$-transverse. The $k$-linear restriction inequality

$$\left\| \left( \prod_{j=1}^{k} E_{U_j} f \right)^{\frac{1}{k}} \right\|_{L^q(B_R)} \lesssim_{\nu,\epsilon} R^\epsilon \|f\|_{L^\infty([-1,1]^{n-1})} \tag{13.23}$$

is a direct consequence (via Hölder's inequality) of (13.22). Conversely, we will next show how such an inequality combined with decoupling implies the corresponding linear restriction inequality.

**Theorem 13.29** *Let $q$ satisfy*

$$\frac{2(2n-k+2)}{2n-k} \le q \le \frac{2k}{k-2}.$$

*If (13.23) holds for $q$, then (13.22) also holds for the same $q$.*

*Proof* We sketch briefly the argument, which is very similar to the one in Section 7.3.

Let $\mathcal{C}_\mu$ denote the partition of $[-1,1]^{n-1}$ into cubes with side length $2/\mu$. Let $K \in 4^{\mathbb{N}}$ be a large parameter whose value (depending on $q,n$) will be determined at the end of the argument.

The starting point in our analysis is the equality

$$Ef(x) = \sum_{\alpha \in \mathcal{C}_K} E_\alpha f(x).$$

When we restrict $x$ to a typical ball $B_K$ of radius $K$, the key dichotomy can be formulated as follows. Either we may find $k$ cubes $\alpha_1, \ldots, \alpha_k \in \mathcal{C}_K$ that are $K^{1-k}$-transverse and produce a significant contribution to $Ef$ on $B_K$, or most of the mass of $Ef$ on $B_K$ is coming from those $\alpha \in \mathcal{C}_K$ near some $k-2$ dimensional affine subspace $V(B_K)$ of $\mathbb{R}^{n-1}$. In more precise terms, if $\mathcal{F}_{V(B_K)}$ denotes the family of those cubes $\beta \in \mathcal{C}_{K^{1/2}}$ intersecting $V(B_K)$, we have

$$\int_{B_K} |Ef|^q \leq C_1 \sum_{\alpha \in \mathcal{C}_K} \int |E_\alpha f|^q w_{B_K} + \int \left| \sum_{\beta \in \mathcal{F}_{V(B_K)}} E_\beta f \right|^q w_{B_K}$$

$$+ C_K \max_{\substack{\alpha_1,\ldots,\alpha_k \in \mathcal{C}_K \\ K^{1-k} - \text{transverse}}} \left( \prod_{i=1}^k \int |E_{\alpha_i} f| w_{B_K} \right)^{\frac{q}{k}}.$$

The region on $\mathbb{P}^{n-1}$ corresponding to $\beta \in \mathcal{F}_{V(B_K)}$ is near the $k-2$ dimensional paraboloid

$$\{(\xi, |\xi|^2) : \xi \in V(B_K)\},$$

so it can be estimated using lower-dimensional decoupling. More precisely, since $q \leq 2k/(k-2)$, Exercise 10.29 combined with Hölder's inequality shows that

$$\int \left| \sum_{\beta \in \mathcal{F}_{V(B_K)}} E_\beta f \right|^q w_{B_K} \lesssim_\epsilon K^{\frac{k-2}{2}(\frac{q}{2}-1)+\epsilon} \sum_{\beta \in \mathcal{F}_{V(B_K)}} \int |E_\beta f|^q w_{B_K}.$$

Summing this inequality over balls $B_K$ in a finitely overlapping cover $\mathcal{B}$ of $B(0, R)$, we may write

$$\int_{B(0,R)} |Ef|^q \leq C_1 \sum_{\alpha \in \mathcal{C}_K} \int |E_\alpha f|^q w_{B(0,R)}$$

$$+ C_\epsilon K^{\frac{k-2}{2}(\frac{q}{2}-1)+\epsilon} \sum_{\beta \in \mathcal{C}_{K^{\frac{1}{2}}}} \int |E_\beta f|^q w_{B(0,R)}$$

$$+ C_K \sum_{B_K \in \mathcal{B}} \max_{\substack{\alpha_1,\ldots,\alpha_k \in \mathcal{C}_K \\ K^{1-k} - \text{transverse}}} \left( \prod_{i=1}^k \int |E_{\alpha_i} f| w_{B_K} \right)^{\frac{q}{k}}.$$

Assume that $\|f\|_\infty = 1$. Combining parabolic rescaling with the $k$-linear restriction estimate (13.23) as in Section 7.3, we arrive at the following inequality:

$$\int_{B(0,R)} |Ef|^q \leq C_{\text{lin}}(R, q)(C_1' K^{2n-q(n-1)} + C_\epsilon' K^{\frac{k-2}{2}(\frac{q}{2}-1)+n-\frac{q(n-1)}{2}+\epsilon})$$

$$+ C_{K,\epsilon} R^\epsilon.$$

If $q > 2(2n - k + 2)/(2n - k)$, the exponents of $K$ are negative for $\epsilon$ small enough. Choosing $K$ large enough leads to the inequality

$$C_{\text{lin}}(q, R) < \frac{1}{2} C_{\text{lin}}(q, R) + C_{K,\epsilon} R^\epsilon.$$

We conclude that $C_{\text{lin}}(q, R) \lesssim_\epsilon R^\epsilon$. The same will hold for $q = 2(2n - k + 2)/(2n - k)$, due to an easy application of Hölder's inequality.  $\square$

This argument can be refined as in [32], by repeating the dichotomy for each subspace $V(B_K)$. Using only multilinear estimates, this approach leads to linear restriction estimates that are superior to those in Corollary 4.7, when $n \geq 4$.

While (13.23) is only known in some partial range if $k \leq n - 1$, a slightly weaker version of the $k$-linear restriction estimates was proved by Guth [56] for a much larger range of exponents $q$. This version of $k$-linear restriction estimates is just as good for applications. Guth's range was slightly improved by the author [44] in $\mathbb{R}^4$, and later by Hickman and Rogers [67] in dimensions greater than three. These improvements combine Guth's method with the restricted Kakeya inequality in Theorem 8.33. The reader is referred to check [67] for the best-known range for the linear restriction problem in all dimensions.

## 13.10 Local Smoothing for the Wave Equation

Let us consider the wave equation on $\mathbb{R}^n \times \mathbb{R}$:

$$\begin{cases} \Delta_x u = u_{tt} \\ u(x,0) = f(x), \quad x \in \mathbb{R}^n \\ u_t(x,0) = 0, \quad x \in \mathbb{R}^n. \end{cases}$$

We assume the initial velocity is zero, but the argument can be easily modified to incorporate nonzero velocities. The solution $u(x,t)$ is given by the linear operator $L$ related to the extension operator for the cone

$$Lf(x,t) = \int \widehat{f}(\xi) \cos(2\pi t |\xi|) e(\xi \cdot x) \, d\xi$$

$$= \frac{1}{2}(E^{\text{Cone}^n}(\widehat{f})(x,t) + E^{\text{Cone}^n}(\widehat{f})(x, -t)).$$

Let us recall the Sobolev norm of $f$:

$$\|f\|_{W^{\sigma,p}(\mathbb{R}^n)} = \|\mathcal{F}^{-1}(\widehat{f}(\xi)(1 + |\xi|^2)^{\frac{\sigma}{2}})\|_{L^p(\mathbb{R}^n)}$$

$$\sim \left\| \left( \sum_{j=0}^\infty 2^{2\sigma j} |\mathcal{P}_j f|^2 \right)^{\frac{1}{2}} \right\|_{L^p(\mathbb{R}^n)}.$$

Here $\mathcal{P}_0$ and $\mathcal{P}_j$ ($j \geq 1$) are radial Littlewood–Paley projections for the ball $|\xi| \leq 1$ and the annuli $|\xi| \sim 2^j$, respectively.

Miyachi [84] and Peral [90] have independently proved that the inequality

$$\|Lf(x,t)\|_{W^{-\sigma,p}(\mathbb{R}^n,dx)} \lesssim \|f\|_{L^p(\mathbb{R}^n)}$$

holds for all $t$ if and only if

$$\sigma \geq (n-1)\left|\frac{1}{2}-\frac{1}{p}\right|.$$

While the solution $u(x,t)$ may be less regular for some values of $t$, it is conjectured that higher regularity holds in an average sense for each $p > 2$.

**Conjecture 13.30** (Local Smoothing Conjecture, [100]) *We have*

$$\left(\int_1^2 \|Lf(x,t)\|^p_{W^{-\sigma,p}(\mathbb{R}^n,dx)}\,dt\right)^{\frac{1}{p}} \lesssim \|f\|_{L^p(\mathbb{R}^n)}$$

*as long as*

$$\begin{cases} \sigma > (n-1)(\frac{1}{2}-\frac{1}{p})-\frac{1}{p}, & p \geq \frac{2n}{n-1} \\ \sigma > 0, & 2 < p < \frac{2n}{n-1} \end{cases}.$$

It is known that this conjecture is stronger than the Restriction Conjecture for the sphere $\mathbb{S}^{n-1}$, see [111]. This gives a very good indication of its extraordinary difficulty when $n \geq 3$. Perhaps surprisingly, the conjecture is also open when $n = 2$.

Some of the papers containing significant progress on the Local Smoothing Conjecture include [66], [81], [85], and [100]. Wolff's paper [119] introduced decoupling mainly for the purpose of establishing new local smoothing estimates. In particular, he was able to verify Conjecture 13.30 for $n = 2$ and $p \geq 74$.

Combining Wolff's approach with the sharp decoupling for the cone proved in the previous chapter leads to the following progress.

**Theorem 13.31** ([29]) *Conjecture 13.30 holds in the range* $p \geq 2(n+1)/(n-1)$, *for all* $n \geq 2$.

*Proof* Let $\rho : \mathbb{R} \to [0,\infty)$ be a Schwartz function such that $1_{[1,2]} \leq \rho$ and $\mathrm{supp}\,(\hat{\rho}) \subset [-1,1]$. Given $f$, let

$$Tf(x,t) = E^{\mathrm{Cone}^n}(\hat{f})(x,t)\rho(t).$$

Fix $p \geq 2(n+1)/(n-1)$. It suffices to prove that for each $j \geq 1$ and each $f$ with the Fourier transform supported on the annulus $\mathbb{A}_j = \{\xi \in \mathbb{R}^n : 2^j \leq |\xi| \leq 2^{j+1}\}$, we have

$$\|Tf\|_{L^p(\mathbb{R}^{n+1})} \lesssim_\epsilon 2^{j[(n-1)(\frac{1}{2}-\frac{1}{p})-\frac{1}{p}+\epsilon]}\|f\|_{L^p(\mathbb{R}^n)}. \tag{13.24}$$

A simple computation shows that for each $(\xi, \xi_n) \in \mathbb{R}^n \times \mathbb{R}$,

$$\widehat{Tf}(\xi, \xi_n) = \widehat{f}(\xi)\widehat{\rho}(\xi_n - |\xi|).$$

Thus, $\widehat{Tf}$ is supported on the neighborhood

$$\mathcal{N}_{\mathbb{A}_j}(1) = \{(\xi, |\xi| + s) : \xi \in \mathbb{A}_j, |s| < 1\}$$

of the cone. Partition the sphere $\mathbb{S}^{n-1}$ into caps $\alpha$ with diameter $\sim 2^{-j/2}$. The sets

$$\mathbb{A}_j(\alpha) = \left\{\xi \in \mathbb{A}_j : \frac{\xi}{|\xi|} \in \alpha\right\}$$

partition $\mathbb{A}_j$, while the sets $\theta = \mathcal{N}_{\mathbb{A}_j(\alpha)}(1)$ partition $\mathcal{N}_{\mathbb{A}_j}(1)$. Rescaling Theorem 12.2 and using Hölder's inequality leads to the $l^p(L^p)$ decoupling for $F = Tf$

$$\|F\|_{L^p(\mathbb{R}^{n+1})} \lesssim_{\epsilon} 2^{j\left[\frac{n-1}{2}\left(\frac{1}{2}-\frac{1}{p}\right)+\frac{n-1}{4}-\frac{n+1}{2p}+\epsilon\right]} \left(\sum_{\theta} \|\mathcal{P}_{\theta}F\|^p_{L^p(\mathbb{R}^{n+1})}\right)^{\frac{1}{p}}$$

$$= 2^{j\left[(n-1)\left(\frac{1}{2}-\frac{1}{p}\right)-\frac{1}{p}+\epsilon\right]} \left(\sum_{\theta} \|\mathcal{P}_{\theta}F\|^p_{L^p(\mathbb{R}^{n+1})}\right)^{\frac{1}{p}}.$$

Note that

$$\mathcal{P}_{\theta}F(x,t) = T(\mathcal{P}_{\mathbb{A}_j(\alpha)}f)(x,t),$$

where $\theta = \mathcal{N}_{\mathbb{A}_j(\alpha)}(1)$. Lemma 13.32 shows that

$$\|\mathcal{P}_{\theta}F\|_{L^p(\mathbb{R}^{n+1})} \lesssim \|\mathcal{P}_{\mathbb{A}_j(\alpha)}f\|_{L^p(\mathbb{R}^n)}.$$

On the other hand, the inequality

$$\left(\sum_{\alpha} \|\mathcal{P}_{\mathbb{A}_j(\alpha)}f\|^p_{L^p(\mathbb{R}^n)}\right)^{\frac{1}{p}} \lesssim \|f\|_{L^p(\mathbb{R}^n)}$$

is an immediate consequence of interpolation (cf. Exercise 9.10). Inequality (13.24) follows by combining the last three inequalities.                $\square$

The following lemma is a standard application of oscillatory integral estimates.

**Lemma 13.32** *Let $\psi_1$ be a smooth function compactly supported on $[1/2, 4]$. Let $\psi$ be a smooth function on $\mathbb{S}^{n-1}$ compactly supported on some cap $\alpha$ on $\mathbb{S}^{n-1}$ with diameter $\sim 2^{-j/2}$. Define $\eta$ on $\mathbb{R}^n$ as follows*

$$\eta(\xi) = \psi_1\left(\frac{|\xi|}{2^j}\right) \psi\left(\frac{\xi}{|\xi|}\right) e(t|\xi|).$$

*Then*

$$\int_{\mathbb{R}^n} |\hat{\eta}(x)| \, dx \lesssim 1$$

*uniformly over $j \geq 0$ and $t \sim 1$.*

*Proof* To simplify notation, we will consider the case $n = 2$. It will be clear from the argument that we may work with $t = 1$. We may also assume that the cap $\alpha$ is an arc containing $(1, 0)$ and that

$$\psi(\cos \phi, \sin \phi) = \psi_2(2^{\frac{j}{2}} \phi)$$

for some smooth $\psi_2$ compactly supported on $[-1, 1]$.

Let $\phi_0$ be the polar angle of $x$, so $x = |x|(\cos \phi_0, \sin \phi_0)$. Then

$$\hat{\eta}(x) = \int_{r=0}^{\infty} \int_{\phi=-\pi}^{\pi} r \psi_1 \left( \frac{r}{2^j} \right) e(r(1 - |x| \cos(\phi - \phi_0))) \psi_2(2^{\frac{j}{2}} \phi) \, dr \, d\phi.$$

Let $\psi_0(r) = r \psi_1(r)$. The relevant features of this function can be summarized as follows: $\psi_0 \in \mathcal{S}(\mathbb{R})$ and $\psi_0^{(l)}(0) = 0$ for all $l \geq 0$.

For $l \geq 1$, let $F_l$ be defined recursively by

$$\begin{cases} F_1(r) = \int_{-\infty}^{r} \hat{\psi}_0(s) \, ds \\ F_{l+1}(r) = \int_{-\infty}^{r} F_l(s) \, ds, \ \ l \geq 1. \end{cases}$$

The properties of $\psi_0$ force each $F_l$ to be a Schwartz function.

We may write

$$\hat{\eta}(x) = 2^{2j} \int_{-\infty}^{\infty} \hat{\psi}_0(2^j(|x| \cos(\phi - \phi_0) - 1)) \psi_2(2^{\frac{j}{2}} \phi) \, d\phi.$$

Let us start by observing that for $|x| \leq 1/2$, we have

$$|\hat{\eta}(x)| \lesssim 1, \tag{13.25}$$

since $2^j(1 - |x| \cos(\phi - \phi_0)) \sim 2^j$ and $\hat{\psi}_0$ has rapid decay.

We distinguish the following regions for $|x| \geq 1/2$. For $0 \leq k \leq j - 1$, the set $S_k$ contains all $|x| \geq 1/2$ satisfying $|\sin(\phi - \phi_0)| \sim 2^{-k/2}$ for all $|\phi| \leq 2^{-j/2}$. The set $S_j$ contains all $|x| \geq 1/2$ such that $|\sin(\phi - \phi_0)| \lesssim 2^{-j/2}$ for all $|\phi| \leq 2^{-j/2}$. Note that the polar angles of all $x \in S_k$ belong to a set $U_k$ of measure $\lesssim 2^{-k/2}$.

We first estimate the contribution from $x \in S_j$, using a change of variables and the fact that $|\cos(\phi - \phi_0)| \gtrsim 1$ for all $|\phi| \leq 2^{-j/2}$:

$$\int_{S_j} |\hat{\eta}(x)| \, dx$$

$$\leq 2^{2j} \int_{\phi_0 \in U_j} \left[ \int_{-\infty}^{\infty} \left[ \int_0^{\infty} r |\hat{\psi}_0(2^j (r \cos(\phi - \phi_0) - 1))| \, dr \right] |\psi_2(2^{\frac{j}{2}} \phi)| \, d\phi \right] d\phi_0$$

$$\leq \int_{\phi_0 \in U_j} \left[ \int_{-\infty}^{\infty} \left| \frac{\psi_2(2^{\frac{j}{2}} \phi)}{\cos(\phi - \phi_0)^2} \right| d\phi \right] d\phi_0 \int_{-\infty}^{\infty} |r \hat{\psi}_0(r - 2^j)| \, dr$$

$$\lesssim |U_j| \int_{-\infty}^{\infty} |\psi_2(2^{\frac{j}{2}} \phi)| \, d\phi \int_{-\infty}^{\infty} |r \hat{\psi}_0(r - 2^j)| \, dr$$

$$\lesssim 1.$$

Assume now that $x$ is in some $S_k$ with $0 \leq k \leq j - 1$. Performing $l$ integrations by parts shows that

$$|\hat{\eta}(x)| = \frac{2^{2j}}{(2^j |x|)^l} \left| \int_{-\infty}^{\infty} F_l(2^j (|x| \cos(\phi - \phi_0) - 1)) D^{(l)}(\psi_2(2^{\frac{j}{2}} \phi)) \, d\phi \right|,$$

$$(13.26)$$

where

$$Dg(\theta) = \frac{d}{d\phi} \left( \frac{g(\phi)}{\sin(\phi - \phi_0)} \right).$$

A simple computation reveals that

$$\int_{-\infty}^{\infty} |D^{(l)}(\psi_2(2^{\frac{j}{2}} \phi))| \, d\phi \lesssim_l 2^{\frac{kl + j(l-1)}{2}}.$$

$$(13.27)$$

In particular, combining this for $l = 3$ with (13.26) shows that for each $x \in S_0$, we have

$$|\hat{\eta}(x)| \lesssim \frac{1}{|x|^3}$$

$$(13.28)$$

and thus

$$\int_{S_0} |\hat{\eta}(x)| \, dx \lesssim 1.$$

When $x \in S_k$ for some $1 \leq k \leq j - 1$, we combine integration by parts with a change of variables. More precisely, using (13.26) with $l = 2$, we get

$$|\hat{\eta}(x)| \leq \frac{2^{2j}}{(2^j |x|)^2} \left| \int_{-\infty}^{\infty} |F_2|(2^j (|x| \cos(\phi - \phi_0)) - 1) |D^{(2)}(\psi_2(2^{\frac{j}{2}} \phi))| \, d\phi \right|.$$

Integrating this over $x \in S_k$ leads to

$$\int_{x \in S_k} |\hat{\eta}(x)| \, dx$$

$$\leq \int_{\phi_0 \in U_k} \left[ \int_{-\infty}^{\infty} \left[ \int_{r \geq \frac{1}{2}} r \frac{2^{2j}}{(2^j r)^2} |F_2|(2^j (r \cos(\phi - \phi_0) - 1)) \, dr \right] \right.$$

$$\left. \times |D^{(2)}(\psi_2(2^{\frac{j}{2}}\phi))| \, d\phi \right] d\phi_0$$

$$\leq \int_{\phi_0 \in U_k} \left[ \int_{|r| \gtrsim 2^j} \frac{1}{r} |F_2|(2^j + r) \, dr \right] \left[ \int_{-\infty}^{\infty} |D^{(2)}(\psi_2(2^{\frac{j}{2}}\phi))| \, d\phi \right] d\phi_0.$$

The integral in $r$ is $O(2^{-j})$. The integral in $\phi$ is $O(2^{k+j/2})$, due to (13.27). Thus

$$\int_{x \in S_k} |\hat{\eta}(x)| \, dx \lesssim 2^{\frac{k-j}{2}}.$$

We conclude that the contribution from $\cup_{1 \leq k \leq j-1} S_k$ is also acceptable:

$$\sum_{1 \leq k \leq j-1} \int_{x \in S_k} |\hat{\eta}(x)| \, dx \lesssim \sum_{k=1}^{j-1} 2^{\frac{k-j}{2}} \lesssim 1. \qquad \square$$

## 13.11 Problems for This Chapter

**Exercise 13.33** Let $S$ be a collection of positive integers in some interval of length $L \geq N$. Assume that

$$\min_{\substack{k,k' \in S \\ k \neq k'}} |k - k'| \geq \frac{L}{N}.$$

Then for each $a_k \in \mathbb{C}$,

$$\left\| \sum_{k \in S} a_k e(k x_1 + k^2 x_2 + \cdots + k^n x_n) \right\|_{L^{n(n+1)}([0,1]^n)} \lesssim_\epsilon N^\epsilon \|a_k\|_{l^2(S)}.$$

The implicit constant is independent of $L$ and $a_k$.

**Exercise 13.34** (Additive energy of $1/\sqrt{R}$-separated $R^{-1}$-balls) Let $\Lambda \subset \mathbb{P}^2$ consist of $1/\sqrt{R}$-separated points.
    Prove that

$$|\{(\lambda_1, \lambda_2, \lambda_3, \lambda_4) \in \Lambda^4 : |\lambda_1 + \lambda_2 - (\lambda_3 + \lambda_4)| \lesssim R^{-1}\}| \lesssim_\epsilon |\Lambda|^2 R^\epsilon.$$

Compare this with Exercise 3.35.

**Exercise 13.35** Let $A$ be a $\delta$-separated set inside $[1,2]$. Prove that

$$|\{(a_1,\ldots,a_6) \in A^6 \colon a_1 + a_2 + a_3 = a_4 + a_5 + a_6 \text{ and } a_1 a_2 a_3 = a_4 a_5 a_6\}|$$
$$\lesssim_\epsilon \delta^{-\epsilon} |A|^3.$$

**Exercise 13.36** Prove that for each $a_k \in \mathbb{C}$

$$\left\| \sum_{k=1}^N a_k e(k^2 x) \right\|_{L^4([0,1])} \lesssim_\epsilon N^\epsilon \|a_k\|_{l^2}$$

and that

$$\left\| \sum_{k=1}^N e(k^2 x) \right\|_{L^4([0,1])} \gtrsim (\log N)^{\frac{1}{4}} N^{\frac{1}{2}}.$$

**Exercise 13.37** Let $l \geq 2$ and $2 \leq p < 2l$. Assume that

$$\left\| \sum_{k=1}^N a_k e(k^l x) \right\|_{L^p([0,1])} \lesssim \|a_k\|_{l^2}$$

holds for each $a_k \in \{0,1\}$, with an implicit constant independent of $N$.
    Recall the definition of $\mathcal{Q}_l(N)$ from (13.20). Prove that

$$\mathcal{Q}_l(N) \lesssim N^{\frac{2}{p}}.$$

Hint: Let $A = \{a + nq \colon 1 \leq n \leq N\}$, $f(x) = \sum_{m \in A} e(mx)$ and $g(x) = \sum_{k \colon k^l \in A} e(k^l x)$. Analyze $\int_0^1 f(x)\overline{g(x)}\,dx$.

**Exercise 13.38** Let $B$ be an arbitrary ball of radius 1 in $\mathbb{R}^3$.

(a) For $1 \leq m \leq M$, consider the functions $w_m \colon B \to \mathbb{C}$ satisfying $\|w_m\|_{L^\infty(B)} \leq 1$ and

$$\sum_{2 \leq m \leq M} \|w_m - w_{m-1}\|_{L^\infty(B)} \leq 1.$$

Prove that for each $a_{m_1,m_2} \in \mathbb{C}$ with $|a_{m_1,m_2}| \leq 1$, we have

$$\left\| \sum_{m_1=1}^M \sum_{m_2=1}^M a_{m_1,m_2} w_{m_1}(x) w_{m_2}(x) e(m_1 x_1 + m_2 x_2 + (m_1^2 + m_2^2) x_3) \right\|_{L^4(B)} \lesssim_\epsilon M^{1+\epsilon}.$$

(b) Let $I_1, I_2 \subset [1, N]$ be intervals with dist $(I_1, I_2) \gtrsim N$. Assume that $I_i$ is the union of pairwise disjoint intervals $J_i$ of length $\sim N^{1/2}$. Prove that

$$\left\| \sum_{k_1 \in I_1} e\left(k_1 x_1 + k_1^2 x_2 + \frac{k_1^3}{N^2} x_3\right) \sum_{k_2 \in I_2} e\left(k_2 x_1 + k_2^2 x_2 + \frac{k_2^3}{N^2} x_3\right) \right\|_{L^4(B)}$$

$$\lesssim_\epsilon N^{\frac{1}{4}+\epsilon} \left( \sum_{J_1 \subset I_1} \sum_{J_2 \subset I_2} \left\| \sum_{k_1 \in J_1} e\left(k_1 x_1 + k_1^2 x_2 + \frac{k_1^3}{N^2} x_3\right) \right. \right.$$

$$\left. \left. \times \sum_{k_2 \in J_2} e\left(k_2 x_1 + k_2^2 x_2 + \frac{k_2^3}{N^2} x_3\right) \right\|_{L^4(B)}^4 \right)^{\frac{1}{4}}.$$

(c) Prove that for each $J_1, J_2$ as in the preceding,

$$\left\| \sum_{k_1 \in J_1} e\left(k_1 x_1 + k_1^2 x_2 + \frac{k_1^3}{N^2} x_3\right) \sum_{k_2 \in J_2} e\left(k_2 x_1 + k_2^2 x_2 + \frac{k_2^3}{N^2} x_3\right) \right\|_{L^4(B)} \lesssim_\epsilon N^{\frac{1}{2}+\epsilon}.$$

Conclude that

$$\left\| \sum_{k_1 \in I_1} e\left(k_1 x_1 + k_1^2 x_2 + \frac{k_1^3}{N^2} x_3\right) \sum_{k_2 \in I_2} e\left(k_2 x_1 + k_2^2 x_2 + \frac{k_2^3}{N^2} x_3\right) \right\|_{L^4(B)} \lesssim_\epsilon N^{1+\epsilon}.$$

(d) Use a bilinear-to-linear reduction to argue that

$$\left\| \sum_{k=1}^N e\left(k x_1 + k^2 x_2 + \frac{k^3}{N^2} x_3\right) \right\|_{L^8(B)} \lesssim_\epsilon N^{\frac{1}{2}+\epsilon},$$

then prove that the exponent 8 cannot be larger. Conclude that for each rectangular box $T_N = [a, a+N] \times [b, b+N^2] \times [c, c+N]$ we have

$$\left\| \sum_{k=1}^N e\left(\frac{k}{N} x_1 + \frac{k^2}{N^2} x_2 + \frac{k^3}{N^3} x_3\right) \right\|_{L^8_\sharp(T_N)} \lesssim_\epsilon N^{\frac{1}{2}+\epsilon}.$$

Compare this with Vinogradov's Mean Value Theorem.

Hint for (a): Use summation by parts.

Hint for (b): Write $k_1 + k_2 = m_1$, $k_1^2 + k_2^2 = m_2$, $k_1^3 + k_2^3 = 3m_1 m_2/2 - m_1^3/2$. Then use periodicity to reduce the desired estimate to an application of decoupling for the surface $\psi(\xi_1, \xi_2) = 3\xi_1\xi_2/2 - \xi_1^3/2$, on balls of radius $\sim N$.

Hint for (c): Change variables $k_i = l_i + m_i$ with $1 \le m_i \lesssim N^{1/2}$, and

$$\begin{cases} y_1 = x_1 + \dfrac{3l_1^2}{N^2}x_3 \\[2mm] y_2 = x_1 + \dfrac{3l_2^2}{N^2}x_3 \\[2mm] y_3 = x_2, \end{cases}$$

then apply (a).

**Exercise 13.39** Given $S \subset \mathbb{Z}^n$, let

$$K(x) = \sum_{\xi \in S} e(\xi \cdot x).$$

Consider a sequence $(a_\xi)_{\xi \in S} \subset \mathbb{C}$ satisfying $\|a_\xi\|_{l^2(S)} = 1$ and let

$$F(x) = \sum_{\xi \in S} a_\xi e(\xi \cdot x).$$

For $\alpha > 0$, define

$$E_\alpha = \{x \in \mathbb{T}^n : |F(x)| > \alpha\}.$$

Prove that for each decomposition $K = K_1 + K_2$, we have

$$\alpha^2 |E_\alpha|^2 \le \|K_1\|_\infty |E_\alpha|^2 + \|\widehat{K_2}\|_\infty |E_\alpha|.$$

Hint: Use a Stein–Tomas-type argument.

**Exercise 13.40** Let

$$K(x) = \sum_{k_1=1}^{N} \cdots \sum_{k_{n-1}=1}^{N} e(k_1 x_1 + \cdots + k_{n-1} x_{n-1} + (k_1^2 + \cdots + k_{n-1}^2)x_n).$$

Let $\eta \colon [-1/10, 1/10] \to \mathbb{R}$ be a smooth function with integral equal to 1. For $N \le Q \le N^2$, define

$$A_Q := \{Q \le q \le 2Q : q \text{ is prime}\},$$

and

$$\mathcal{F}_Q := \left\{ \frac{a}{q} : q \in A_Q, \ 1 \le a \le q-1 \right\}.$$

Let

$$K^Q(x) = K(x) \frac{Q^2}{|\mathcal{F}_Q|} \sum_{\frac{a}{q} \in \mathcal{F}_Q} \eta \left( Q^2 \left( x_n - \frac{a}{q} \right) \right).$$

(a) Prove that $\|K^Q\|_\infty \lesssim Q^{(n-1)/2} \log N$ and $\|\widehat{K - K_Q}\|_\infty \lesssim \log N/Q$.

(b) Combine these estimates with Exercise 13.39 to argue that for each $\alpha \gtrsim N^{n-1/4}(\log N)^{1/2}$ and each $l^2$ normalized sequence $a_{k_1,\dots,k_{n-1}} \in \mathbb{C}$, we have

$$|\{x \in \mathbb{T}^n : |F(x)| > \alpha\}| \lesssim \frac{(\log N)^{\frac{n+1}{n-1}}}{\alpha^{\frac{2(n+1)}{n-1}}},$$

where

$$F(x) = \sum_{k_1=1}^{N} \cdots \sum_{k_{n-1}=1}^{N} a_{k_1,\dots,k_{n-1}} e(k_1 x_1 + \cdots + k_{n-1}x_{n-1} + (k_1^2 + \cdots + k_{n-1}^2)x_n).$$

Hint for (a): Use the Prime Number Theorem

$$\left|\pi(Q) - \frac{Q}{\log Q}\right| \lesssim \frac{Q}{(\log Q)^2}$$

to argue that $|\mathcal{F}_Q| \sim Q^2/\log Q$. Prove that

$$\left|\sum_{k=1}^{N} e(kt + k^2 x_n)\right| \lesssim Q^{\frac{1}{2}}$$

whenever $|x_n - a/q| \lesssim 1/Q^2$ for some $a/q \in \mathcal{F}_Q$.

# References

[1] Arkhipov, G. I., Chubarikov, V. N. and Karatsuba, A. A. *Trigonometric sums in number theory and analysis*, Translated from the 1987 Russian original. De Gruyter Expositions in Mathematics, 39. Walter de Gruyter GmbH & Co. KG, Berlin, 2004

[2] Bak, Jong-Guk and Seeger, A. *Extensions of the Stein–Tomas theorem*, Math. Res. Lett. 18 (2011), no. 4, 767–81

[3] Barcelo, B. *On the restriction of the Fourier transform to a conical surface*, Trans. Amer. Math. Soc. 292 (1985), 321–33

[4] Bejenaru, I. *Optimal bilinear restriction estimates for general hypersurfaces and the role of the shape operator*, Int. Math. Res. Not. IMRN (2017), no. 23, 7109–47

[5] Bejenaru, I. *Optimal multilinear restriction estimates for a class of surfaces with curvature*, Anal. PDE 12 (2019), no. 4, 1115–48

[6] Bejenaru, I. *The multilinear restriction estimate: a short proof and a refinement*, Math. Res. Lett. 24 (2017), no. 6, 1585–1603

[7] Bejenaru, I. *The optimal trilinear restriction estimate for a class of hypersurfaces with curvature*, Adv. Math. 307 (2017), 1151–83

[8] Bennett, J. *Aspects of multilinear harmonic analysis related to transversality*, Harmonic analysis and partial differential equations, Edited by Patricio Cifuentes, José García-Cuerva, Gustavo Garrigós, et al., 1–28, Contemp. Math., 612, Amer. Math. Soc., Providence, RI, 2014

[9] Bennett, J. *A trilinear restriction problem for the paraboloid in $\mathbb{R}^3$*, Electron. Res. Announc. Amer. Math. Soc. 10 (2004), 97–102

[10] Bennett, J., Bez, N., Flock, T. and Lee, S. *Stability of Brascamp–Lieb constant and applications*, Amer. J. Math. 140 (2018), no. 2, 543–69

[11] Bennett, J., Carbery, A., Christ, M. and Tao, T. *Finite bounds for Hölder–Brascamp–Lieb multilinear inequalities*, Math. Res. Lett. 17 (2010), no. 4, 647–66

[12] Bennett, J., Carbery, A., Christ, M. and Tao, T. *The Brascamp–Lieb inequalities: finiteness, structure and extremals*, Geom. Funct. Anal. 17 (2008), no. 5, 1343–1415

[13] Bennett, J., Carbery, A. and Tao, T. *On the multilinear restriction and Kakeya conjectures*, Acta Math. 196 (2006), no. 2, 261–302

322

[14] Bennett, J., Carbery, A. and Wright, J. *A non-linear generalisation of the Loomis–Whitney inequality and applications*, Math. Res. Lett. 12 (2005), no. 4, 443–57

[15] Biswas, C., Gilula, M., Li, L., Schwend, J. and Xi, Y. $l^2$ *decoupling in* $\mathbb{R}^2$ *for curves with vanishing curvature*, available on arXiv

[16] Blomer, V. and Brüdern, J. *The number of integer points on Vinogradov's quadric*, Monatsh. Math. 160 (2010), no. 3, 243–56

[17] Bombieri, E. and Bourgain, J. *A problem on sums of two squares*, Int. Math. Res. Not. IMRN (2015), no. 11, 3343–407

[18] Bombieri, E. and Zannier U. *Note on squares in arithmetic progressions II*, Atti Accad. Naz. Lincei Cl. Sci. Fis. Mat. Natur. Rend. Lincei (9), Mat. Appl. 13 (2002), no. 2, 69–75

[19] Bourgain, J. *On the Vinogradov integral* (Russian) English version published in Proc. Steklov Inst. Math. 296 (2017), no. 1, 30–40. Tr. Mat. Inst. Steklova 296 (2017)

[20] Bourgain, J. *Moment inequalities for trigonometric polynomials with spectrum in curved hypersurfaces*, Israel J. Math. 193 (2013), no. 1, 441–58

[21] Bourgain, J. *Fourier transform restriction phenomena for certain lattice subsets and applications to nonlinear evolution equations. I. Schrödinger equations*, Geom. Funct. Anal. 3 (1993), no. 2, 107–56

[22] Bourgain, J. *Besicovitch type maximal operators and applications to Fourier analysis*, Geom. Funct. Anal. 1 (1991), no. 2, 147–87

[23] Bourgain, J. *Eigenfunction bounds for the Laplacian on the n-torus*, Internat. Math. Res. Not. (1993), no. 3, 61–6

[24] Bourgain, J. *On* $\Lambda(p)$*-subsets of squares*, Israel J. Math. 67 (1989), no. 3, 291–311

[25] Bourgain, J., Demeter, C. and Kemp, D. *Decouplings for real analytic surfaces of revolution*, to appear in GAFA seminar notes

[26] Bourgain, J., Demeter, C. and Guo, S. *Sharp bounds for the cubic Parsell–Vinogradov system in two dimensions*, Adv. Math. 320 (2017), 827–75

[27] Bourgain, J. and Demeter, C. *Decouplings for curves and hypersurfaces with nonzero Gaussian curvature*, Journal d'Analyse Mathematique 133 (2017), 279–311

[28] Bourgain, J. and Demeter, C. *Mean value estimates for Weyl sums in two dimensions*, J. Lond. Math. Soc. (2) 94 (2016), no. 3, 814–38

[29] Bourgain, J. and Demeter, C. *The proof of the* $l^2$ *decoupling conjecture*, Ann. of Math. 182 (2015), no. 1, 351–89

[30] Bourgain, J. and Demeter, C. *New bounds for the discrete Fourier restriction to the sphere in 4D and 5D*, Int. Math. Res. Not. IMRN (2015), no. 11, 3150–84

[31] Bourgain, J. and Demeter, C. *Improved estimates for the discrete Fourier restriction to the higher dimensional sphere*, Illinois J. Math. 57 (2013), no. 1, 213–27

[32] Bourgain, J. and Guth, L. *Bounds on oscillatory integral operators based on multilinear estimates*, Geom. Funct. Anal. 21 (2011), no. 6, 1239–95

[33] Bourgain, J., Demeter, C. and Guth, L., *Proof of the main conjecture in Vinogradov's mean value theorem for degrees higher than three*, Ann. of Math. (2) 184 (2016), no. 2, 633–82

[34] Brandolini, L., Gigante, G., Greenleaf, A., Iosevich, A., Seeger, A. and Travaglini, G. *Average decay estimates for Fourier transforms of measures supported on curves*, J. Geom. Anal. 17 (2007), no. 1, 15–40

[35] Carbery, A. *A remark on reverse Littlewood–Paley, restriction and Kakeya*, available on arXiv at https://arxiv.org/pdf/1507.02515.pdf

[36] Carleson, L. *On the Littlewood–Paley theorem*, Report, Mittag-Leffler Inst. (1967)

[37] Christ, M. *On the restriction of the Fourier transform to curves: endpoint results and the degenerate case*, Trans. Amer. Math. Soc. 287 (1985), no. 1, 223–38

[38] Cilleruelo, J. and Granville, A. *Lattice points on circles, squares in arithmetic progressions and sumsets of squares*, Additive combinatorics, Edited by Andrew Granville, Melvyn B. Nathanson, and József Solymosi, 241–62, CRM Proc. Lecture Notes, 43, Amer. Math. Soc., Providence, RI, 2007

[39] Córdoba, A. *Vector valued inequalities for multipliers*, Conference on harmonic analysis in honor of Antoni Zygmund, Vol. I, II, 295–305 (Chicago, Il, 1981), Wadsworth Math. Ser., Wadsworth, Belmont, CA, 1983

[40] Córdoba, A. *Geometric Fourier analysis*, Ann. Inst. Fourier (Grenoble) 32 (1982), no. 3, vii, 215–26

[41] Córdoba, A. *The Kakeya maximal function and the spherical summation multipliers*, Amer. J. Math. 99 (1977), no. 1, 1–22

[42] Davies, R. O. *Some remarks on the Kakeya problem*, Proc. Cambridge Philos. Soc. 69 (1971), 417–21

[43] Demeter, C. *A decoupling for Cantor-like sets*, Proc. Amer. Math. Soc. 147 (2019), no. 3, 1037–50

[44] Demeter, C. *On the restriction theorem for paraboloid in* $\mathbb{R}^4$, Colloq. Math. 156 (2019), no. 2, 301–11

[45] Drury, S. W. *Restrictions of Fourier transforms to curves*, Ann. Inst. Fourier (Grenoble) 35 (1985), no. 1, 117–23

[46] Dvir, Z. *On the size of Kakeya sets in finite fields*, J. Amer. Math. Soc. 22 (2009), no. 4, 1093–7

[47] Fefferman, C. *A note on spherical summation multipliers*, Israel J. Math. 15 (1973), 44–52

[48] Fefferman, C. *The multiplier problem for the ball*, Ann. of Math. (2) 94 (1971), 330–6

[49] Fefferman, C. *Inequalities for strongly singular convolution operators*, Acta Math. 124 (1970), 9–36

[50] Foschi, D. *Global maximizers for the sphere adjoint Fourier restriction inequality*, J. Funct. Anal. 268 (2015), no. 3, 690–702

[51] Garrigós, G. and Seeger, A. *A mixed norm variant of Wolff's inequality for paraboloids*, Harmonic analysis and partial differential equations, Contemporary Mathematics, Vol. 505, Edited by Patricio Cifuentes, José García-Cuerva, Gustavo Garrigós, et al., Amer. Math. Soc., Providence, RI, 2010, 179–97

[52] Garrigós, G. and Seeger, A. *On plate decompositions of cone multipliers*, Proc. Edinb. Math. Soc. (2) 52 (2009), no. 3, 631–51

[53] Greenleaf, A. *Principal curvature and harmonic analysis*, Indiana Univ. Math. J. 30 (1981), no. 4, 519–37

[54] Grosswald, E. *Representations of integers as sums of squares*, Springer-Verlag, New York, NY, 1985

[55] Guo, S. and Zhang, R. *On integer solutions of Parsell–Vinogradov systems*, available on arXiv

[56] Guth, L. *Restriction estimates using polynomial partitioning II*, available on arXiv

[57] Guth, L. *A restriction estimate using polynomial partitioning*, J. Amer. Math. Soc. 29 (2016), no. 2, 371–413

[58] Guth, L. *Polynomial methods in combinatorics*, University Lecture Series, 64. Amer. Math. Soc., Providence, RI, 2016

[59] Guth, L. *A short proof of the multilinear Kakeya inequality*, Math. Proc. Cambridge Philos. Soc. 158 (2015), no. 1, 147–53

[60] Guth, L. *The endpoint case of the Bennett–Carbery–Tao multilinear Kakeya conjecture*, Acta Math. 205 (2010), no. 2, 263–86

[61] Guth, L. and Katz, N. H. *On the Erdös distinct distances problem in the plane*, Ann. of Math. (2) 181 (2015), no. 1, 155–90

[62] Guth, L. and Katz, N. H. *Algebraic methods in discrete analogs of the Kakeya problem,* Adv. Math. 225 (2010), no. 5, 2828–39

[63] Guth, L. and Zahl, J. *Polynomial Wolff axioms and Kakeya-type estimates in* $\mathbb{R}^4$, available on arXiv

[64] Guth, L., Iosevich, A., Ou, Y. and Wang, H. *On Falconer's distance set problem in the plane*, available on arXiv

[65] Hardy, G. H. and Littlewood, J. E. *Some problems of diophantine approximation*, Acta Math. 37 (1914), no. 1, 193–239

[66] Heo, Y., Nazarov, F. and Seeger, A. *Radial Fourier multipliers in high dimensions*, Acta Math. 206 (2011), no. 1, 55–92

[67] Hickman, J. and Rogers, K., M. *Improved Fourier restriction estimates in higher dimensions*, available on arXiv

[68] Hörmander, L. *Oscillatory integrals and multipliers on* $FL^p$, Ark. Mat. 11, (1973), 1–11

[69] Hua, L. K. *On Waring's problem*, Q. J. Math 9. 199–202

[70] Kaplan, H., Sharir, M. and Shustin, E. *On lines and joints*, Discrete Comput. Geom. 44 (2010), no. 4, 838–43

[71] Katznelson, Y., *An introduction to harmonic analysis*. Third edition. Cambridge Mathematical Library. Cambridge University Press, Cambridge, 2004

[72] Katz, N. H. and Rogers, K. *On the polynomial Wolff axioms*, available on arXiv

[73] Katz, N. H. and Zahl, J. *An improved bound on the Hausdorff dimension of Besicovitch sets in* $\mathbb{R}^3$, available on arXiv

[74] Katz, N. H. and Tao, T. *New bounds for Kakeya problems. Dedicated to the memory of Thomas H. Wolff*, J. Anal. Math. 87 (2002), 231–63

[75] Killip, R. and Visan, M. *Scale invariant Strichartz estimates on tori and applications*, Math. Res. Lett. 23 (2016), no. 2, 445–72

[76] Łaba, I. and Pramanik, M. *Wolff's inequality for hypersurfaces*, Collect. Math. 2006, Vol. Extra, 293–326

[77] Łaba, I. and Tao, T. *An improved bound for the Minkowski dimension of Besicovitch sets in medium dimension*, Geom. Funct. Anal. 11 (2001), no. 4, 773–806

[78] Łaba, I. and Tao, T. *An X-ray transform estimate in* $\mathbb{R}^3$, Rev. Mat. Iberoamericana 17 (2001), no. 2, 375–407

[79] Łaba, I. and Wolff, T. *A local smoothing estimate in higher dimensions, dedicated to the memory of Tom Wolff.* J. Anal. Math. 88 (2002), 149–71

[80] Lacey, M. T. *Issues related to Rubio de Francia's Littlewood–Paley inequality*, New York Journal of Mathematics. NYJM Monographs, 2. State University of New York, University at Albany (2007)

[81] Lee, S. and Vargas, A. *On the cone multiplier in* $R^3$, J. Funct. Anal. 263 (2012), no. 4, 925–40

[82] Lee, S. and Vargas, A. *Restriction estimates for some surfaces with vanishing curvatures*, J. Funct. Anal. 258 (2010), no. 9, 2884–909

[83] Mattila, P. *Fourier analysis and Hausdorff dimension*, Cambridge Studies in Advanced Mathematics, 150. Cambridge University Press, Cambridge, 2015

[84] Miyachi, A. *On some estimates for the wave equation in* $L^p$ *and* $H^p$, J. Fac. Sci. Univ. Tokyo Sect. IA Math. 27 (1980), no. 2, 331–54

[85] Mockenhaupt, G., Seeger, A. and Sogge, D. *Wave front sets, local smoothing and Bourgain's circular maximal theorem*, Ann. of Math. (2) 136 (1992), no. 1, 207–2018

[86] Müller, D., Ricci, F. and Wright, J. *A maximal restriction theorem and Lebesgue points of functions in* $F(L^p)$, available on arXiv

[87] Nicola, F. *Slicing surfaces and the Fourier restriction conjecture*, Proc. Edinb. Math. Soc. (2) 52 (2009), no. 2, 515–27

[88] Ou, Y. and Wang, H. *A cone restriction estimate using polynomial partitioning*, available on arXiv

[89] Pach, J. and Sharir, M. *Repeated angles in the plane and related problems*, J. Combin. Theory Ser. A, 59 (1992), 12–22

[90] Peral, Juan, C. $L^p$ *estimates for the wave equation*, J. Funct. Anal. 36 (1980), no. 1, 114–45

[91] Pramanik, M. and Seeger, A. $L^p$ *regularity of averages over curves and bounds for associated maximal operators*, Amer. J. Math. 129 (2007), no. 1, 61–103

[92] Prestini, E. *A restriction theorem for space curves*, Proc. Amer. Math. Soc. 70 (1978), no. 1, 8–10

[93] Prestini, E. *Restriction theorems for the Fourier transform to some manifolds in* $\mathbb{R}^n$, Harmonic analysis in Euclidean spaces (Proc. Sympos. Pure Math., Williams Coll., Williamstown, MA, 1978), Part 1, 101–9, Proc. Sympos. Pure Math., XXXV, Part I, Amer. Math. Soc., Providence, RI, 1979

[94] Quilodrán, R. *The joints problem in* $\mathbb{R}^3$, SIAM J. Discrete Math. 23 (2009/10), no. 4, 2211–13

[95] Ramos, J. *The trilinear restriction estimate with sharp dependence on the transversality*, available on arXiv

[96] Rubio de Francia, J, L. *A Littlewood–Paley inequality for arbitrary intervals*, Rev. Mat. Iberoamericana 1 (1985), no. 2, 114

[97] Rudin, W. *Trigonometric series with gaps*, Journal of Mathematics and Mechanics, 9 (1960), no. 2, 203–27

[98] Sanders, T. *The structure theory of set addition revisited*, Bull. Amer. Math. Soc. (N.S.) 50 (2013), no. 1, 93–127

[99] Sogge, C., D. *Improved critical eigenfunction estimates on manifolds of nonpositive curvature*, Math. Res. Lett. 24 (2017), no. 2, 549–70

[100] Sogge, C., D. *Propagation of singularities and maximal functions in the plane*, Invent. Math. 104 (1991), no. 2, 349–76

[101] Sogge, C., D. *Concerning the $L^p$ norm of spectral clusters for second-order elliptic operators on compact manifolds*, J. Funct. Anal. 77 (1988), no. 1, 123–38

[102] Stein, E. M. *Harmonic analysis: real-variable methods, orthogonality, and oscillatory integrals. With the assistance of Timothy S. Murphy*, Princeton Mathematical Series, 43. Monographs in Harmonic Analysis, III. Princeton University Press, Princeton, NJ, 1993

[103] Stein, E. M. *Oscillatory integrals in Fourier analysis. Beijing lectures in harmonic analysis (Beijing, 1984)*, 307–355, Ann. of Math. Stud., 112, Princeton University Press, Princeton, NJ, 1986

[104] A. Stone, and J. Tukey, *Generalized sandwich theorems*, Duke Math. J. 9, (1942), 356–59

[105] Strichartz, R. S. *Restrictions of Fourier transforms to quadratic surfaces and decay of solutions of wave equations*, Duke Math. J. 44 (1977), no. 3, 705–14

[106] Székely, L., A. *Crossing numbers and hard Erdös problems in discrete geometry*, Combin. Probab. Comput. 6 (1997), no. 3, 353–8

[107] Tao, T. *Lectures Notes 5*, www.math.ucla.edu/~tao/254b.1.99s/

[108] Tao, T. *A sharp bilinear restrictions estimate for paraboloids*, Geom. Funct. Anal. 13 (2003), no. 6, 1359–84

[109] Tao, T. *Endpoint bilinear restriction theorems for the cone, and some sharp null form estimates*, Math. Z. 238 (2001), no. 2, 215–68

[110] Tao, T. *The Bochner–Riesz conjecture implies the restriction conjecture*, Duke Math. J. 96 (1999), no. 2, 363–75

[111] Tao, T. *The weak-type endpoint Bochner–Riesz conjecture and related topics*, Indiana Univ. Math. J. 47 (1998), no. 3, 1097–1124

[112] Tao, T. and Vargas, A. *A bilinear approach to cone multipliers I. Restriction estimates*, Geom. Funct. Anal. 10 (2000), no. 1, 185–215

[113] Tao, T. and Vu, V. *Additive combinatorics*, Cambridge Studies in Advanced Mathematics, 105. Cambridge University Press, Cambridge, 2006

[114] Tao, T., Vargas, A. and Vega, L. *A bilinear approach to the restriction and Kakeya conjectures*, J. Amer. Math. Soc. 11 (1998), no. 4, 967–1000

[115] Tomas, P. A. *A restriction theorem for the Fourier transform*, Bull. Amer. Math. Soc. 81 (1975), 477–8

[116] Wang, H. *A restriction estimate in $\mathbb{R}^3$ using brooms*, available on arXiv

[117]  Wolff, T. *Lectures on harmonic analysis. With a foreword by Charles Fefferman and preface by Izabella Łaba*, Edited by Łaba and Carol Shubin. University Lecture Series, 29, Amer. Math. Soc., Providence, RI, 2003

[118]  Wolff, T. *A sharp bilinear cone restriction estimate*, Ann. of Math. (2) 153 (2001), no. 3, 661–98

[119]  Wolff, T. *Local smoothing type estimates on $L^p$ for large $p$*, Geom. Funct. Anal. 10 (2000), no. 5, 1237–88

[120]  Wolff, T. *A mixed norm estimate for the X-ray transform*, Rev. Mat. Iberoamericana 14 (1998), no. 3, 561–600

[121]  Wolff, T. *An improved bound for Kakeya type maximal functions*, Rev. Mat. Iberoamericana 11 (1995), no. 3, 651–74

[122]  Wongkew, R. *Volumes of tubular neighbourhoods of real algebraic varieties*, Pacific J. Math. 159 (1993), no. 1, 177–84

[123]  Wooley, T. *Nested efficient congruencing and relatives of Vinogradov's mean value theorem*, available on arXiv

[124]  Wooley, T. D., *The cubic case of the main conjecture in Vinogradov's mean value theorem*, Adv. Math. 294 (2016), 532–61

[125]  Wooley, T. D., *Translation invariance, exponential sums, and Warings problem*, Proceedings of the International Congress of Mathematicians–Seoul 2014, Vol. II, 505–529, Kyung Moon Sa, Seoul, 2014

[126]  Zahl, J. *A discretized Severi-type theorem with applications to harmonic analysis*, available on arXiv

[127]  Zhang, R. *The endpoint perturbed Brascamp–Lieb inequality with examples*, available on arXiv

[128]  Zygmund, A. *On Fourier coefficients and transforms of functions of two variables*, Studia Math. 50 (1974), 189–201

# Index

Printed in the United States
by Baker & Taylor Publisher Services